6th International Conference on Advanced Manufacturing Engineering and Technologies (NEWTECH 2020)

AA007032

IOP Conference Series: Materials Science and Engineering
Volume 968

Galati, Romania
9 - 11 September 2020

Editors:

Viorel Paunoiu
Nicusor Baroiu
Virgil Teodor

ISBN: 978-1-7138-3203-4
ISSN: 1757-8981

Printed from e-media with permission by:

Curran Associates, Inc.
57 Morehouse Lane
Red Hook, NY 12571

Some format issues inherent in the e-media version may also appear in this print version.

No changes have been made to the content of these proceedings. There may be changes to pagination and minor adjustments for aesthetics.

Printed with permission by Curran Associates, Inc. (2021)

For permission requests, please contact the Institute of Physics
at the address below.

Institute of Physics
Dirac House, Temple Back
Bristol BS1 6BE UK

Phone: 44 1 17 929 7481
Fax: 44 1 17 920 0979

techtracking@iop.org

Additional copies of this publication are available from:

Curran Associates, Inc.
57 Morehouse Lane
Red Hook, NY 12571 USA
Phone: 845-758-0400
Fax: 845-758-2633
Email: curran@proceedings.com
Web: www.proceedings.com

TABLE OF CONTENTS

IOP Conf. Series: Materials Science and Engineering **968** (2020) 011001 doi:10.1088/1757-899X/968/1/011001

PREFACE

The 6th International Scientific Conference: NEWTECH 2020 – Advanced Manufacturing Engineering and Technologies was organized by Dunarea de Jos University of Galati, Romania. It was held between 9th-11th of September 2020 in Galati, Romania and was supported by the The Executive Unit for Financing Higher Education, Research, Development and Innovation (UEFISCDI).

The Conference: NEWTECH 2020 – Advanced Manufacturing Engineering and Technologies has a tradition of over 10 years, the inaugural edition having been organized in 2009 by the Dunarea de Jos University of Galati. This was followed by the 2011 edition in Brno, Czech Republic; 2013 at the KHT Institute of Technology, Stockholm; 2015 at Wroclaw University of Technology and 2017 at the University of Belgrade.

One achievement we can be proud of is that the current edition, the sixth, has returned to the place where this scientific event originated, and takes place in the year in which the Manufacturing Engineering Department from Faculty of Engineering, Dunarea de Jos University of Galati, Romania, celebrates 60 years of existence.

The situation with COVID-19 imposed restrictions on holding mass events and travelling. The health of the participants is key and so, the organizing committee, encouraged by the eagerness of the participants to share their scientific work despite the circumstances, decided to hold the conference online.

The 6th International Scientific Conference: NEWTECH 2020 – Advanced Manufacturing Engineering and Technologies provides a forum where recent advances and future directions in the manufacturing processes and forming are discussed by scientists and engineers from academia and industry worldwide. The topics covered are of great interest for academics and for professionals (engineers and researchers) from industry involved in traditional and novel forming and manufacturing technologies for conventional and emerging materials.

A total of 40 oral presentations from 10 worldwide countries, Algeria, Cech Republic, France, Greece, Portugal, Serbia, Spain, Romania, Russia, and United States of America are registered including:

• 10 plenary lectures given by renowned experts from academia;

• 30 oral presentations organized in 4 general sessions. The 30 oral presentations are composed of 30 peer-reviewed full papers (between 6 and 14 pages each one).

All the papers have been reviewed by two expert referees in their relevant fields. The papers selected for the volume depended on their quality and relevancy to the conference.

Free access to all papers will be granted to all the registered attendees. However, the peer-reviewed full papers and a part of the plenary lectures will be published after the conference, in the open access proceedings by IOP's Conference Series journals, IOP Conference Series: Materials Science and Engineering (https://iopscience.iop.org/journal/1757-899X).

Many ideas have been delivered, covering the different aspects of advanced manufacturing as seen from the perspective of academics and industry alike. I would briefly like to review some of the main points tackled during the conference:

IOP Conf. Series: Materials Science and Engineering **968** (2020) 011001 doi:10.1088/1757-899X/968/1/011001

1. The adoption of AI techniques, based on stochastic analysis, machine learning, neural network, to industrial applications must continue for the development of the new manufacturing technologies
2. Engineering models derived, based on the fundamental understanding of the engineering design and analysis, are an important topic which will contribute to design new tools for cutting and forming and new processes
3. Macro, Micro and nano modeling are beginning to receive more attention and will remain a focal point of interest both in cutting and metal forming
4. Incremental forming and 3d Printing are high topics in research with large application
5. New systems based on Industrial Internet of Things (IIoT) for the monitoring of the manufacturing parameters could lead to future progress in this domain
6. Smart metrology models for Industry 4.0 must continue to be developed
7. The experimental work together with modern metrological systems and control are the key factor for a quality research
8. Application of Product lifecycle management (PLM) it's important to be used for reducing the time to market and save resources

The sessions took the form of video conferences, with the scientific community showing great interest all throughout. More than 80 researchers took part online. Each of the authors used Microsoft Teams to present their report, after which the participants in the forum were given the opportunity for discussions.

The main goal of NEWTECH 2020, under the given conditions, was successfully achieved - to bring together researchers from academia and industry, scientists and experts in engineering and technology to address new challenges, share solutions and discuss future research directions.

We would like to thank all those researchers who have trusted the conference and who recognized NEWTECH 2020 as an appropriate forum to share their recent developments and learn about the new trends in the field of technical sciences and engineering. I would like to express my gratitude to Professors Miroslav Piska, Andreas Archenti and Vidosav Majstorovic for encouraging us to organize this conference.

We thank all the authors for their contribution to this valuable volume. We must also thank all the reviewers for their effort, which ensured not only excellent feedbacks for the authors but also a high-quality scientific research.

Last but not least, we sincerely thank the whole team involved in organizing the NEWTECH 2020 conference. I kindly thank our organizing committee, especially Nicusor Baroiu, Assoc. Professor, Virgil Teodor, Professor, Cezarina Afteni, PhD Researcher Assistant and Georgiana Costin, PhD candidate. Without their support this conference would not have been possible to arrange.

We plan to continue this conference. As such, the steering committee has decided to organize the next conference in Rennes, France, in 2022. We wish you good luck to the future president of the event, Adinel Gavrus, Assoc. Professor.

Prof. dr. eng. Viorel PAUNOIU
NEWTECH 2020 - Conference Chairman

IOP Conf. Series: Materials Science and Engineering **968** (2020) 011001 doi:10.1088/1757-899X/968/1/011001

EDITORS NEWTECH 2020

Prof. dr. eng. Viorel PAUNOIU
Director of Research Centre in Manufacturing (ITCM)
Department of Manufacturing Engineering (DME)
Dunarea de Jos University of Galati, Faculty of Engineering,
Galati, cod 800201, Romania

Assoc. prof. dr. eng. Nicusor BAROIU
Department of Manufacturing Engineering (DME)
Dunarea de Jos University of Galati, Faculty of Engineering,
Galati, cod 800201, Romania

Prof. dr. eng. Virgil TEODOR
Department of Manufacturing Engineering (DME)
Dunarea de Jos University of Galati, Faculty of Engineering,
Galati, cod 800201, Romania

ORGANIZERS

UNIVERSITAS GALATIENSIS

UNIVERSITAS GALATIENSIS FACULTATEA DE INGINERIE

ITCM

UEFISCDI

UNIVERSITY PARTNERS

BRNO UNIVERSITY OF TECHNOLOGY FACULTY OF MECHANICAL ENGINEERING

Wrocław University of Science and Technology

KTH VETENSKAP OCH KONST
KTH
Royal Institute of Technology

University of Belgrade
Faculty of Mechanical Engineering

IOP Conf. Series: Materials Science and Engineering **968** (2020) 011001 doi:10.1088/1757-899X/968/1/011001

CONFERENCE COMMITTEES

Conference Chairman
Professor V. PĂUNOIU (Romania)

Steering Committee
Professor M. PÍŠKA (Czech Republic)
Professor C. NICOLESCU (Sweden)
Professor A. ARCHENTI (Sweden)
Professor J. JEDRZEJEWSKI (Poland)
Professor W. GRZESIK (Poland)
Professor V. MAJSTOROVIC (Serbia)

Scientific Committee
Professor S. ADJERID (Algeria)
Professor P. ARRAZOLA (Spain)
Professor D. BANABIC (Romania)
Associate Professor N. BAROIU (Romania)
Professor M. BANU (USA)
Professor N. BÂLC (Romania)
Professor O. BOLOGA (Romania)
Professor I. BONDREA (Romania)
Professor D. BRABAZON (Ireland)
Professor GHE. BRABIE (Romania)
Professor S. CHIKH (Algeria)
Professor M. CACHO (Spain)
Professor R. ČEP (Czech Republic)
Professor S. CHATTI (Tunisie)
Associate Professor B. CHIRITA (Romania)
Professor M. CHOUCHANE (Tunisie)
Professor G. CONSTANTIN (Romania)
Professor C. CROITORU (Romania)
Professor J. P. DAVIM (Portugal)
Professor C. DOICIN (Romania)
Professor A. EPUREANU (Romania)
Professor C. FETECAU (Romania)
Professor P. FRANCO (Spain)
Professor G. R. FRUMUSANU (Romania)
Associate Professor A. GAVRUS (France)
Professor M. GHEORGHE (Romania)
Associate Professor GHE. GURAU (Romania)
Professor T. HAMA (Japan)
Professor H. HOCHENG (Taiwan)
Professor M. IORDACHE (Romania)

IOP Conf. Series: Materials Science and Engineering **968** (2020) 011001 doi:10.1088/1757-899X/968/1/011001

Organizing Committee

IOP Conf. Series: Materials Science and Engineering **968** (2020) 011001 doi:10.1088/1757-899X/968/1/011001

PLENARY LECTURES

Prof. Dr. Dorel BANABIC
Technical University of Cluj-Napoca, ROMANIA

From Micro to Macro Scale Modeling in Sheet Metal Forming

Prof. Dr. Jun NI
University of Michigan, USA

Industrial AI and Its Applications

Prof. Dr. Pedro-José ARRAZOLA
Mondragon University, SPAIN

Surface Integrity of Machined Surfaces

Prof. Dr. Vidosav D. MAJSTOROVIC
University of Belgrade, SERBIA

Cyber Manufacturing Metrology

Assoc. Prof. Hab. Dr. Adinel GAVRUS
INSA Rennes, FRANCE

Computational Framework Concerning the Formulation of Maximum Work Principle used in Plasticity, Materials Forming and Tribology as a Consequence of a Variational Optimization Problem Defined From the Constructal Law

IOP Conf. Series: Materials Science and Engineering **968** (2020) 011001 doi:10.1088/1757-899X/968/1/011001

Prof. Dr. Miroslav PÍŠKA
Brno University of Technology, CZECH REPUBLIC

Tribological Performance of Graphene and Graphene Oxide Films as Solid Lubricant Layers on Tool Steel Surfaces

Prof. Dr. Andreas ARCHENTI
KTH Royal Institute of Technology, SWEDEN

Precision engineering's impact on society

Prof. Dr. Hilde Pérez GARCÍA
University of Leon, SPAIN

Collaborative Robots is Smart Manufacturing

Prof. Dr. Dumitru NEDELCU
Gheorghe Asachi Technical University of Iasi, ROMANIA

Mechanical and Thermal Analyses of Biodegradable Materials for Injection Moulding and Additive Manufacturing

Prof. Dr. Gabriel FRUMUSANU
Dunarea de Jos University of Galati, Romania

A new approach in manufacturing process optimization: the holistic optimization

CONFERENCE CHAIRS:

1. Prof. dr. eng. Eugen RUSU, Dunarea de Jos University of Galati, Romania
2. Prof. dr. eng. Viorel PAUNOIU, Dunarea de Jos University of Galati, Romania
3. Prof. dr. eng. Virgil TEODOR, Dunarea de Jos University of Galati, Romania
4. Prof. dr. eng. Miroslav PÍŠKA, Brno University of Technology, Czech Republic
5. Prof. dr. eng. Hilde PEREZ GARCÍA, University of Leon, SPAIN
6. Prof. dr. eng. Andreas ARCHENTI, KTH Royal Institute of Technology, Sweeden
7. Prof. dr. eng. Nikolaos VAXEVANIDIS, School of Pedagogical and Technological Education (ASPETE), Greece
8. Prof. dr. eng. Dumitru NEDELCU, Gheorghe Asachi Technical University of Iasi, Romania
9. Prof. dr. eng. Elena SCUTELNICU, Dunarea de Jos, University of Galati, Romania
10. Prof. dr. eng. Catalina MAIER, Dunarea de Jos, University of Galati, Romania

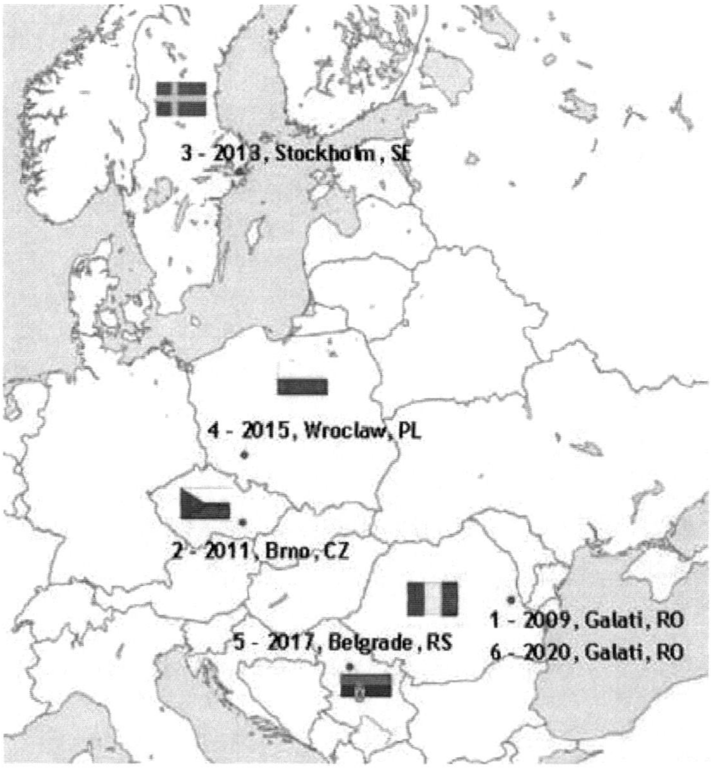

IOP Conf. Series: Materials Science and Engineering **968** (2020) 011001 doi:10.1088/1757-899X/968/1/011001

REVIEWERS LIST

Professor A. Archenti (Sweden)
Professor P. Arrazola (Spain)
Professor M. Banu (USA)
Associate Professor N. Baroiu (Romania)
Professor I. Bondrea (Romania)
Professor M. Cacho (Spain)
Professor M. Chikh (Algeria)
Associate Professor B. Chirita (Romania)
Professor M. Chouchane (Tunisia)
Professor C. Croitoru (Romania)
Professor G. R. Frumusanu (Romania)
Associate Professor A. Gavrus (France)
Associate Professor Ghe. Gurau (Romania)
Professor T. Hama (Japan)
Professor M. Iordache (Romania)
Professor Ž. Jakovljević (Serbia)
Professor P. Kyratsis (Greece)
Professor Y. Korkolis (USA)
Professor L. N. López de Lacalle (Spain)
Professor C. Maier (Romania)
Professor C. Mironeasa (Romania)
Professor D. Nedelcu (Romania)
Professor C. Nicolescu (Sweden)
Professor J. Ni (USA)
Professor E. Nitu (Romania)
Professor Ghe. Oancea (Romania)
Professor V. Paunoiu (Romania)
Professor M. Píška (Czech Republic)
Professor G. Racz (Romania)
Professor S. Radzevich (USA)
Professor J. Rech (France)
Professor D. Rodrigues (Portugal)
Professor A. Rosochowski (England)
Assistant Professor C. C. Rusu (Romania)
Professor T. Savu (Romania)
Professor E. Scutelnicu (Romania)
Professor L. Slatineanu (Romania)
Associate Professor M.D. Stanciu (Romania)
Professor V. G. Teodor (Romania)
Professor N. Vaxevanidis (Greece)
Professor I. Voiculescu (Romania)
Professor Đ. Vukelić (Serbia)

IOP Conf. Series: Materials Science and Engineering **968** (2020) 011001 doi:10.1088/1757-899X/968/1/011001

MEDIA SUPPORT

ALL CONFERENCE ALERT
Academics conferences at a glance

CONFERENCE IN EUROPE

INTERNATIONAL CONFERENCE ALERTS

Conference Alerts
Global Academic Conferences

IOPscience

with the support of

ACADEMIA DE ŞTIINŢE TEHNICE ROMÂNIA

PROECONOMICA ASSOCIATION

SOCIETY OF AUTOMOTIVE ENGINEERS OF ROMANIA
SIAR
SOCIETATEA INGINERILOR DE AUTOMOBILE DIN ROMANIA

URB

Spectromas

AUIF

DUCTIL

LINCOLN® ELECTRIC

IRISROBOTICS

INAS
power to change

ZONA LIBERA GALATI

NEPTUN

UZINSIDER
ENGINEERING S.A.
Galati - Romania

IHP

IOP Conf. Series: Materials Science and Engineering **968** (2020) 011002 doi:10.1088/1757-899X/968/1/011002

Peer review declaration

NEWTECH 2020 International Conference, September 09-11, 2020, Galati, Romania Online edition

All papers published in this volume of IOP Conference Series: Materials Science and Engineering have been peer reviewed through processes administered by the Editors. Reviews were conducted by expert referees to the professional and scientific standards expected of a proceedings journal published by IOP Publishing.

- **Type of peer review:**
  The type of peer review was **Single-blind.** The review forms were sent to reviewers and their identities was hidden from authors. In the attached document you will find the documents which certified this action, including a list with the reviewers. In the document, the review forms are presented before being sent to authors.
  Each paper was sent to the two independent reviewers by the conference president. At least one more reviewer was used if the evaluation between the two reviewers was much different,.
  The review reports were received by the organizing committee and sent to the authors.
  The revised papers by the authors were retransmitted to the reviewers in order to receive the final answer.

- **Conference submission management system:**
  The conference management system was done through the conference platform (www.newtech2020.ugal.ro). Each author, after registering could upload the paper.
  Initially, the abstract of the paper was uploaded. If the abstract concerning the conference topics was accepted, the paper was sent by email to the organizing committee.
  In the first step, the work was checked to avoid plagiarism. If the similarity percentage was adequate, following the analysis performed by the organizing committee, the revision process was started, according to those presented above.
  Only papers that were presented to the conference were accepted.
  If during the papers presentations were observations, the authors were obliged to resubmit the corrected paper.

- **Number of submissions received:**
  Total papers received were 32 (for IOP Conference Series: Materials Science and Engineering) from which 2 were rejected by the reviewers. The documents regarding these rejects, were not included.

- **Number of submissions sent for review:**
  32 papers were sent to the reviewers.

IOP Conf. Series: Materials Science and Engineering **968** (2020) 011002 doi:10.1088/1757-899X/968/1/011002

- **Number of submissions accepted:**
  30 papers were accepted.

- **Acceptance Rate (Number of Submissions Accepted / Number of Submissions Received X 100):**
  Number of Submissions Accepted / Number of Submissions Received X 100 = 30/32X100 = 93,75%

- **Average number of reviews per paper:**
  2 reviews per paper.

- **Total number of reviewers involved:**
  42 reviewers from which 21 from abroad.

- **Contact person for queries:**
  Viorel PAUNOIU, viorel.paunoiu@ugal.ro

IOP Conf. Series: Materials Science and Engineering **968** (2020) 012001 doi:10.1088/1757-899X/968/1/012001

Cyber Physical Manufacturing Metrology

V D Majstorovic and S Stojadinovic
University of Belgrade, Faculty of Mechanical Engineering, Belgrade, Serbia

E-mail: vidosav.majstorovic@sbb.rs

Abstract: The Cyber Physical Manufacturing Metrology (CP2M) is based on integration of the Cyber Physical Manufacturing (CPM) and connection between Internet of Things (IoT) and Cloud technology (CT). These are high-level methodologies for development of new generation manufacturing metrology systems, which are more intelligent, flexible and self-adaptable. CP2M generates Big Data, horizontally by integration (network of machines/CMMs, processes and sensors) and vertically by control (usually defined over five levels) which should be analytically processed and managed by the CP2M. In this paper was given, a detailed analysis of the current framework of development the CP2M. A brief overview of the concept CP2M research, particularly in Serbia is given as well.

1. Introduction

Industry 4.0 is the digital transformation of manufacturing and related industries and value chain creation processes, based on intelligent of machines and processes for industry by information and communication technology. Industry 4.0 is used interchangeably with the fourth industrial revolution. This is a new era in the organization and control of the industrial value chain, common call Industry 4.0.

The original definition of Industry 4.0, as proposed in 2011 year [1,2]: "Industry 4.0 is a German-government-sponsored vision for advanced manufacturing. Smart industry or Industry 4.0 refers to the technological evolution from embedded systems to cyber-physical systems" ... "Industry 4.0 represents the coming fourth industrial revolution on the way to an Internet of Things, Data and Services. Decentralized intelligence helps create intelligent object networking and independent process management, with the interaction of the real and virtual worlds representing a crucial new aspect of the manufacturing and production process".

This definition contains several key words, and the most important is – advanced manufacturing including advanced metrology. This means that advanced manufacturing/metrology is the basis for the fourth industrial revolution, with industrial manufacturing being integrated into digital technologies on the Internet. Industry 4.0 is the original German term. In the same context, the following terms are used worldwide: a smart factory (metrology), and intelligent manufacturing (metrology) [3].

If we perform a comparison between the development of technological systems and manufacturing metrology, we will also to come the stages of development of metrology systems, as shown in figure 1.

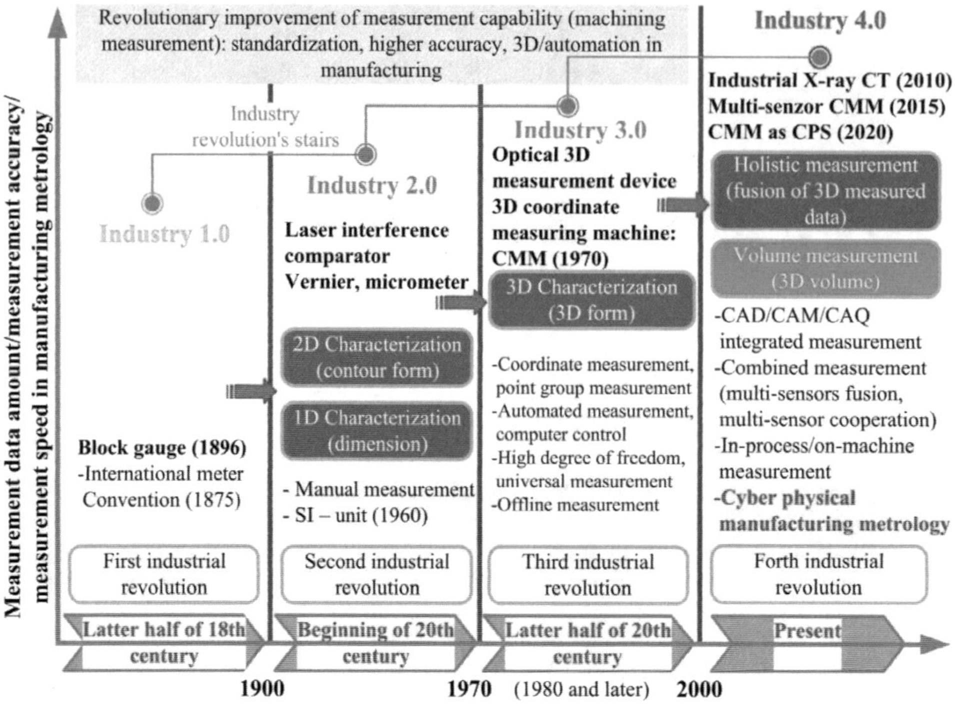

Figure 1. Correlation between era industry models and manufacturing measurement (adopted according [4]).

We can define the following facts: (i) holistic measurement model applied as a smart system (workshop, machine and part (3D measured data)), and (ii) coordinate measuring machine (CMM) as multi sensor and cyber physical system (CPS), make up today the basic framework smart metrology model for Industry 4.0.

In concept of smart manufacturing, model for CP2M is integral part, figure 2 [4]. This model has two parts: virtual and physical. They are connected by the infrastructure of the Industrial Internet of Things (IIoT). In this concept, the basic element that connects of those worlds is the CMM agent, and its function is to control CMM as a CPS. The specific units that are particularly important for the application of this model are: measuring part as a smart product, big data of metrological analysis and cloud metrology model.

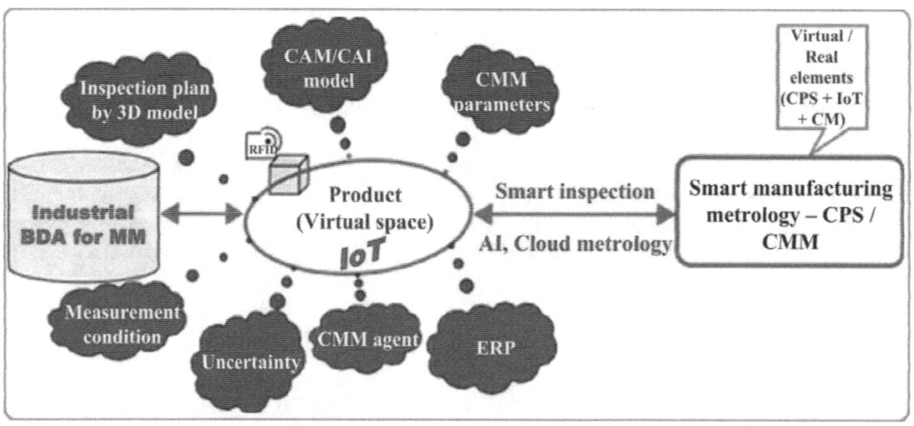

Figure 2. Model of cyber physical manufacturing metrology (CMM) (adopted according [4]).

This paper has several sections: (i) the definitions of concept Industry 4.0 and CP2M is initially given, (ii) then the CP2M model from four corner is explained in detail: metrology processes, devices for metrology, software and engineering approach, (iii) some results of the research and development of our CP2M are shown as a next, and (iv) conclusions and future research.

2. Manufacturing metrology and Industry 4.0

According [5] manufacturing metrology consist following wholes: (i) shape (size, position, form and surface texture), (ii) material (fissure, texture, hardness, and E/G modulus), (iii) function (force, moment, speed, noise), and (iv) electricity (current, voltage, resistor and magnetic size). If we look at the Industry 4.0 model framework, we can conclude that it has all types of measurements, with the first component being the most represented. From Industry 4.0 perspective, the CP2M concept can be defined through four wholes: metrology processes, devices for metrology, software for information exchange and modeling for engineering approach, which is given below.

2.1. Metrology Processes

They are issues related to the manufacturing process, would be mapped and recorded as in the virtual information system with real time in smart workshop. On figure 3 is given a link between metrological processes and elements of Industry 4.0 that support it: big data analytics (BDA, AI) and industrial internet of things (IIoT).

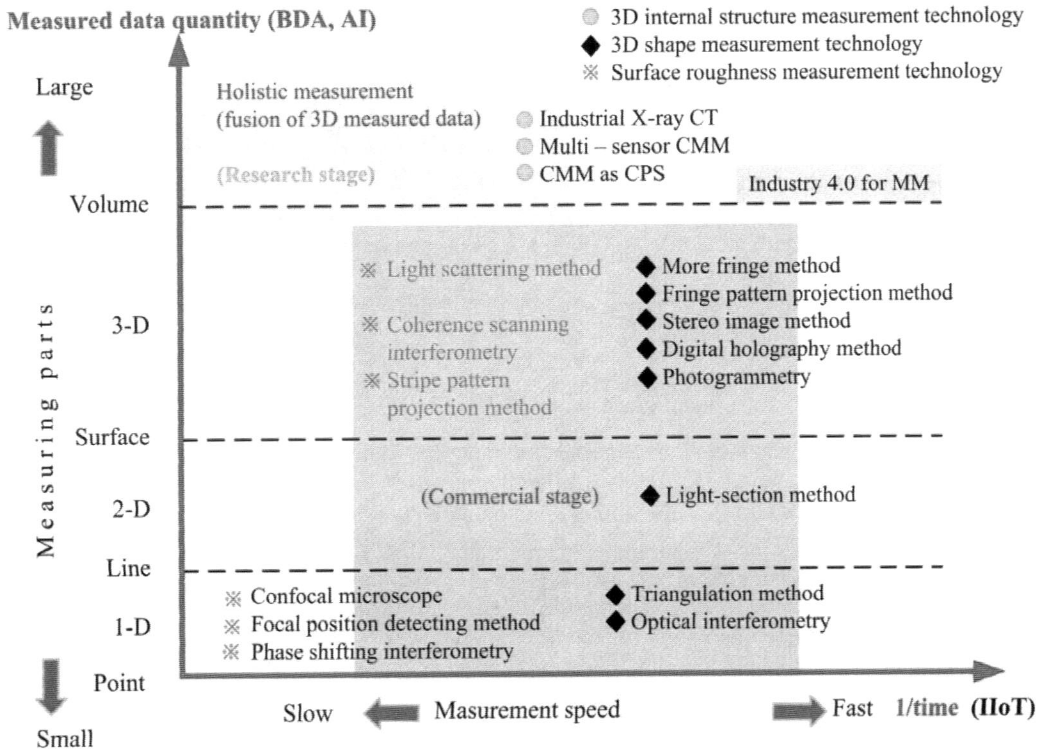

Figure 3. Map of manufacturing metrological processes and link to Industry 4.0 model (adopted according [4]).

The holistic concept of measurement generates large data sets, which are analyzed by BDA using AI techniques, and key metrological characteristics are generated on the intelligent product itself. The mapping of this information is done using IIoT, and its elements make up the infrastructure of the overall CP2M model.

2.2. Devices for Metrology

Metrological devices within the manufacturing system in Industry 4.0 model would be considered as "Things" or "objects". These include (smart) automation devices (like as CMM), field devices (measuring robot), operating devices, mobile devices, field - buses, programmable logic controllers, servers, web access devices, etc. Relative embedded technologies will also be used to merge these physical devices with virtual systems, and also the conditions of the physical devices would be recorded automatically to the information system of smart manufacturing, figure 4.

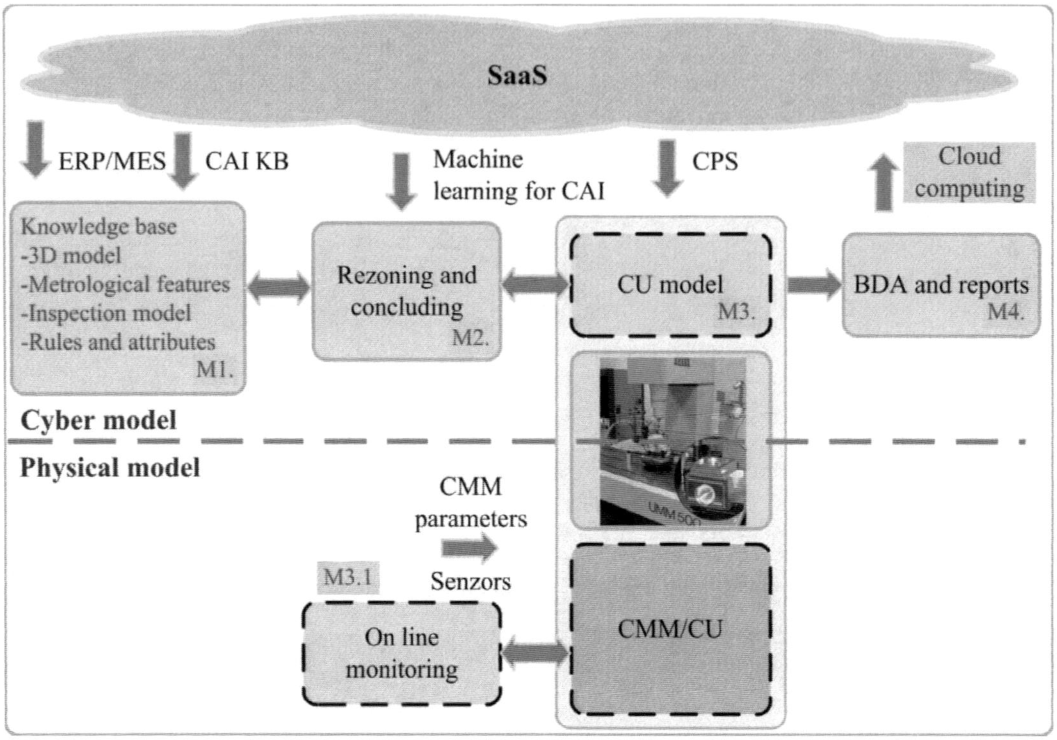

Figure 4. Cyber physical model CP2M.

On virtual enterprise level the CP2M model is connected to smart workshop (manufacturing) via ERP (enterprise resource planning) and MES (manufacturing execution system), as a rule using SaaS cloud computing model, and by IIoT this connection is realized at the physical level.

2.3. Software for information exchange

The software perspective emphasizes on the software realization for the interfaces and integration among items (IoT) from the physical-, cyber- and automation-levels in Industry 4.0. Feature-based technology and STEP standard could be considered as a main integrator in terms of linking the engineering and manufacturing domain within various CAx systems, including Industry 4.0 model. To specify the part data representation for a specific application, STEP (ISO 10303) uses Application Protocols (AP). Beside STEP APs, the following standards and interfaces are important for CAI. A vendor-independent Dimensional Measuring Interface Standard (DMIS) provides the bidirectional communication of inspection data between systems and inspection equipment, and is frequently used with CMMs. It is intermediate format between a CAD system and a CMM's native proprietary language. Dimensional Markup Language (DML) translates the measurement data from CMMs into a standardized file that could be used for data analysis and reporting. I++ DME-Interface provides communications protocol, syntax and semantics for command and response across the interface, providing low level inspection instructions for driving CMMs, figure 6 [6].

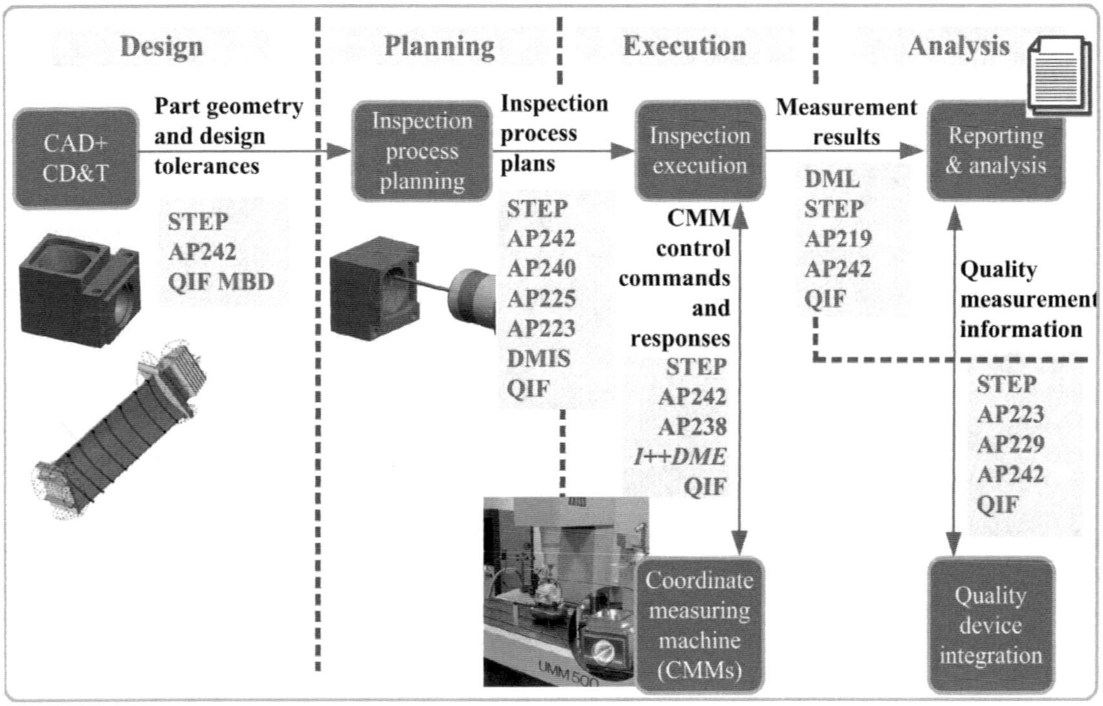

Figure 5. CMM interoperability model for Industry 4.0 / CP2M model.

The inspection planning is the most important step in the development of CAI framework a freeform surface, for CP2M model as a part of Industry 4.0 concept. In a context of the research presented in this paper (section 3) [3,12,13,14], the most important approach is STEP and STEPNC enabled inspection. Starting from STEP AP219, AP224, AP238 and AP242, CMM generates an inspection plan in DMIS (Dimensional Measuring Interface Standard - ISO 22093) format [6]. The Quality Information Framework (QIF) is an ANSI (the American National Standards Institute) standard that defines an integrated set of XML information models to enable the productive exchange of metrology data throughout the entire manufacturing quality measurement process [7]. OPC UA is next open standard for information exchange on industrial communication: devices within machines, between machines and from machines to systems. OPC UA was the recommended industrial communication standard in the Reference Architecture Model Industry 4.0 (RAMI 4.0) [8]. It provides both communication protocol and information modeling method (IIoT), allowing easy modeling and development of digital twins of manufacturing equipment in the cyberspace for Industry 4.0 model.

2.4. Modeling for engineering approaches

Industry 4.0 is building smart factories in deployment that have four dimensions, figure 6 [9,10,11].

Elements of manufacturing metrology include all part on smart factory: (i) smart manufacturing (smart metrology) based on advanced digital-oriented technologies (additive manufacturing, cloud computing and internet of things). It has automated flexible lines that adapt production processes with changed conditions to the type of product, while sometime maintaining high quality, high productivity and flexibility, as well as production volume, with optimal consumption of resources; (ii) smart products (advanced production mode and new characteristics, including also of metrological parameters). They generate and send to the manufacturer exploitation feedback information, which is used primarily in the field of customer service. On this way increases the value added of the product, and the manufacturer develops a new business model (product + service); (iii) new ways workers perform their activities, based on advanced digital-oriented technologies (smart working), and (iv) smart supply-chain (procurement of raw materials and delivery of finished products). Bidirectional exchange of information in collaborative production, using it exchange also for digital platforms of design of the innovative products.

Figure 6. Framework of Industry 4.0 model for smart factory (adopted according [9]).

3. Cyber-physical manufacturing metrology model (CP3M) – some research results

The framework of CP3M model consists of the following sub – modules [3,12,13,14]: (a) module for definition and recognition of geometrical features (GF) from CAD/GD&T model of the measurement part, where we used them for definition of metrological feature (MF) (b) module for building of intelligent inspection process planning (IIPP), that contains methods for prismatic parts presented, and method for freeform surfaces application, (c) CMM – generation of control data list for CMM that is transferred to CMM using cloud technology, and (d) module for analysis of results and generation of the final measuring reports. Cloud services within the organization provide the necessary information for integration of knowledge and data from various phases in product design and manufacturing/metrology into inspection planning, and make available information about inspection results to all interested parties in product lifecycle.

In CP3M is a feature-based model for probe path planning for sculptured surfaces using an example of turbine blades [12, 13,14]. The probe path planning model represents a part of CP3M for free form surfaces. In this model the geometrical information for feature description is taken from CAD model of the part in IGES format, figure 7.

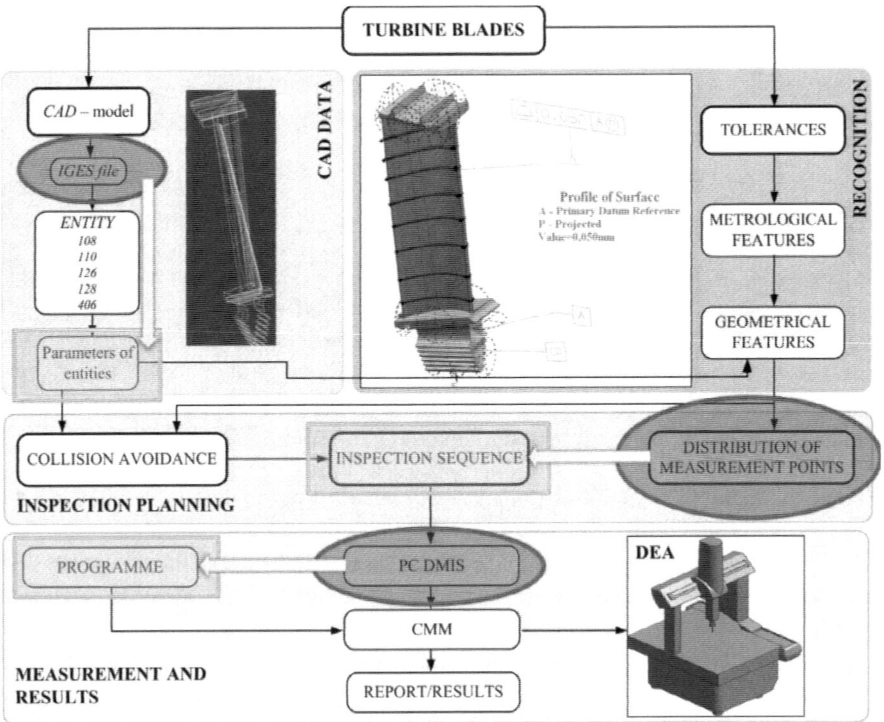

Figure 7. Inspection planning of turbine blades CP3M model.

The inspection plan consists of input CAD data, recognition of metrological and geometrical primitives, definition of inspection sequences, distribution of measuring points, collision avoidance principle, and measurement and analysis of the results. Realization of the CPM3 model in practice is being investigated for the DEA CMM, which is based on the IoT Cloud technology. Connection of the physical and the virtual world is provided using an IIoT device - Raspberry Pi3, which, via industrial router, has access to the Internet and realizes direct communication with the CMM control unit, using Ethernet or Wi-Fi connection. After each measurement, report is generated and automatically sent to the Cloud where it is stored. The system sends the report to a user or a predefined group of users, via email or SMS service: start/end of measurement, various alarms, environment sensor readings and other important information.

4. Conclusions

Cyber-physical manufacturing metrology play very important role in Industry 4.0 model for manufacturing. Already today, manufacturers of metrology equipment, in particular CMM, produce them as CPS, which can be used in Industry 4.0 concept in industry. On the other hand, researchers around of the world have already presented their results from the field research of CP2M. Some of them have already been presented in this paper. In the coming period, rapid diffusion of these models (Industry 4.0 and CP2M) into industrial applications should be expected, as soon as possible.

IOP Conf. Series: Materials Science and Engineering **968** (2020) 012001 doi:10.1088/1757-899X/968/1/012001

5. References

[1] https://i40mc.de/en/, Accessed, Apr. 2020.

[2] https://www.i-scoop.eu/industry-4-0/, Accessed, Apr. 2020.

[3] Majstorovic V D, Durakbasa N, Takaya Y and Stojadinovic S 2019 Advanced Manufacturing Metrology in Context of Industry 4.0 Model, *Proceedings of the 12th International Conference on Measurement and Quality Control - Cyber Physical Issue. IMEKO TC14 2019* (Belgrade: Springer) p1

[4] Yasuhiro T 2018 Strategic Vision for Smart Machining Tool and Measuring Instrument, *NTN Technical Review* **86**, pp 20-41

[5] Berthold J and Imkamp D, Looking at the future of manufacturing metrology: roadmap document of the German VDI/VDE Society for Measurement and Automatic Control, *J. Sens. Sens. Syst.* 2 pp 1–7

[6] Majstorović V D, Mačužić J, Šibalija T and Živković S 2015 Cyber-Physical Manufacturing Systems – Manufacturing Metrology Aspects *Journal of Proceedings in Manufacturing Systems* 1 **10** pp 9−14

[7] DMSC. Home - QIF Standard. http://www.qifstandards.org/ Accessed April 2020

[8] Schleipen M, Gilani S S, Bischoff T and Pfrommer J 2016 OPC UA - Industrie 4.0-enabling technology with high diversity and variability *Proceedings of 49th CIRP Conference on Manufacturing Systems* **57** (Stuttgard: Elsevier) pp 315-320

[9] Frank A, Dalenogare L and Ayala N 2019 Industry 4.0 technologies: Implementation patterns in manufacturing *International Journal of Production Economics* **210** pp 15–26

[10] Emmanouilidis C, Pistofidis P, Bertoncelj L, Katsouros V, Fournaris A, Koulamas C and Ruiz-Carcel C 2019 Enabling the human in the loop: Linked data and knowledge in industrial cyber-physical systems *Annual Reviews in Control* **47** pp 249–265

[11] Dassisti M, Giovannini A, Merla P, Chimienti M and Panetto H 2019 An approach to support Industry 4.0 adoption in SMEs using a core-metamodel *Annual Reviews in Control* **47** pp 266-274

[12] Majstorović V et al., Cyber-Physical Manufacturing Metrology Model (CP3M) for Sculptured Surfaces – Turbine Blade Application *Procedia CIRP* **63** ed Tseng M M et al (Taichung: Elsevier) pp 658–663

[13] Majstorovic V D et al. 2018 Cyber-Physical Manufacturing Metrology Model (CP3M) – Big Data Analytics Issue *Procedia CIRP* **72** de Wang L (Stockholm: Elsevier) pp 503–508

[14] Majstorovic V D et al., Building of Internet of Things Model for Cyber-Physical Manufacturing Metrology Model (CP3M) *Procedia CIRP* **81** ed Butala P et al (Ljubljana: Elsevier) pp 862-867

IOP Conf. Series: Materials Science and Engineering **968** (2020) 012002 doi:10.1088/1757-899X/968/1/012002

Condition monitoring of rolling element bearings: benchmarking of data-driven methods

E Iunusova[1], K Szipka[1] and A Archenti[2]
[1] Sustainable Production Development Department, KTH Royal Institute of Technology Kvarnbergagatan 12, 15136 Södertälje, Sweden
[2] Production Engineering Department, KTK Royal Institute of Technology Brinellvägen 68, 100 44 Stockholm, Sweden

E-mail: iunusova@kth.se

Abstract. Condition-based maintenance (CBM) is a maintenance strategy used to gain updated information about equipment condition and is today considered a natural part of the engineering field. The replacement of the traditional scheduled maintenance strategy in favor of CBM has the potential to significantly improve the safety of the system operating in harsh environments of the operation and increase in productivity by prolonging the life of an asset and preventing costly breakdowns. For many years CBM remained the subject of vigorous research and discussions. Increasing the automation level and the number of sensors in industries allowed obtaining and collecting data in large amounts. The current level of computational power allows us to process and analyse this massive amount of data, which has given a new leap in the development of industrial analytics. Rather than in the case of classical knowledge-based modelling tools, data-driven methods propose modelling and forecasting frameworks based on data analysis. Consequently, the transition to the data-driven modelling gave a leap in CBM research and has recently drawn increasing attention, providing new case studies, algorithms, and results. However, technical challenges remain. Despite great flexibility and good forecasting performances, there are several limitations of data-driven algorithms. This paper provides an overview of the data-driven failure algorithms for rolling element bearings monitoring. Bearings have played a pivotal role in industrial machinery to operate with high effciency and safety. They are considered to be one of the most common machine elements of precision rotating machinery. A benchmarking of various predictive and descriptive algorithms was performed. The analysis was carried out on a dataset from the run-to-failure experiments on bearings from NASAs Data Repository. This paper also summarizes the current trends and highlights the limitations with respect to traditional knowledge-based modelling. Special attention is paid to identifying research gaps and promising research directions.

1. Introduction

The Fourth Industrial Revolution is strongly associated with the integration between physical and digital systems in production environments. The proliferation of sensing technologies exponentially increases the amount of data extracted from the manufacturing process, production system, and equipment. This data can bring out valuable information and knowledge that could enhance the effectiveness of the whole manufacturing process, as a valuable asset for decision support in various areas, particularly in health monitoring and condition-based maintenance. Driving by the aspiration to increase equipment uptime and reduce unnecessary maintenance work, more and more attention has been put into predictive maintenance, which necessitates advanced prognosis tools to focus on performance degradation monitoring and assessment so failure can be predicted and prevented [1].

The various methods and tools regarding failure prognostic can be grouped into the knowledge-based and data-driven approaches. Knowledge-based approaches rely on the use of a set of algebraic and differential equations obtained by using traditional laws of physics. Data-driven approaches aim to transform raw data into relevant information and behavior models by using only the data provided by the monitoring system. In this case, the target is not to establish an analytical model of the system (physical and mathematical knowledge of the process) neither to provide a deep understanding of its physical parameters [2].

The design of knowledge-based process monitoring and fault diagnosis systems has been a remarkable research topic for several decades. The well-established knowledge-based techniques have been successfully applied to plenty of processes for industrial electronics, automatic control systems, etc. Recently, to ensure the reliability and safety of modern large-scale industrial processes, data-driven methods have been receiving considerably increasing attention, particularly for the purpose of process monitoring [3]. The malfunction of manufacturing machines can be generally attributed to various faults of different categories, including drive inverter failures, stator winding insulation breakdown, bearing faults, and air gap eccentricity. Several surveys regarding the likelihood of induction machine failures conducted by the IEEE Industry Application Society and the Japan Electrical Manufacturers' Association reveal that bearing fault is one of the foremost causes of breakdowns and is responsible for 30 to 40 percent of all machine failures [4]. Such failure can be catastrophic, resulting in costly downtime.

Bearings have played a pivotal role in industrial machinery to operate with high efficiency and safety, and they are considered to be one of the most common machine elements of precision rotating machinery. Furthermore, bearings are critical to almost all forms of rotating machinery and are among the most common machine elements. Therefore, rolling element bearing condition monitoring and diagnostics has been a research frontier for engineers and scientists for many years. In order to prevent unexpected bearing failure, vibration analysis has been used extensively for various bearing condition monitoring techniques [5]. A rolling-element bearing is illustrated in figure 1a, which contains the outer race typically mounted on the motor cap, the inner race to hold the motor shaft, the balls or the rolling elements, and the cage for restraining the relative distances between adjacent rolling elements [4].

This paper provides an overview of the most widely used data-driven failure algorithms for rolling element bearing monitoring. The results show that the improvement of existing Machine Learning (ML) algorithms and the creation of new ones are gathering more significant momentum from year to year. Modern algorithms allow reaching good prediction results even for those who don't understand the fundamentals of the investigation subject. Complete disregard of the physical foundations in pursuit of a more accurate forecast leads to the creation of untraceable models that are adapted to the particular case with particular data from particular equipment. They do not rely on understanding physical and mathematical laws, and in many cases, trained models do not generalize well.

This paper aims to assess the current state and the development direction of failure prognostics approaches, investigate modern data-driven trends, and highlight their limitations. The remaining sections of this paper are organized as follows: In Section 2, the description of the run-to-failure test dataset and fault types are provided. In Section 3, the workflow of the used methodology is shown and briefly described. The comparison of the results of implementing the proposed methodology is presented in Section 4. Finally, Section 5 presents a summary of the work with the highlighting limitations and future investigation plans.

IOP Conf. Series: Materials Science and Engineering **968** (2020) 012002 doi:10.1088/1757-899X/968/1/012002

2. Overview

Data is the foundation for data-driven modelling. It is necessary to have a good collection of datasets to be able to develop effective ML algorithms for bearing fault detection. Natural bearing degradation is a gradual process and may develop throughout months or years. Bearing faults are investigated trough three mayor approaches for monitoring purposes. First, with the application of simulations after the detailed modelling of bearing behavior, including the simulation of the sensor signals. Second, artificially induced faults, for instance by scratching or drilling the bearing surface or created by exerting a shaft current for accelerated life testing. Third, bearings are run under given conditions with performing run-to-failure tests. While the data collection (especially in the third case) can be expensive and time-consuming, fortunately, a few organizations have made an effort and published bearing fault datasets for investigations of various ML algorithms. Their widespread prevalence has another great advantage. Researches from all over the world could compare the results of implementing different algorithms to the same dataset. The data contains a complete record of natural bearing defect evolution generated by the NSF I/UCR Center for Intelligent Maintenance Systems with support from Rexnord Corp. in Milwaukee was used in this paper [6].

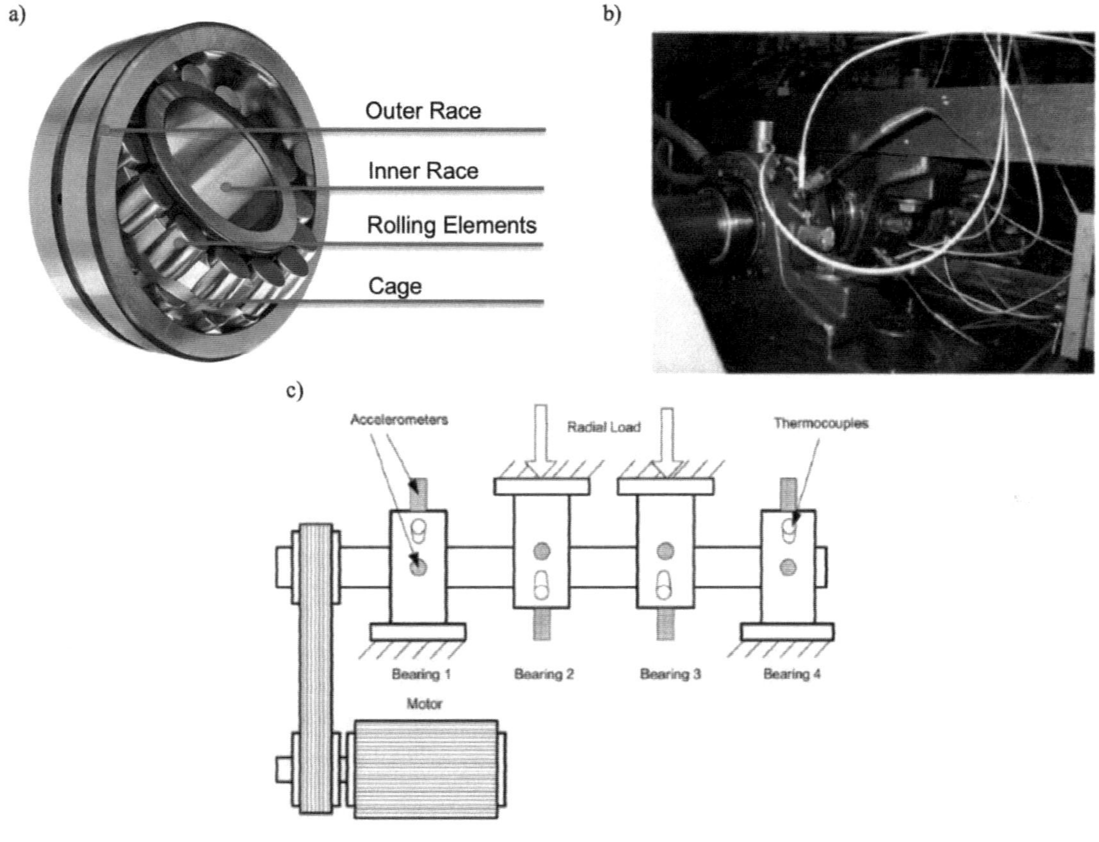

Figure 1. The test rig: a) the main components of the rolling element bearing [7], b) the photo of the test rig [6], c) the test rig setup scheme [6].

2.1. Test Rig Setup

In the test rig, four Rexnord ZA-2115 double row bearings were installed on the shaft (figure 2). The test rig element's characteristics are presented in table 1 [6].

Table 1. The test rig element's characteristics.

Element	Characteristics
Test Rig	• The rotation speed was kept constant at 2000 RPM • A radial load of 6000 lbs is applied onto the shaft and bearing • All bearings are force lubricated
Rolling elements bearings	• Bearings have 16 rollers in each row • The pitch diameter of 2.815 in. • The roller diameter of 0.331 in. • The tapered contact angle of 15.17 degrees
Accelerometers	• PCB 353B33 High Sensitivity Quartz ICP accelerometers • Installed on the bearing housing (two for each bearing [x- and y-axes] for data set 1, one for each bearing for data sets 2 and 3) • All failures occurred after exceeding the designed lifetime of the bearing, which is more than 100 million revolutions

2.2. Data Structure

During the three independent run-to-failure bearing experiments the acquisition of the vibration signals were collected from start to failure with explicit time stamps. Each test dataset consists of individual files that are 1-second signals recorded every 10 minutes. Each file consists of 20,480 points, with the sampling rate set at 20 kHz. This sampling frequency can assure that valuable higher frequency ranges are also captured, which are commonly used for early bearing degradation identification. Data was acquired by DAQCard-6062E NI. Inspections confirmed that test 1 ended up with an Inner Race (IR) failure in bearing 3 and a Rolling Elements (RE) failure in bearing 4. Test 2 and 3 ended with an Outer Race (OR) failure in bearing 1 and 3, respectively [6].

2.3. Equipment life cycle stages

The typical life cycle of the bearings from the beginning of operation until the defect condition could be divided into three stages: normal, degradation, and defect. At the beginning of the operation, the equipment is in a normal stage which can be understood as the operational condition within functional specifications. After any fault occurs, the equipment goes into the degradation stage. Depending on the type of fault, the duration and speed of captured health degradation will vary. This behavior, together with large amount of noise in the input data can result in potential challenges for abnormality identification.

Usually, data collection is not interrupted immediately after a breakdown. Due to this reason, data is still captured from the defected part. In this research, the defective part will be excluded from the investigation after its identification. The target of the first part of the proposed workflow is to identify the fault and failure points and divide the data into phases, which are shown in figure 2.

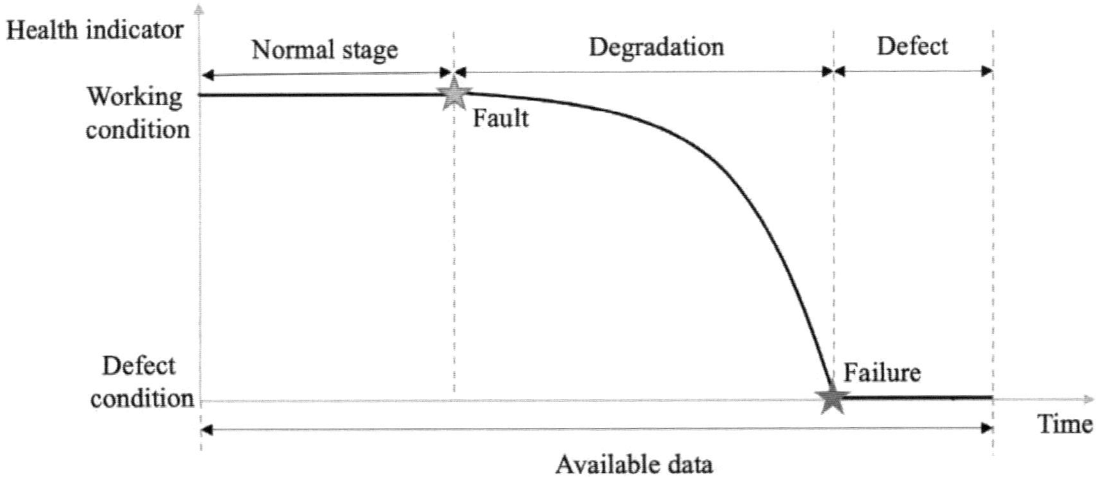

Figure 2. The typical life cycle of the equipment.

3. Methodology

In order to create prediction models for root cause identification, supervised learning algorithms will be used. A labelling process, including clustering approaches, will support the creation of this prediction model. This is necessary since the run-to-failure data collection does not contain labels for training a model based on various supervised learning. Thus, the overall workflow was divided into two parts: labelling and modelling. In this research, the unsupervised learning algorithms were used in the labelling part to support the supervised algorithms in the modelling part. The most important parts of the workflow are demonstrated in figure 3.

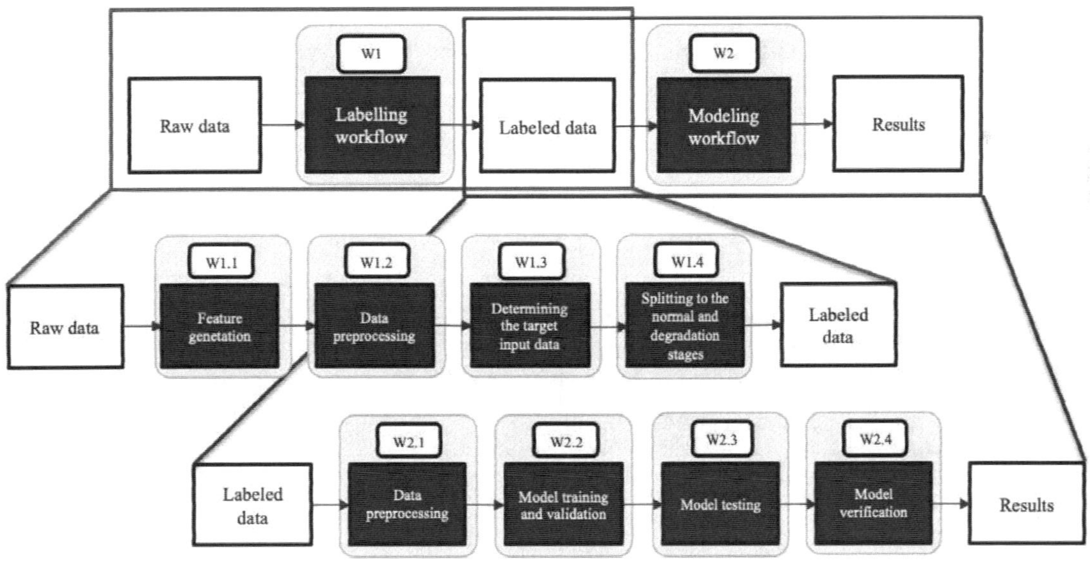

Figure 3. The workflow used for data investigation.

3.1. Labelling workflow

The first step in the workflow is the time-domain feature generation. More than 390 features were created by TSFEL library [8]. Furthermore, several statistic features as crest factor, shape factor were added. In total 406 features were created for each sensor. The examples of several different features are shown in figure 4a. It can be seen that the variation of the values for each feature becomes bigger after crossing some timestamp point. Figure 4b shows a 2-dimensional representation of all the features after implementing the Principal Component Analysis (PCA). It has to be highlighted that all calculations were made in higher dimensions, and the visualizations are only provided for two dimensions, after the application of PCA. However, even in this simplified case, cases that potentially can be abnormal seem to be detectable. No filtering is implemented at this stage to reduce the effect of noise or focus on certain bandwidths of the input signal.

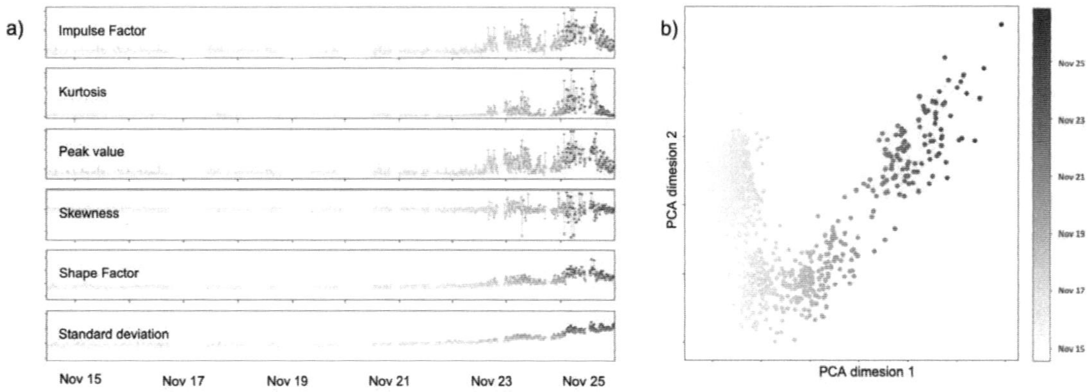

Figure 4. a) the examples of the created features, b) the two-dimentional representation of all the features after implementing PCA.

After performing data preprocessing, the failure point was identified in order to distinguish the combined normal and degradation stages from the defect stages for each test case. This was implemented using unsupervised learning to divide into two clusters using four different clustering algorithms, such as K-Means, Mean Shift, DBSCAN, and Birch. The final clusters were determined by the implementation of the voting technique. To find a particular split point, the logistic regressions sigmoid were used (figure 5).

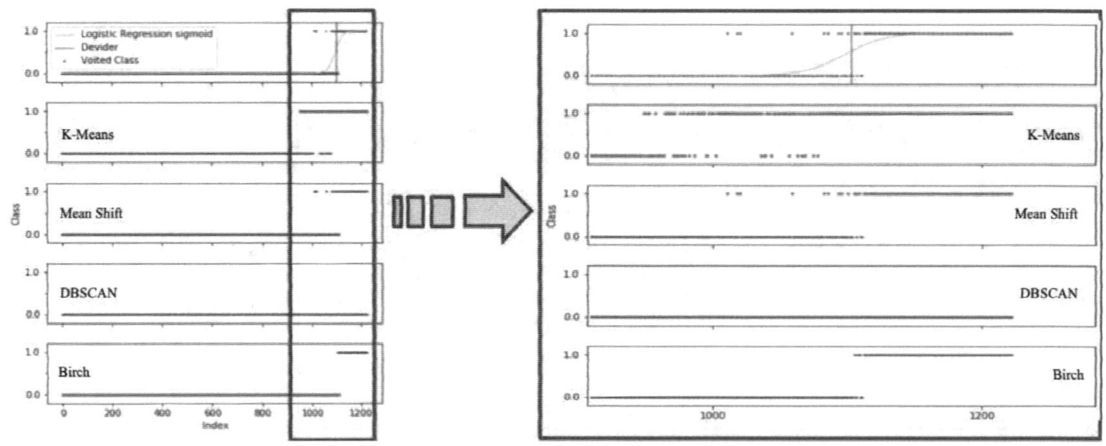

Figure 5. The clustering process.

A similar approach was implemented in the next step to determine the point of splitting the normal and degradation stages (fault point). The first cluster accepted as a normal stage, and the second was accepted as a degradation stage. The data above 0.5 of probability were accepted as degradation cluster, under 0.5 - as a normal cluster.

3.2. Modelling for abnormality and root cause identification

The labeled data were used as input for the supervised learning algorithms to create separate abnormality detection models for each type of failure and create a general predictive model for all failures to identify the Root Cause Analysis (RCA). In the modelling part were used such algorithms as Random Forest (RF), Logistic Regression (LR), Naive Bayes (NB), XGBoost (XGB). To improve results, the model hyperparameters were optimized through cross-validation. In figure 6 the results from the RF model are presented.

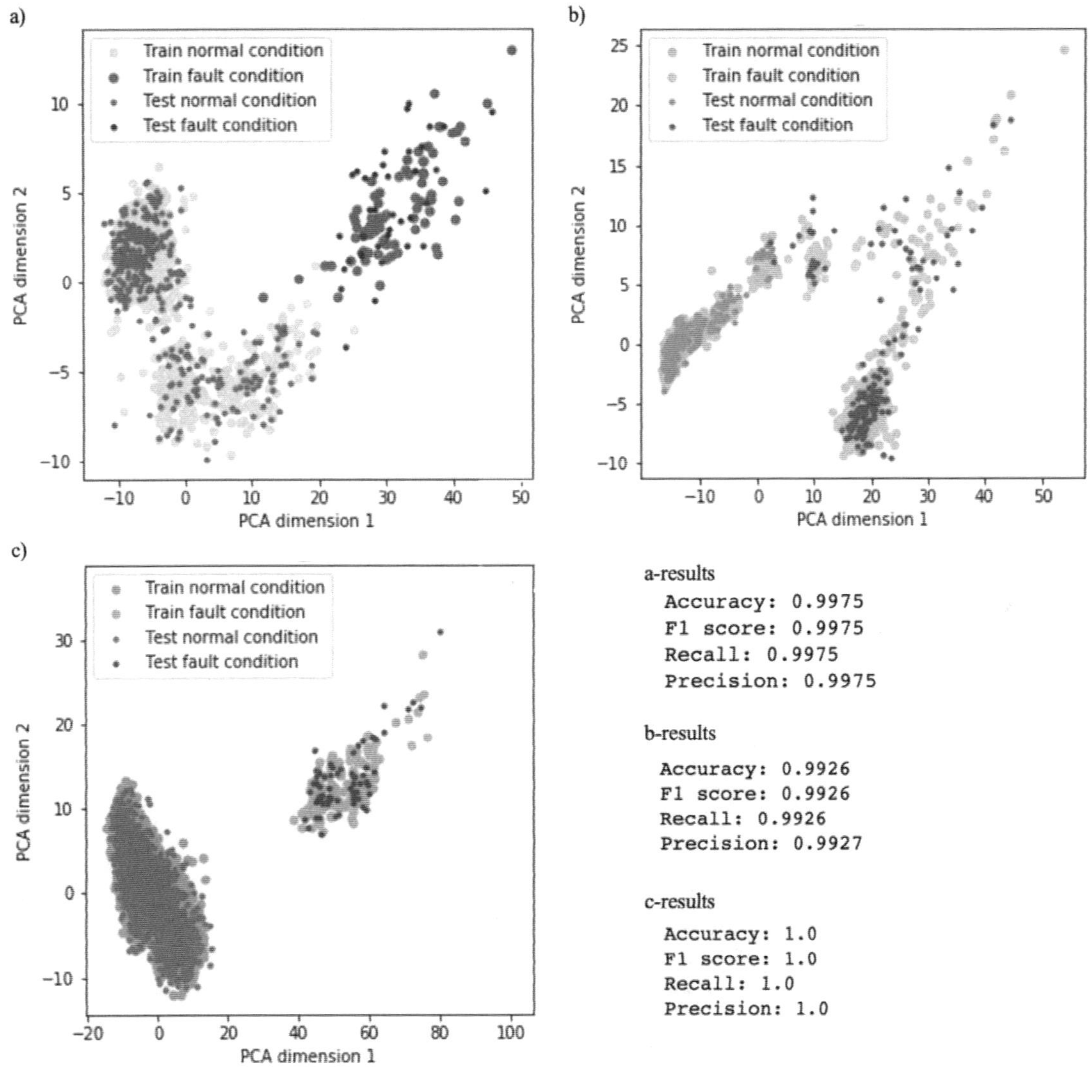

Figure 6. Modelling results of RF algorithm for: a) IR failure, b) RE failure, C) OR failure.

The first model (figure 6a) detected the IR failure in bearing 3 from test 1. It can be seen that the results show a high model performance. All calculations were made in higher dimensions. This figure is only a two-dimensional representation of this higher-order solution. However, in this case, the decision boundaries can be already easily determined. The same is true for the second model (figure 6b) that detects the failure of the RE in bearing 4 from test 1 and, in particular, for model three (figure 6c) that detect the OR failure in bearing 1 from test 2. In OR determining case, the failure resulted in a significant sudden change of system behavior, which made it easy to detect an abnormality.

The confusion matrix of a general model trained for RCA also shows high performance (figure 7a). The separate results of predictive models for each failure are shown on the three subfigures on the left side of the figure including IR, RE, and OR failure detection. In the middle, all various conditions and failures are represented.

4. Results and benchmarking

The results for the rest of the algorithms are shown in table 2. Splitting of the sensor data was carried out in order to implement training and testing. For testing, the one-third part of the data was used, including each third timestamp. The rest of the data was used for training purposes. Cross-validation technique was applied to enhance training performance. In order to evaluate if the trained models generalize well, the OR failure was verified on a bearing dataset, which was acquired from a completely independent test. The IR and RE failures were verified using data from the duplicated sensors from the same bearings. A limitation in those verifications was that for the IR and RE cases, the test conditions were not changed as the setup was the same. Only different sensors were used for the same setup.

Table 2. Benchmarking of the abnormality detection models.

Model	Parameter	LR		RF		NB		XGB	
		test	verif.	test	verif.	test	verif.	test	verif.
IR failure detection	Accuracy	0,9975	0,9950	0,9975	0,9959	0,9508	0,9779	0,9950	0,9959
	Recall	0,9975	0,9950	0,9975	0,9959	0,9508	0,9779	0,9950	0,9959
	Precision	0,9975	0,9951	0,9975	0,9959	0,9675	0,9813	0,9950	0,9959
RE failure detection	Accuracy	0,9975	0,9926	0,9926	0,9975	0,9606	0,9991	1,0000	0,9991
	Recall	0,9975	0,9926	0,9926	0,9975	0,9606	0,9991	1,0000	0,9991
	Precision	0,9975	0,9927	0,9926	0,9975	0,9644	0,9991	1,0000	0,9991
OR failure detection	Accuracy	1,0000	0,9122	1,0000	0,9200	0,9970	0,8566	0,9985	0,8977
	Recall	1,0000	0,9122	1,0000	0,9200	0,9970	0,8252	0,9985	0,8977
	Precision	1,0000	0,9106	1,0000	0,9183	0,9973	0,8728	0,9986	0,8956

This verification process was applied to the general model, where the same failures were aimed to be reproduced. The results of model verification can be seen in figure 7b through a confusion matrix.

Even with the same type of bearings, at the same test rig, the results worsened by repeating the same test. This highlights that the models are over fitted on the given dataset and show overconfidence. This over fit is already the result of an optimization, highlighting significant limitations of using datasets without proper considerations of bearing frequencies and filtering of the input data. The generalization is also affected by the implemented labelling, using unsupervised learning, which is not investigated during the benchmarking.

IOP Conf. Series: Materials Science and Engineering **968** (2020) 012002 doi:10.1088/1757-899X/968/1/012002

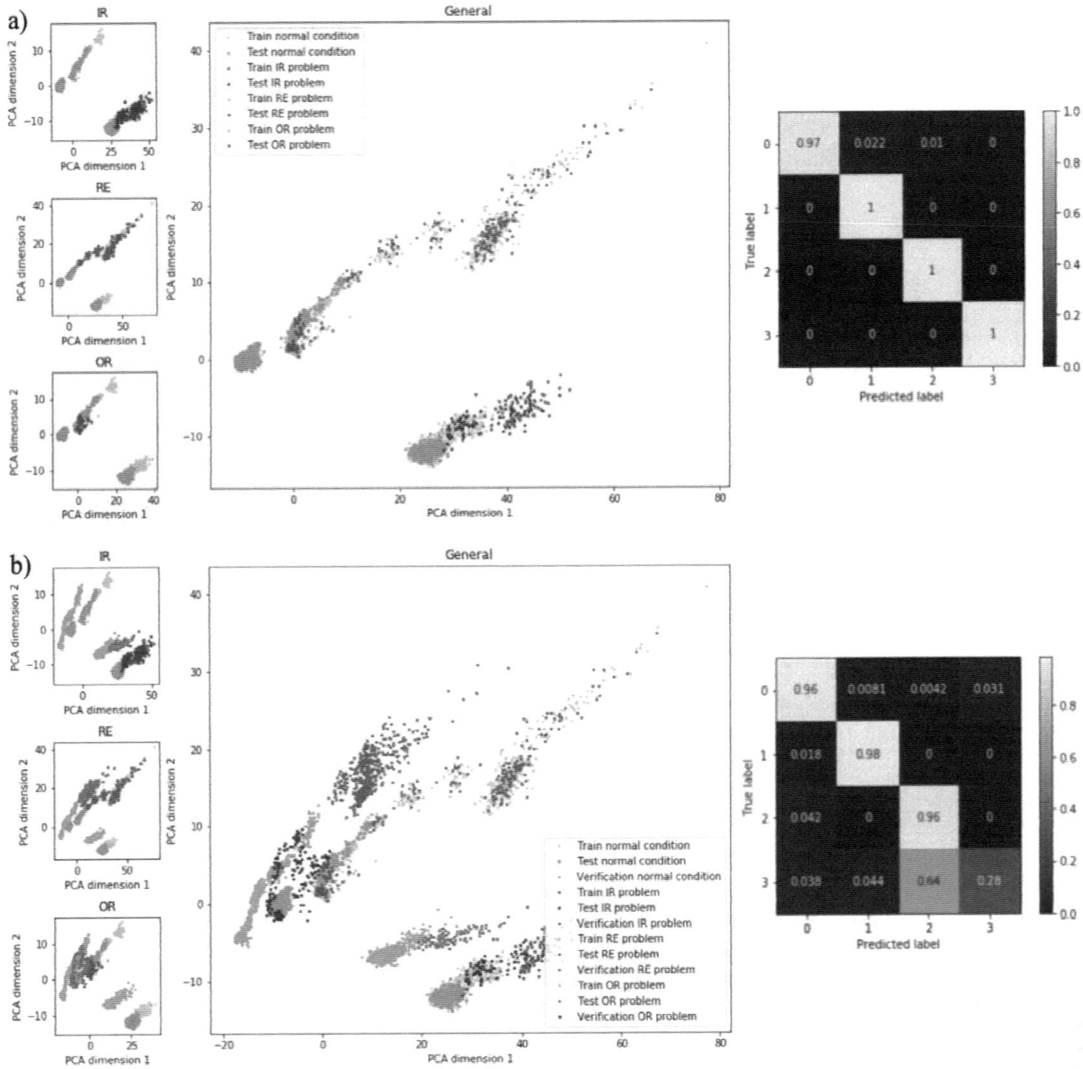

Figure 7. Modelling results of RF algorithm for RCA: a) test, b) verification

5. Conclusion

During the last decades, a data-driven approach seemed like a powerful tool for condition monitoring. Widespread use has led to its rapid growth. However, it has significant limitations. Using the data-driven approaches results in illusive reliability on the trained model, which can happen due to limitations in the conditions mapped into the studied dataset. Even though the testing results show high performance, the models do not generalize well. It is crucial to keep in mind that in the case of using a data-driven approach, the quality of the model cannot get better than the quality of the used data.

After the discussion of the main limitations of the data-based approach and the performed benchmarking of algorithms, an important conclusion can be drawn. To improve the quality of the input data and the results of trained models, more elements of the knowledge-based approach have to be involved. A hybrid approach, which is the combination of knowledge-based and data- driven approaches, can effectively address the limitations. In such a hybrid approach, knowledge of the physical behaviour of the test case will be involved. The effect of filtering, taking into account the bearing frequencies, will be considered. Furthermore, a detailed study of feature generation tools will be performed, which uses the physical parameters of the experimental setup. In the examination of the feature selection algorithms, the focus must be on keeping the traceability of the acquired data. The investigation of labelling techniques based on domain knowledge should have a significant impact on the results. The aim of that research will be to create a well-generalized model for different types of machine components, equipment and other industrial machinery.

6. References

[1] Lughofer E and Sayed-Mouchaweh M 2019 Predictive Maintenance in Dynamic Systems Advanced Methods, Decision Support Tools and Real-World Applications (1st ed. 2019. ed.)

[2] Tobon-Mejia D A, Medjaher K, Zerhouni N and Tripot G 2012 A Data-Driven Failure Prognostics Method Based on Mixture of Gaussians Hidden Markov Models *IEEE Transactions on Reliability* **61**(2) pp. 491-503

[3] Carvalho T P, Soares F A A M N, Vita R, Francisco R da P, Basto J P and Alcalá S G S 2019 A systematic literature review of machine learning methods applied to predictive maintenance *Computers and Industrial Engineering* **137** p. 106024

[4] Zhang S, Zhang S, Wang B and Habetler T G 2020 Deep Learning Algorithms for Bearing Fault Diagnostics—A Comprehensive Review *IEEE Access* **8** pp. 29857-29881

[5] Qiu H, Lee J, Lin J and Yu G 2003 Robust performance degradation assessment methods for enhanced rolling element bearing prognostics *Advanced Engineering Informatics* **17**(3-4) pp. 127-140

[6] Qiu H, Lee J, Lin J and Yu G 2006 Wavelet filter-based weak signature detection method and its application on rolling element bearing prognostics. *Journal of Sound and Vibration* **289**(4-5) pp. 1066-1090

[7] Amin Almasi 2017 Vibration, noise and failure in rolling element bearings Turbomachinery *International Magazine*

[8] Barandas M, Folgado D, Fernandes L, Santos S, Abreu M, Bota P, Liu H, Schultz T and Gamboa H 2020 TSFEL: Time Series Feature Extraction Library. SoftwareX **11** p. 100456

IOP Conf. Series: Materials Science and Engineering **968** (2020) 012003 doi:10.1088/1757-899X/968/1/012003

Computational Framework Concerning the Formulation of Maximum Work Principle used in Plasticity, Materials Forming and Tribology as a Consequence of a Variational Optimization Problem Defined From the Constructal Law

Adinel Gavrus
National Institute of Applied Sciences of Rennes – INSA Rennes, Civil and Mechanical Eng. Laboratory – LGCGM (EA 3913), 20 Av. des Buttes de Coësmes, 35708 Rennes, France

E-mail : adinel.gavrus@insa-rennes.fr

Abstract. This research work proposes a fundamental application of the Constructal Theory developed by Prof. Adrian BEJAN of Duke University (USA) to prove on a mathematical point of view the "Principle" of Maximum Work, used by the theory of plasticity, material forming and tribology, as a consequence of the global solution defining a constrained variational optimization problem. According to the first and second law of thermodynamics, the principal law of Constructal Theory try to complete them with a quantitative prediction of the natural tendency of any finite size system to evolve towards an optimal space-time configuration minimizing the losses and the entropy generation simultaneously with a required maximum of the global entropy. In this sense, regarding a material plastic deformation characterising the forming processes, among all possible and admissible flow undergoing well-specified boundary conditions and loadings, the real one is the one who minimizes the sum of the dissipated power of volume deformation and surfaces friction. Thus, all the mechanical variables defining the real mechanical state (velocities, stresses, strain and strain rate) are those ones which minimize the total dissipated power. A variational minimization problem under a lot of defined constraints is then obtained. Using the Principle of Virtual Powers it can be shown finally that the "Principle" of Maximum Work, used particularly in metals plasticity, is obtained as a consequence of a minimization problem under constraints based on the Constructal Theory. This generalizes its application to any type of continuous media (metals, polymers, fluid, mushy state) and allows proving an equivalent form for the friction stresses occurring on contact interfaces. The convexity properties of both the plastic and the friction potential together with their normal rule properties can be also proven using the proposed mathematical framework. It is concluded that only the rheological and tribological flow laws associated with a potential become to satisfy the second thermodynamics principle completed with the Constructal Law. Analytical computations concerning plane and cylindrical crushing show the feasibility of the proposed minimization problem formulation to define material flow giving accurate approximate solutions. In order to valid the whole presented theory, comparisons are made using classical analyses based on the upper and lower bound theorems (obtained as consequences of the proposed optimization variational problem), the well-known slices method and a finite element modelling (FEM). A second application concerning the anisotropic formulation of a Coulomb friction law from a quadratic convex tribologic potential will be presented to define contact evolution during rectilinear sliding of a circular pion on a plane laminated thick plate surface along different orientations.

1. Introduction

During the last twenty years Prof. Adrian BEJAN of Duke University has developed theoretical and applied researches concerning the thermodynamic optimization of complex systems based on the principle called as constructal defining any system evolution by the maximization of entropy, minimization of generated entropy and minimization of resistance and losses related to local or global flows searching to maximize the speed to reach a stable equilibrium state. As mentioned in previous scientific works [1-2], the Constructal Theory of Professor A. BEJAN can be seen as a definition and resolution of a general problem of optimization under constraints taking into account the postulate than: *"for a finite-size system to persists in time (to live or to be able to survive), it must evolve in such a way that it provides easier access to the imposed (global) currents that flow through it"* as to facilitate access as much as possible under the constraints flows or which cross them minimizing the corresponding losses [1]. In addition, it is also observed that all system search to optimize the distribution of their imperfections in order to facilitate the flow and to minimize the local resistance or the required transformation powers. Numerous studies concerning thermal problems, fluid or porous media flow, nature, economic or societal behavior confirm this principle [2]. Starting from recent author's scientific works, this article proposes to consolidate a general mathematical proof of the "Principle" of Maximum Work used in materials plasticity through its definition as a theorem which results from the application of the Contructal Principle seen as being a more general postulate from a thermodynamic point of view and valid for any type of continuum media [3-5]. It is well known that the Maximum Work Principle (MWP) is used by plasticity theory [6] to obtain associated flow laws or to define constitutive equations related to describe the material's strength and to develop analytical or numerical computations in particular during forming processes. A lot of previous works [7-8] try only to explain and to postulate the Maximum Working Principle by a purely phenomenological approach regarding polycrystalline metals starting from local slips of atomic particles planes having a maximum density. By exploiting the Virtual Powers Principle (VPP) defining the mechanic dynamic equilibrium if is applied the Constructal Theory, the Maximum Work Principle can be obtained as a direct consequence of a general variational optimization problem which search to minimize the losses of material flow, see to minimize the sum of all the dissipations energies. After theoretical backgrounds accompanied by their consequences concerning the convexity and the normality rule law of rheological and tribological potentials describing the bulk material behavior and contact interfaces, will be recalled the theorems of the upper and lower bound together with applications concerning a plane compression and a cylindrical crushing taking into account a rigid-plastic behavior with a Tresca plastic friction law. A validation of the proposed analyzes together with all the obtained analytical solutions will be made through comparisons between the estimation of the compression loads evolution and the corresponding numerical results given by Finite Element Modeling using Forge2®.

2. Theoretical framework

According to the theory of continuum mechanics for all material flow in a well-defined body volume Ω, the dynamic mechanical equilibrium can be written from the Virtual Powers Principle (VPP):

$$\int_{\Omega}[\sigma]:[\dot{\varepsilon}^*]dV+\int_{\partial\Omega}-\vec{\tau}\cdot\Delta\vec{v}^*\,dS=\int_{\partial\Omega'}\vec{T}\cdot\vec{v}^d dS'+\int_{\partial\Omega''}\vec{T}^d\cdot\vec{v}^* dS''+\int_{\Omega}\rho\vec{f}\cdot\vec{v}^* dV-\int_{\Omega}\rho\frac{d\vec{v}^*}{dt}\cdot v^*dV \quad (1)$$

It is known that this principle, equivalent to Newton's fundamental law, is valid for any field of admissible virtual speeds $\vec{v}^*$ respecting the boundary conditions taking into account the Cauchy stress tensor $[\sigma]$, the virtual strain rate tensor $[\dot{\varepsilon}^*]=\frac{1}{2}\{[grad(\vec{v}^*)]+[grad(\vec{v}^*)]^T\}$, the friction stress vector $\vec{\tau}$ acting on the contact surface or on the surface of velocities discontinuities $\partial\Omega$, the specific loading $\vec{T}$, the imposed velocities $\vec{v}^d$ on the body border part $\partial\Omega'$, the specific imposed loads $\vec{T}^d$ on the body border part $\partial\Omega''$, the specific mass forces $\vec{f}$ and the specific density ρ. For the real velocity field $\vec{v}$ of material flow the dynamic equilibrium is described by the Power Work Principle (PWP):

$$\int_{\Omega}[\sigma]:[\dot{\varepsilon}]dV+\int_{\partial\Omega}-\vec{\tau}\cdot\Delta\vec{v}\,dS=\int_{\partial\Omega'}\vec{T}\cdot\vec{v}^d dS'+\int_{\partial\Omega''}\vec{T}^d\cdot\vec{v}dS''+\int_{\Omega}\rho\vec{f}\cdot\vec{v}dV-\int_{\Omega}\rho\frac{d\vec{v}}{dt}\cdot\vec{v}dV \quad (2)$$

2.1. Theorem of « Maximal Work Principle »

Using the constructal law, in the case of a deformable material or during a forming process, we can postulate that the flow of the material takes place such that it minimizes the sum of the dissipated powers corresponding to the flow of the material, to the friction at the contact interfaces or velocities discontinuities and to the imposed loads. Consequently, the real values of all the kinematic and mechanical variables (velocities, stresses, strains, strain rates) are those which minimize the total power of dissipation. Thus, for any state of material defined by virtual mechanics variables, the real ones must minimize the functional of total dissipation power. It can be then concluded that, for all plastic materials (metallic or non-metallic, fluids or polymers) and for any virtual state, the real strain rate tensor $[\dot{\varepsilon}]$ and the real Cauchy stress tensor $[\sigma]$ corresponding to the material flow can be obtained by minimizing the sum of the dissipated plastic power, of the friction power or of the velocities discontinuities and of the imposed loads power. In addition, the plasticity of the material is generally governed in terms of the stress tensor and the friction vector through the definition of a plastic criterion defined by a multi-variables scalar function $\Phi_p([\sigma])=0$ together with a similar form regarding the friction term i.e. $\Psi_f(\vec{\tau}) = 0$. In this case, from all other admissible virtual speed fields v^* (different with respect to the real one) characterized by a virtual strain rate tensor $[\dot{\varepsilon}^*] \neq [\dot{\varepsilon}]$ and a virtual plastic constraints tensor $[\sigma^*]$ ($\Phi_p([\sigma^*]) = 0$ et $\Psi_f(\vec{\tau}^*) = 0$), the real material plastic flow state can be obtained by minimizing the functional defined through the virtual total dissipated power:

$$\dot{W}_d = Min(\dot{W}_d^*) \; with \; \dot{W}_d^* = \int_{\Omega}[\sigma^*]:[\dot{\varepsilon}^*]dV + \int_{\partial\Omega} -\vec{\tau}^* \cdot \Delta\vec{v}^* \, dS + \int_{\partial\Omega''} -\vec{T}^d \cdot \vec{v}^* \, dS'' \tag{3}$$

Using the boundary conditions concerning the kinematics and the loading of the material, a variational minimization problem under constraints is then obtained. As shown by the Constructal Theory the optimization solution is obtained for the real flow state. So for all virtual states it is required to have the following inequality:

$$\int_{\Omega}[\sigma]:[\dot{\varepsilon}]dV + \int_{\partial\Omega} -\vec{\tau} \cdot \Delta\vec{v} \, dS + \int_{\partial\Omega''} -\vec{T}^d \cdot \vec{v} \, dS'' \leq \int_{\Omega}[\sigma^*]:[\dot{\varepsilon}^*]dV + \int_{\partial\Omega} -\vec{\tau}^* \cdot \Delta\vec{v}^* \, dS + \int_{\partial\Omega''} -\vec{T}^d \cdot \vec{v}^* \, dS'' \tag{4}$$

For consistent materials and quasi-static conditions, the mass and the inertial forces can be neglected, so the VPP principle can be then written in the following simplified form:

$$\int_{\Omega}[\sigma]:[\dot{\varepsilon}]dV + \int_{\partial\Omega} -\vec{\tau} \cdot \Delta\vec{v} \, dS + \int_{\partial\Omega''} -\vec{T}^d \cdot \vec{v} \, dS'' = \int_{\Omega}[\sigma]:[\dot{\varepsilon}^*]dV + \int_{\partial\Omega} -\vec{\tau} \cdot \Delta\vec{v}^* \, dS + \int_{\partial\Omega''} -\vec{T}^d \cdot \vec{v}^* \, dS'' \tag{5}$$

Starting from (4) and (5), it can be concluded that for any virtual stress state $[\sigma^*]$ ($\Phi_p([\sigma^*]) = 0$) or any virtual friction $\vec{\tau}^*$ ($\Psi_f(\vec{\tau}^*) = 0$) and for any body Ω is obtained the below equivalent form:

$$\int_{\Omega}([\sigma^*]-[\sigma]):[\dot{\varepsilon}^*]dV + \int_{\partial\Omega'} (\vec{\tau}-\vec{\tau}^*) \cdot \Delta\vec{v}^* \, dS' \geq 0 \tag{6}$$

This condition requires having positive values for each term i.e.:

$$([\sigma^*]-[\sigma]):[\dot{\varepsilon}^*] \geq 0, \forall [\sigma^*], \Phi_p([\sigma^*])=0 \text{ and } -(\vec{\tau}^*-\vec{\tau}) \cdot \Delta\vec{v}^* \geq 0, \forall \vec{\tau}^*, \Psi_f(\vec{\tau}^*)=0 \tag{7}$$

Indeed if one considers that there exists a virtual state of the stresses or friction for which the terms are negative, one could build another identical virtual state with real state of stresses outside the field for which the terms of the integrals are negative (therefore with zero integrals values outside this domain) and one would obtain a negative value for the inequality (6) which contradicts the positive value to be respected. In the opposite sense, taking into account the real plastic flow characterized by the velocity field v, the strain rate tensor $[\dot{\varepsilon}]$ and the Cauchy stress tensor $[\sigma]$, any other state of the admissible stresses must also verify the following inequalities:

$$\left([\sigma]-\left[\sigma^*\right]\right):[\dot{\varepsilon}]\geq 0,\forall\left[\sigma^*\right],\Phi_p\left(\left[\sigma^*\right]\right)=0 \text{ and } -\left(\vec{\tau}-\vec{\tau}^*\right)\cdot\Delta\vec{v}\geq 0,\forall\vec{\tau}^*,\Psi_f\left(\vec{\tau}^*\right)=0 \quad (8)$$

Starting from the above obtained relationships (7) and (8), it is possible to prove mathematically the convex shape of the potential functions defining the plastic and friction criteria and subsequently the known property of normality to the criterion. Consequently the strain rate should be proportional to the gradient of the plastic criterion with respect to each components of the stress i.e .:

$$[\dot{\varepsilon}]=\lambda_p\partial\Phi_p/\partial[\sigma],\lambda_p\geq 0 \quad (9)$$

In the same sense the associated friction law can be written in the form:

$$\Delta\vec{v}=-\lambda_f\partial\Psi_f/\partial\vec{\tau},\lambda_f\geq 0 \quad (10)$$

In this case, based on the property of the convexity of plastic criteria $\Phi_p\left(\left[\sigma^*\right]\right)$ and $\Psi_f\left(\vec{\tau}^*\right)$, the two inequalities expressed by (8) can be extended to the virtual stress and friction shear respecting $\Phi_p\left(\left[\sigma^*\right]\right)\leq 0$ and $\Psi_f\left(\vec{\tau}^*\right)\leq 0$. It is then possible to conclude that for any state of virtual stresses is obtained:

$$\left([\sigma]-\left[\sigma^*\right]\right):[\dot{\varepsilon}]\geq 0,\forall\left[\sigma^*\right],\Phi_p\left(\left[\sigma^*\right]\right)\leq 0 \text{ and } -\left(\vec{\tau}-\vec{\tau}^*\right)\cdot\Delta\vec{v}\geq 0,\forall\vec{\tau}^*,\Psi_f\left(\vec{\tau}^*\right)\leq 0 \quad (11)$$

The first inequality is known in the plasticity theory as the Maximum Work Principle [4-8]. It is proved here that this can be obtained as a consequence of Constructal Theory [1], [3] and of the Principle of Virtual Powers. The second inequality reflects the same principle applied to the friction stress state.

2.2. Practical Consequences and Synthesis

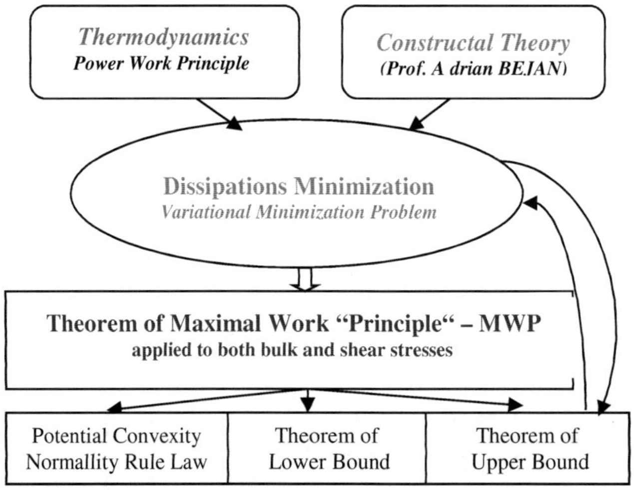

Figure 1. Flowchart concerning the formulation of the dissipation variational minimization problem based on the Constructal Theory and the proof of the "Principle" of Maximum Work (MWP) together with the characteristic properties and theorems [3-5]

Using the theoretical proofs presented above, it can be concluded that the Maximum Work Principle can be applied for any type of continuous media: fluid, solid or pasty material, metallic or non-metallic, as well as to define in a similar form the principle of maximum power concerning the friction stresses. Specific expressions concerning the convex criterion of plasticity or of friction $\Phi_p([\sigma])=0$, $\Psi_f(\vec{\tau})=0$ can be proposed and used to define the isotropic or anisotropic plastic behavior of materials, respectively of the constitutive laws defining sliding at the contact interfaces. It is also possible to obtain and prove the theorems of the Lower and Upper Bound (Figure 1) [6-8].

These can also be considered as variational optimization formulations and are frequently used to obtain analytical estimations of the loads applied during a material forming process under particular assumptions. They are also used to compare and validate the results obtained by Finite Element Modeling (FEM). Regarding the Upper Bound Theorem [8], it is observed that it is indeed obtained an equivalent form of the variational minimization problem (3). One can thus conclude on the coherence and especially on the equivalence between the Maximum Work Principle and the Principle of Minimization of the total dissipations or losses defined by the Constructal Theory.

3. Applications: plane compression and cylindrical crushing

In order to validate the variational minimization principle of dissipations expressed by the optimization problem (3) postulating the real state of a material plastic deformation, two cases of simple forming are considered: plane compression and cylindrical crushing (Figure 2a and Figure 2b).

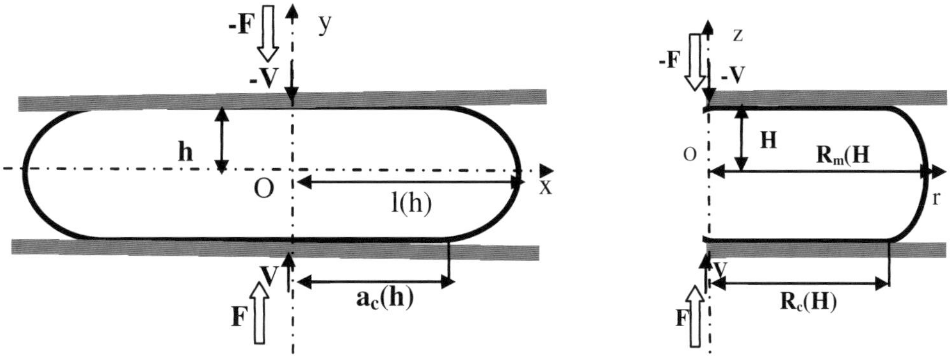

Figure 2. Simple forming operations [5]: a) plane compression, b) cylindrical crushing

It is assuming a material with a perfect rigid-plastic behaviour defined by a constant value of the equivalent Von-Mises stress σ_0 (i.e. $\Phi_p\left(\left[\sigma^*\right]\right)=\bar{\sigma}^*-\sigma_0=0$). In view of large plastic deformations generated during a plane compression or crushing process on a double-action press at constant speed V, a Tresca friction is taken into account on the specimen-tool surface contacts $\tau=\bar{m}\sigma_0/\sqrt{3}$ with $\bar{m}\in[0,1]$. Corresponding to the coordinate systems of Figure 2, taking into account the axial, respectively revolution symmetry, for each time instant t the geometry of the plane compression parallelepiped sample is defined by a transverse surface S_t = $2a$ x $2h$ and a longitudinal surface S_p = $2a$ x L considering $a\approx l(h)\approx a_c(h)$, while in the case of cylindrical crushing the specimen is defined by a cylindrical surface S_c of radius $R\approx R_m(H)\approx R_c(H)$ and a height $2H$. To be able to exploit the variational principle expressing the real state of the flow through the minimization problem (3), it can be defined a virtual kinematics through an incompressible velocity field ($div(\vec{v}^*)=0$) and kinematically admissible (in accordance with the boundaries conditions defined in Figure 2) by expressing the velocities components using the below expressions proposed by Avitzur in the case of a plane compression respectively of a cylindrical crushing [9]:

$$
\begin{aligned}
& v_x^*(x,y)=AV\frac{x}{h}\exp(-\beta y/h) && v_r^*(r,z)=A'V\frac{r}{H}\exp(-\beta z/2H) \\
& v_y^*(y)=AV\left[\exp(-\beta y/h)-1\right]/\beta && v_z^*(z)=4A'V\left[\exp(-\beta z/2H)-1\right]/\beta \\
& v_x^*(0,y)=0,\ v_y^*(0)=0,\ v_y^*(h)=-V && v_r^*(0,z)=0, v_\theta(r,z)=0,\ v_z^*(0)=0,\ v_z^*(H)=-V \\
& A=\beta/[1-\exp(-\beta],\ x\in[0,a],\ y\in[0,h] && A'=\beta/4[1-\exp(-\beta/2)],\ r\in[0,R],\ z\in[0,H]
\end{aligned}
\tag{12}
$$

Here β represents a dimensionless parameter varying the shape of the virtual velocity field. Based on the variational problem formulated by (3), the real total dissipations power $\dot{W}_d$ must verify:

$$\dot{W}_d = Min\ \dot{W}_d^* \le Min\ \dot{W}_d^*(\beta) \tag{13}$$

Since we have $\vec{T}^d = 0$ the subfamily of the virtual total power of the dissipations is defined by:

$$\dot{W}_d^*(\beta) = \int_\Omega [\sigma^*]:[\dot{\varepsilon}^*]dV + \int_{\partial\Omega} -\vec{\tau}^* \cdot \Delta\vec{v}^*\, dS = \int_\Omega \sigma_0 \dot{\bar{\varepsilon}}^* dV + \int_{\partial\Omega} (\bar{m}\sigma_0/\sqrt{3}) \cdot |\Delta\vec{v}^*| dS \tag{14}$$

where:

$$\dot{\bar{\varepsilon}}^* = \sqrt{(2/3)Trace\{[\dot{\varepsilon}^*]^2\}} \text{ and } [\dot{\varepsilon}^*] = \frac{1}{2}\{[grad(\vec{v}*)] + [grad(\vec{v}*)]^T\} \tag{15}$$

The summary of the analytical computations of all the terms involved by the evaluation of $\dot{W}_d^*$ is presented in Table 1 using the assessments detailed by the author in previous works [3-5] on the basis of design principles and material forming theory presented in [7-11]. Given the complexity of analytical evaluation the β_{opt} which minimize $\dot{W}_d^*(\beta)$, this study is limited to the necessary condition $\dot{W}_d \le Min\ \dot{W}_d^*(\beta) \le \dot{W}_d^*(0)$.

Table 1. Definition of the slenderness factor e, of the shape functions with respect to the dimensionless kinematic parameter β, of the total reference power P and of the functional expressing the virtual power of dissipations $\dot{W}_d^*$ [5]

		Plane Compression	Cylindrical Crushing
e (slenderness)		a/h	R/H
V' (volume)		$2ahL$	$\pi R^2 H$
Sp, Sc (area)		$2aL$	πR^2
$\alpha(\beta)$	$\beta \ne 0$	$\beta e/2$	$\lvert\beta\rvert e/2$
	$\beta \to 0$	0	0
$\phi[\alpha(\beta)]$	$\beta \ne 0$	$\sqrt{1+\alpha(\beta)^2} + \frac{\ln\left[\alpha(\beta)+\sqrt{1+\alpha(\beta)^2}\right]}{\alpha(\beta)}$	$8\alpha(\beta)\left[\left(\frac{1}{12}+\alpha(\beta)^{-2}\right)^{3/2} - \left(\alpha(\beta)^{-2}\right)^{3/2}\right]$
	$\beta \to 0$	2	1
$\varphi(\beta)$	$\beta \ne 0$	$\beta\exp(-\beta)/[1-\exp(-\beta]$	$2(\beta/2)\exp(-\beta/2)/[1-\exp(-\beta/2]$
	$\beta \to 0$	1	2
β_{opt}		Condition nécessaire $2\frac{d\phi}{d\beta} + \bar{m}e\frac{d\varphi}{d\beta} = 0 \Leftrightarrow \frac{d\phi}{d\alpha} + \bar{m}\frac{d\varphi}{d\beta} = 0$	Condition nécessaire $2\frac{d\phi}{d\beta} + \frac{\bar{m}e}{3\sqrt{3}}\frac{d\varphi}{d\beta} = 0 \Leftrightarrow \frac{\beta}{\lvert\beta\rvert}\frac{d\phi}{d\alpha} + \frac{\bar{m}}{3\sqrt{3}}\frac{d\varphi}{d\beta} = 0$
P (Ref. Power)		$\frac{4}{\sqrt{3}}\sigma_0 aLV = \frac{2}{\sqrt{3}}\sigma_0 S_p V'$	$\sigma_0 \pi R^2 V = \sigma_0 S_c V'$
$\dot{W}_d^*(\beta)$		$P[\phi[\alpha(\beta)] + \bar{m}e\varphi(\beta)/2]$	$P[2\phi[\alpha(\beta)] + \bar{m}e\varphi(\beta)/3\sqrt{3}]$
$\dot{W}_d^*(0)$		$\dot{W}_d^*(0) = 2P\left[1+\bar{m}\frac{e}{4}\right]$	$\dot{W}_d^*(0) = 2P\left[1+\bar{m}\frac{e}{3\sqrt{3}}\right]$

Thus concerning the plane compression the real load F must verify:

$$F \le \tilde{F} = \frac{\dot{W}_d^*(0)}{2V} = \frac{4aL\sigma_0}{\sqrt{3}}[1+\bar{m}e/4] = \frac{2\sigma_0 S_p}{\sqrt{3}}[1+\bar{m}e/4] \tag{16}$$

Starting from the lower bound theorem, using a specific statically admissible virtual stress state $Div([\sigma]^*) = 0)$ [3], while relying on the virtual power principle and on the MWP as a consequence of the minimization problem (3), it can be written:

$$F \geq \tilde{\tilde{F}} = \frac{4aL\sigma_0}{\sqrt{3}}\left[\sqrt{1-\bar{m}^2} + \bar{m}e/4\right] = \frac{2\sigma_0 S_p}{\sqrt{3}}\left[\sqrt{1-\bar{m}^2} + \bar{m}e/4\right] \quad (17)$$

It is easy to see that the real compressive load could be approximated by the mean value $F \approx (\tilde{F} + \tilde{\tilde{F}})/2$ with a maximal estimation error of $(1-\sqrt{1-\bar{m}^2})/(1+\bar{m}e/4)$ with $\bar{m} \in [0,1]$.

In the particular case of a compression without friction ($\bar{m}=0$) $\tilde{F} = \tilde{\tilde{F}}$ and $F = \dfrac{4aL\sigma_0}{\sqrt{3}} = \dfrac{2}{\sqrt{3}}\sigma_0 S_p$.

This expression is the same with that obtained by an analytical study based on the resolution of the differential incompressibility equation (defining the real velocities by $v_x(x) = Vx/h$, $v_y(y) = -Vy/h$), on the static equilibrium equation and of the law of normality of the potential rigid-plastic (giving the stresses defined by $\sigma_{xx} = 0, \sigma_{yy} = -2\sigma_0/\sqrt{3}, \sigma_{zz} = -\sigma_0/\sqrt{3}$) [6]. It is important to note that in this case, by choosing $\beta = 0$ (because as can be seen in the Table 1 $\beta_{opt} = 0$ if $\bar{m} = 0$), one finds practically the overall minimum of the expression (13), which proves well that the real flow is the solution of the variational optimization problem (3). A similar analysis for the cylindrical crushing gives:

$$F \leq \tilde{F} = \frac{\dot{W}_d^*(0)}{2V} = \pi R^2 \sigma_0\left[1 + \bar{m}e/3\sqrt{3}\right] = \sigma_0 S_c\left[1 + \bar{m}e/3\sqrt{3}\right] \quad (18)$$

Starting from the detailed results in [4] it is obtained:

$$F \geq \tilde{\tilde{F}} = \pi R^2 \sigma_0\left[\sqrt{1-\bar{m}^2} + \bar{m}e/3\sqrt{3}\right] = \sigma_0 S_c\left[\sqrt{1-\bar{m}^2} + \bar{m}e/3\sqrt{3}\right] \quad (19)$$

The average crushing load may be defined by $F \approx (\tilde{F} + \tilde{\tilde{F}})/2$ with a maximal estimation error around $(1-\sqrt{1-\bar{m}^2})/\left(1+\bar{m}e/3\sqrt{3}\right)$. For a crushing without friction ($\bar{m}=0$) $F = \tilde{F} = \tilde{\tilde{F}} = \pi R^2 \sigma_0$ i.e $F = \sigma_0 S_c$. Similarly, it is found that this coincides with the analytical solution (real velocity defined by $v_r(r) = Vr/H$, $v_z(z) = -Vz/H$ and diagonal stress tensor $\sigma_{rr} = 0, \sigma_{\theta\theta} = 0, \sigma_{zz} = -\sigma_0$) [6]. The same conclusion is reached i.e. we have practically $\beta_{opt} = 0$ what again justifies the achievement of the overall minimum of the optimization problem (3), see the condition expressed by (13).

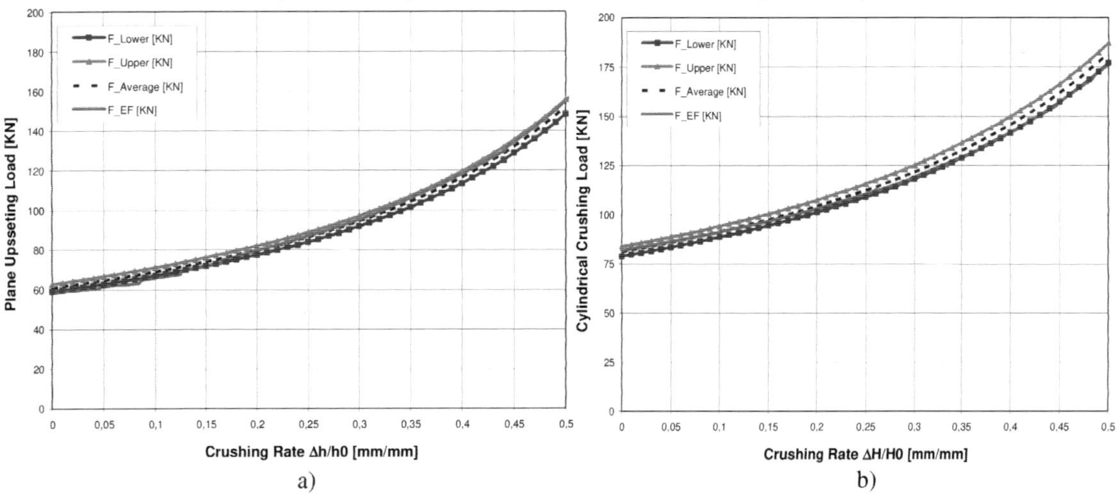

Figure 3. Comparisons of the compression load-crushing rate evolutions obtained by analytical and numerical estimations: a) plane compression, b) cylindrical crushing

To illustrate the degree of accuracy of the analytical estimates of the forming forces compared to a Finite Element (EF) modeling using the Forge2® software, a parallelepiped sample of dimensions 20 mm x 20 mm x20 mm with an initial slenderness e_0 = 1 and a cylindrical specimen of equivalent volume defined by a radius of 10 mm, a height of 20 mm and the same initial slenderness $e_0 = 1$. Assuming $\sigma_0 = 250$ MPa and an average friction defined by $\bar{m} = 0.35$ the load-stroke evolution curves are shown in Figure 3a and Figure 3b. It is easy to see that in both cases (plane compression and cylindrical crushing) the EF solution of the evolution of the load is well between the upper $\tilde{F}$ and lower $\tilde{\tilde{F}}$ theoretical limits with values very close to those given by the estimated average $F \approx (\tilde{F} + \tilde{\tilde{F}})/2$ (error less than 1%) while the load estimation error is maximum 6.5%. for a crushing rate of 0,5 with a logarithmic deformation of about 70%, seen by the difference of the variation between the lower and upper limit.

In the ideal case without friction the analytical curves are superposed and have values practically equal to the numerical finite elements ones. Furthermore, for an identical initial slenderness and an almost equivalent material volume (with a ratio of $4/\pi$) the cylindrical crushing force have values approximately 1.15-1.25 times greater than in the case of a plane compression, thus regaining the ratio $2/\sqrt{3}$ between the two modes of deformation.

4. Conclusions

During the content of this article, it was proved that for a plastic flow of all continuous media, starting from the general principle defined by the Constructal Theory concerning the minimization of the total dissipated energy characterizing the evolution of a finite size system, in particular a deformable material body, one obtains and one proven in a more general form the "Principle" of Maximum Work both with regard to the state of material bulk stresses and that of friction ones. The application for a plane compression and a cylindrical crushing of a rigid-plastic material shows the feasibility of the proposed formulation through the solve of a corresponding variational minimization problem expressed in terms of total dissipated energy, with possibilities to obtain analytical estimations of the required material forming loads evolution. Thus the expressions and the optimal analytical values of the upper and lower limits corresponding to the evolution of the compression load must be seen as a consequence of the proposed variational optimization problem obtained as a consequence of the Constructal Theory. The comparisons with the simulation results obtained by finite element modeling show the feasibility and especially the high degree precision of obtained analytical values.

5. References

[1] A. Bejan 2000 *Shape and Structure: From Engineering to Nature* (UK: Cambridge Univ.)
[2] A. Bejan, S. Lorente 2008 *Design with Constructal Theory* (New Jersey: John Willey & Sons)
[3] A. Gavrus 2017 The Maximum Work Principle Regarded as a Consequence of an Optimization Problem Based on Mechanical Virtual Power Principle and Application of Constructal Theory, *AIP Conf. Proc.* 1896, 100009-1-6
[4] A. Gavrus 2018 Application of Constructal Theory to Write Mechanical Maximum Work Principle and Equilibrium State of Continuum Media Flow as a Solution of a Variational Optimization Problem *Int. J. of Modeling and Optimization* vol. 8(4) p 227-231
[5] A. Gavrus 2019 Principe du Travail Maximal en Mécanique des Milieux Continus et Plasticité Regardé Comme Conséquence et Solution d'un Problème d'Optimisation Variationelle en Utilisant la Théorie Constructale *24ème Congrès Française de Mécanique* (Brest : France).
[6] G. A. Maugin 1992 *The thermodynamics of plasticity and fracture* (UK: Cambridge Univ.)
[7] M. Bellet, J-L. Chenot, L. Fourment, E. Massoni, P. Montmitonnet 1994 *Plasticity Seminary: Elements Finis et Mise en Forme des Métaux*, (Sophia-Antipolis : ENSMP – Mines Paris'Tech)
[8] E. Felder 2007 *Plasticité en mise en forme – Comportement rigide-plastique* (Paris : Tech. Ing.)
[9] B. Avitzur 1968, *Metal Forming: Processes and Analysis* (New-York: McGraw-Hill)
[10] A. Gavrus, H. Francillette 2003, Identification of Anisotropic Hill Criteria from the Channel Die Compression Test. Application to a Normal Anisotropy of Zirconium 702α *J. de Phys. IV (European Phys. J.)* vol. 105 (Paris: EDP Sciences) p 11-18
[11] R. Ebrahimi and A. Najafizadech 2004, A new method for evaluation of friction in bulk metal forming *J. of Mat. Proc. Tech.* vol. 152 (Amsterdam: Elsevier) p 136-143

IOP Conf. Series: Materials Science and Engineering **968** (2020) 012004 doi:10.1088/1757-899X/968/1/012004

Profiling of the flat-tool for manufacturing worms with circular eccentric profile by cold forming

G Frumuşanu, V Teodor and N Oancea
"Dunărea de Jos" University, Manufacturing Engineering Department,
Domnească Street 111, 800201 – Galaţi, Romania

E-mail: gabriel.frumusanu@ugal.ro

Abstract. The incremental pumps are encountered in pharmaceutical and cosmetics industries. One of the most representative categories of these pumps is formed by the progressing cavity pumps. The helical surfaces with circular profile, eccentrically positioned relative to worm axis, are frequently used in their rotors construction. In this specific case, the constructive dimensions of the worms allow their manufacturing by cold forming, with flat-tools similar to screwing dies. In this paper, an analytical algorithm for profiling such flat-tool is proposed. The algorithm starts from the analytical definition of the helical cylindrical surface with constant pitch of the generated worm, and it lays on the "Minimum distance method". The equations of the gearing surface between worm and flat-tool are found, and, based on this, the active surface of the tool is determined. A MatLab application developed for implementing the profiling algorithm is presented. The results obtained after running the application in the case of a particular generated surface are also included.

1. Introduction

Progressing cavity pumps belong to the group of rotating positive displacement pumps. They are self-priming, valve-less, and due to their high process liability and suction capacity they are often used for the continuous, gentle conveyance and precise dosing - in proportion to speed - of difficult media [1].

Among their principal benefits, one can mention:

- Flow in proportion to speed;
- Precise dosing;
- Continuous, gentle and low-pulsation flow;
- Large free ball passage (up to 150 mm);
- High suction capacity;
- Low life-cycle costs due to long lifetime and meantime between failure;
- Reverse rotation and flow.

In what concerns their functioning, the delivery principle is based on a rotor, 1, which turns in an oscillating motion within a fixed stator, 2, see figure 1. The coordinated, spiral geometry of both components creates delivery chambers in which the medium is transported from the suction to the pressure side. Due to the design of rotor and stator, hardly any pulsation or shearing forces are affecting the fluid. Instead, the medium is transported gently and continuously. Consistency and especially the viscosity are not a factor for the delivered flow with this displacement technology; the transported quantity is determined by the speed alone and – in combination with a frequency inverter –

can be regulated conveniently and with precision. This enables an accuracy of between five and three percent to be achieved, while small dispensers even achieve one percent.

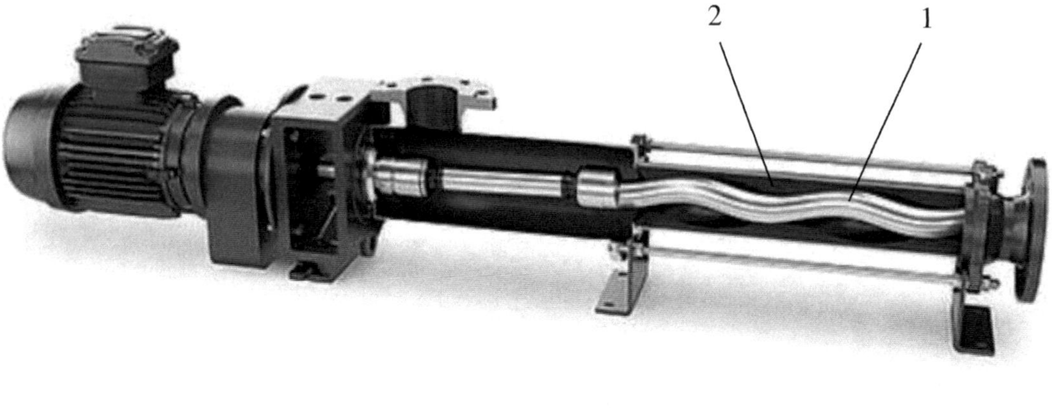

Figure 1. Progressing cavity pump [1].

The cylindrical helical surface with constant pitch of the progressing cavity pump rotor can be generated by milling, with disc-tools having specific profile, or by slotting, with cutters having their active surface delimited by cylindrical surfaces. The profiles of the mentioned types of tools can be found on the base of Nikolaev theorem [2-5].

Another solution for machining the progressing cavity pump rotor is turning with rotating cutter, which is a highly productive process. In this case, the tool specific profile can be determined by "Minimum distance" method [5], applied for finding profiles associated to rolling centrodes.

The "Minimum distance" method resulted from a new approach of well-known Willis theorem [3], applied in the case of profiles associated to a couple of rolling centrodes (C_1, C_2, see figure 2). According to the method, the envelop of a profile associated to a couple of rolling centrodes is the locus of the profile points for which, in the successive rolling positions, the distance *d* to gearing pole *P* (meaning the point of tangency between the centrodes) is minimum.

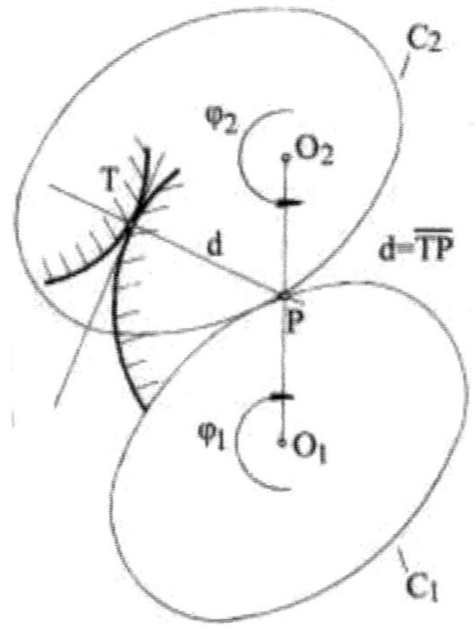

Figure 2. Conjugated profiles & rolling centrodes [5].

The threads manufacturing by cold forming [6, 7] presents some notable benefits relative to their cutting, such as smoother flanks, higher resistance to wear and fatigue, higher productivity. Flat or

IOP Conf. Series: Materials Science and Engineering **968** (2020) 012004 doi:10.1088/1757-899X/968/1/012004

cylindrical tools can be used in this purpose. In both cases, the active surface of the tool is the envelop of thread cylindrical helical surface, in the specific generating motion.

In this paper, an analytical algorithm for profiling such flat-tool, enabling the manufacturing of progressing cavity pump rotor is proposed. The algorithm starts from the analytical definition of the helical cylindrical surface with constant pitch of the generated worm, and it lays on the "Minimum distance method". In the next section, the equations of the helical surface of mentioned pump rotor are deducted. The third section is dedicated to profiling of the flat-tool active surface. The fourth section presents a numerical application, based on a MatLab application, developed for implementing the profiling algorithm. The last section is for conclusion.

2. Equations of the pump rotor surface

The equations defining the helical surface of the pump rotor can be obtained after defining the following reference systems (figure 3):

- *xyz*, meaning the global system, fix and having its origin *O* onto rotor axis,
- $X_1Y_1Z_1$ – local system, having the origin O_1 into the center of the eccentrical circle, and
- *XYZ* – local system, having the origin *O* onto rotor axis and executing a helical motion of φ – angular and *p* – helical parameters, along rotor axis.

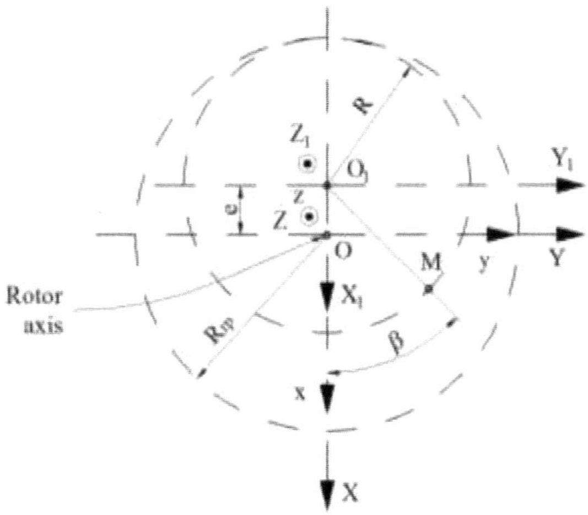

Figure 3. The eccentrical circle generating the helical surface of the rotor.

The co-ordinates of the generic point from the eccentrical circle are:

$$\begin{vmatrix} X_1 = -e + R\cos\beta\,; \\ Y_1 = R\sin\beta\,, \end{vmatrix} \tag{1}$$

where β means an angular variable, $\beta \in [0, 2\pi]$.

The helical motion of XYZ reference system has the following equation, written with matrices:

$$\begin{pmatrix} X \\ Y \\ Z \end{pmatrix} = \begin{pmatrix} \cos\varphi & -\sin\varphi & 0 \\ \sin\varphi & \cos\varphi & 0 \\ 0 & 0 & 1 \end{pmatrix} \begin{pmatrix} -e + R\cos\beta \\ R\sin\beta \\ 0 \end{pmatrix} + \begin{pmatrix} 0 \\ 0 \\ p\cdot\varphi \end{pmatrix}. \tag{2}$$

After development and mathematical calculus, relation (2) becomes:

$$\Sigma \begin{vmatrix} X = -e\cos\varphi + R\cos(\beta+\varphi)\,; \\ Y = -e\sin\varphi + R\sin(\beta+\varphi)\,; \\ Z = p\cdot\varphi\,, \end{vmatrix} \tag{3}$$

meaning the equations of rotor helical surface, Σ.

3. Profiling of the flat tool active surface

The manufacturing of progressing cavity pump rotor by cold forming with flat-tools is accomplished by rolling the cylindrical workpiece between a couple of profiled flat tools (see figure 4). Corresponding to this situation, a cylindrical axoid A_1, of R_{rp} radius is considered for the workpiece, while each tool has a plane as axoid, A_2 and A_2', respectively. The A_1 axoid executes a rotation motion of φ_1 parameter (figures 4 and 5), while A_2 and A_2' have translation motions, of λ parameter. The rolling motion requires to keep always the rolling condition:

$$\lambda = R_{rp} \cdot \varphi_1 \tag{4}$$

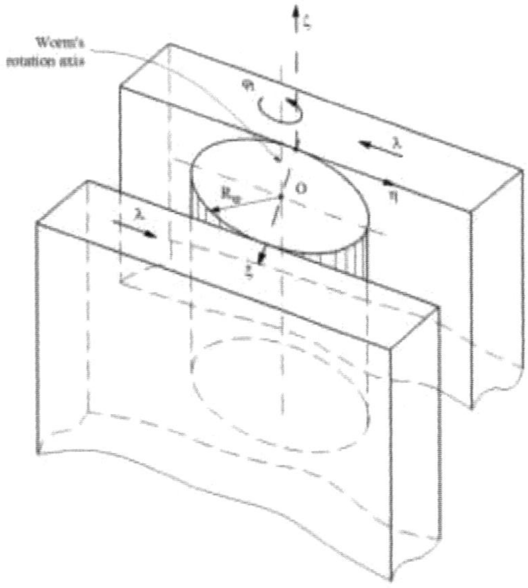

Figure 4. Manufacturing scheme.

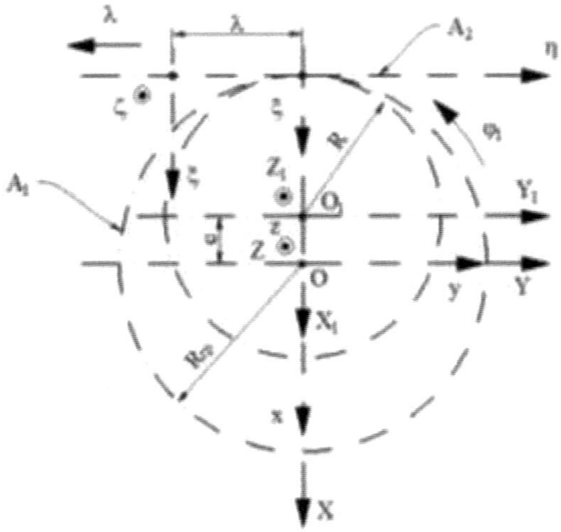

Figure 5. Transversal section of rolling axoids.

Because both flat-tools have identical profile, only one of them is further considered, together with its axoid, A_2, as depicted in figure 5. The active surface of the flat-tool results as envelop of the surfaces family generated in the motion of rotor surface Σ relative to the local system $\xi\eta\varsigma$, attached to A_2 axoid.

The equation of workpiece motion relative to considered flat-tool is:

$$\xi = \omega_3{}^T(\varphi) \cdot X - a\,, \quad \text{with } a = \begin{pmatrix} -R_{rp} \\ -R_{rp} \cdot \varphi \\ 0 \end{pmatrix}. \tag{5}$$

In relation (5) ω_3 is the well-known matrix of coordinates transform for rotation around Oz axis. If here X means the vector formed by the coordinates of the generic point from Σ surface, calculated with (3), then (5) gives the equations of the surfaces family generated in the motion of rotor surface, depending on the parameters β, φ and φ_1. The envelop of this family results by associating to (5) the enveloping condition, which, according to Gohman theorem [8] has the expression:

$$\begin{vmatrix} \xi_\beta{}' & \eta_\beta{}' & \zeta_\beta{}' \\ \xi_\varphi{}' & \eta_\varphi{}' & \zeta_\varphi{}' \\ \xi_{\varphi_1}{}' & \eta_{\varphi_1}{}' & \zeta_{\varphi_1}{}' \end{vmatrix} = 0. \tag{6}$$

In relation (6), ξ, η and ζ mean the components of the vector ξ calculated with (5). The ensemble formed by equation (5) and condition (6) defines the flat-tool active profile. As it can be noticed, the finding of profile in this manner is relative complicated from calculus point of view. Instead, we suggest a simplified profiling algorithm, consisting in two steps:

- The flat-tool profile is determined, at first, in a plane section, corresponding to $\varphi = 0$, in equation (5), by using a numerical method (e.g. the "Minimum distance" method, [5]) for finding the generated profile envelop. Thus, after imposing the $\varphi = 0$ condition, (5) becomes:

$$C_{\varphi_1}\begin{vmatrix}\xi = -e\cos\varphi_1 + R\cos(\varphi_1+\beta) + R_{rp};\\ \eta = -e\sin\varphi_1 + R\sin(\varphi_1+\beta) + R_{rp}\cdot\varphi_1.\end{vmatrix} \tag{7}$$

According to "Minimum distance" method, in the case of generating with the rack-tool (applicable when profiling the flat-tool for manufacturing by rolling) the enveloping condition has the shape:

$$\xi\cdot\xi'_\beta + (\eta - R_{rp}\cdot\varphi_1)\eta'_\beta = 0. \tag{8}$$

After replacing the expressions for ξ and η and some mathematical calculus, relation (8) becomes:

$$\varphi_1 = \sin^{-1}(e\sin\beta/R_{rp}) - \beta. \tag{9}$$

- The flat-tool 3D profile is generated by extruding the plane profile from above after a direction inclined with the angle α of the generatrix line of the tool cylindrical surface (see figure 6).

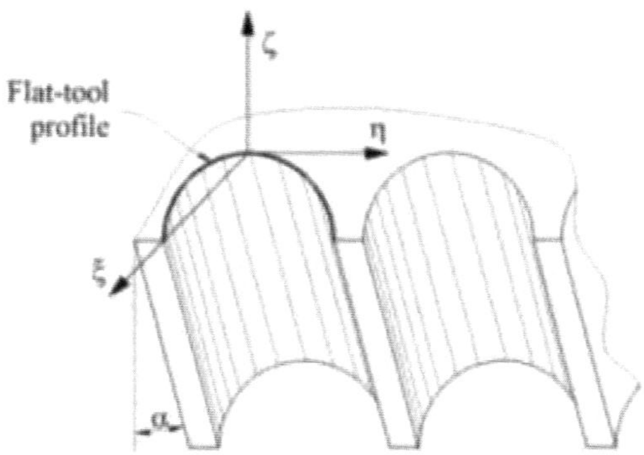

Figure 6. The generating of flat-tool 3D profile.

The angle α can be easily calculated with:

$$\alpha = \tan^{-1}(R_{rp}/p). \tag{10}$$

4. Numerical application

The feasibility of the algorithm for profiling the flat-tool that can be used for generating the rotor of progressing cavity pump was tested by running a numerical application, further presented. The application addresses the case of the rotor having $R = 10$ mm, $e = 3$ mm and $p = 10$ mm.

Table 1. Calculated coordinates of the points defining rotor surface (excerpt).

Crt. no.	X [mm]	Y [mm]	Z [mm]	Crt. no.	X [mm]	Y [mm]	Z [mm]
1	7	0	0	431	13	0	31.4159
2	6.6574	2.1631	3.1416	432	12.3637	4.0172	34.5575
3	5.6631	4.1145	6.2832		....		
4	4.1145	5.6631	9.4248	858	4.1145	-5.6631	53.4071
	....			859	5.6631	-4.1145	56.5487
429	10.5172	-7.6412	25.1327	860	6.6574	-2.1631	59.6903
430	12.3637	-4.0172	28.2743	861	7	0	62.8319

The coordinates of the points defining rotor surface were calculated with the help of a dedicated MatLab application, on the base of relation (3). Because the rotor profile is periodic, it was determined and drawn only for a length equal to its helical surface pitch. The value of β angle was discretized between 0 and 2π, with the increment of $\pi/20$, while the value of φ, in the same interval, with the increment of $\pi/10$. Hereby, the coordinates of 861 points from rotor surface were found, some of them sampled in Table 1. The shape of the corresponding rotor surface is depicted in figure 7.

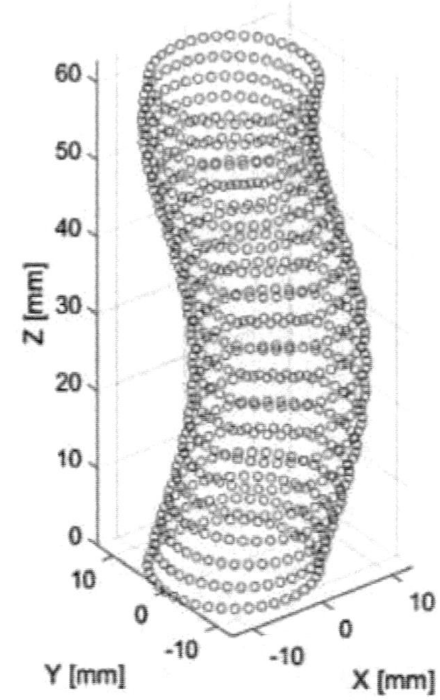

Figure 7. Profile of the pump helical rotor.

The flat-tool profile in the plane section corresponding to $\varphi = 0$ was then determined, with the help of another dedicated MatLab application, on the base of relations (7) and (9), by discretizing the values of β angle in 101 equidistant points between 0 and 2π. Some of the points belonging to flat-tool profile are sampled in Table 2, while the profile resulted by joining these points is depicted in figure 8.

Table 2. Calculated coordinates of the points defining flat-tool profile (excerpt).

Crt. no.	ξ [mm]	η [mm]	Crt. no.	ξ [mm]	η [mm]	Crt. no.	ξ [mm]	η [mm]
1	6	-40.8407	46	0.2457	-3.1706	91	5.7362	31.6670
2	5.9975	-39.9224	47	0.1583	-2.5282	92	5.7890	32.5825
3	5.9901	-39.0041	48	0.0895	-1.8912	93	5.8352	33.4986
4	5.9777	-38.0859	49	0.0399	-1.2585	94	5.8752	34.4153
5	5.9602	-37.1679	50	0.0100	-0.6285	95	5.9092	35.3325
6	5.9374	-36.2501	51	0	0	96	5.9374	36.2501
7	5.9092	-35.3325	52	0.0100	0.6285	97	5.9602	37.1679
8	5.8752	-34.4153	53	0.0399	1.2585	98	5.9777	38.0859
9	5.8352	-33.4986	54	0.0896	1.8912	99	5.9901	39.0041
10	5.7890	-32.5825	55	0.1583	2.5282	100	5.9975	39.9224
	....			....		101	6	40.8407

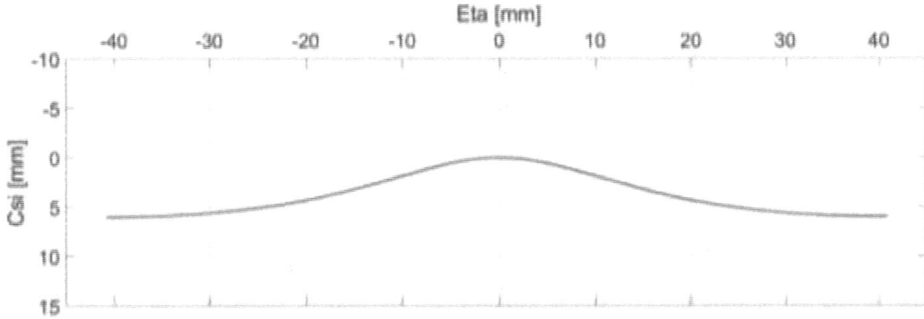

Figure 8. The flat-tool profile in the plane section corresponding to $\varphi = 0$.

The solid model of the flat-tool (depicted in figure 9) has been generated in Catia graphical environment after calculating with relation (10) the value of angle α, which, in the addressed case is of 52.4314 degrees.

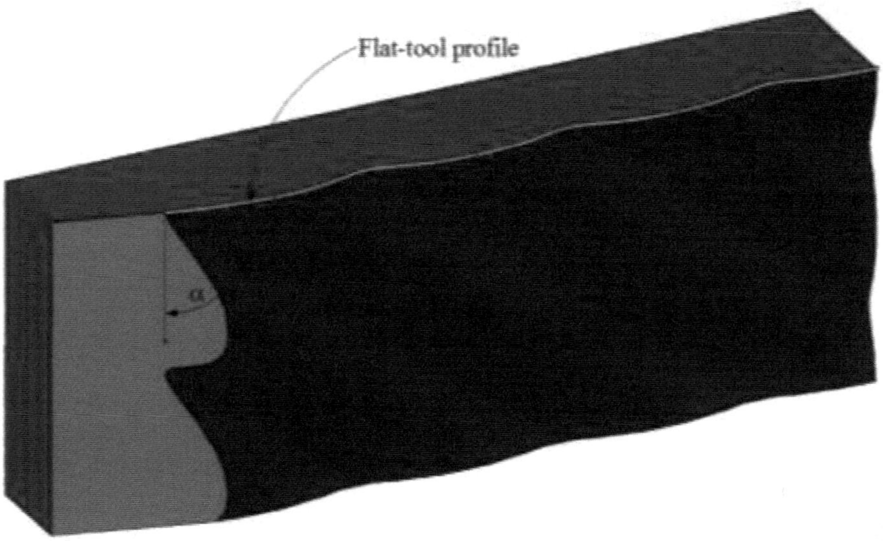

Figure 9. The flat-tool 3-D model.

The flat-tool is provided with an engagement surface, having plane shape and inclined with 30° relative to feed direction (depicted in grey, figure 9). According to [6, 7], its length, measured into feed direction is equal to πR_{rp}, which means 31.4 mm, while the active surface length is equal to $6\pi R_{rp}$ (188.4 m), measured into the same direction.

5. Conclusion

The paper presents an algorithm that can be used in order to profile the flat-tool used for manufacturing the helical rotor of progressing cavity pumps by cold forming. In this purpose, the geometrical model of the rotor has been built, at first. Then the equations of the surfaces family described by the rotor in its motion relative to the tool reference system have been found. By associating to these the enveloping condition, written on the base of the "Minimum distance" method, the flat-tool active profile has been obtained. The profiling algorithm feasibility was tested by running a numerical application, with the help of MatLab soft. The results are showing that the profiling algorithm is fast, robust and precise, which recommend its application in solving actual profiling problems industrial practice.

6. References

[1] https://www.starpumpalliance.com
[2] Luksin V S 1968 *Teoria vintovih poverhnosti v proiectirovanie instrumentov* (eng. *Helical Surfaces Theory Applied in Cutting Tools Design*), (Moscow: Maşinostroienie)
[3] Litvin F L 1994 *Gear Geometriy and Applied Theory*, (New York: Prentince Hall)
[4] Brîndaşu P D and Beju L D 2002 *Bazele prelucrării suprafeţelor şi scule aşchietoare* (eng. *Basics of Surfaces Machining and Cutting Tools*), (Sibiu, Editura Universităţii "Lucian Blaga")
[5] Oancea N 2004 *Generarea suprafeţelor prin înfăşurare, vol.II* (eng. *Surfaces generation through winding, vol. II*) (Galaţi: University Foundation Press)
[6] Moraru V, Plăhteanu B, Velicu S and Aurite T 1982, *Maşini-unelte speciale* (eng. *Special Machine Tools*) (Bucharest: EDP)
[7] Frumuşanu G, Mitu Ş, Fetecău C and Stoian C 1999 *Maşini-unelte speciale şi neconvenţionale*, (eng. *Special and Unconventional Machine Tools*) (Galaţi: Mongabit)
[8] Oancea N 2004 *Generarea suprafeţelor prin înfăşurare, vol.I* (eng. *Surfaces generation through winding, vol. I*), (Galaţi: University Foundation Press)

Acknowledgement
This work was supported by the Romanian Ministry of Research and Innovation, CCCDI – UEFISCDI, project number PN-III-P1-1.2-PCCDI-2017-0446 / *Intelligent manufacturing technologies for advanced production of parts from automobiles and aeronautics industries* (TFI PMAIAA) - 82 PCCDI/2018, within PNCDI III.

IOP Conf. Series: Materials Science and Engineering **968** (2020) 012005 doi:10.1088/1757-899X/968/1/012005

Choosing the optimal order within reconfigurable manufacturing systems based on the Earning Power value

V Marinescu, F Teodor, C Maier, V Paunoiu and A Epureanu
"Dunărea de Jos" University of Galaţi, Faculty of Engineering, Manufacturing Engineering Department

E-mail: catalina.maier@ugal.ro

Abstract: In this paper will be presented the method of choosing an optimal order within a Reconfigurable Manufacturing System (RMS). The reconfigurable manufacturing systems for which we perform the analysis of optimal management belong to make-to-order (MTO) production companies. These companies start the manufacturing process based on the needs of the customers. After the content of the customers orders has been known and accepted they will deliver personalized products, provided that the cost of the production is minimum. A Virtual Workshop will be created associated with each manufacturing order with well-specified requirements (DD - DeadDate, TTF - TimeToFinish, EP - Earning Power). The workshops and orders will be transposed into a Petri net based on the new created three-dimensional model RPD3D [2]. The 3D network will be used to simulate the products manufacturing included in the orders in the virtual workshops and then the optimal order that will actually be produced on RMS from several possible orders based on the evaluation of the indicator called specific profit rate "earning power "- EP will be choosen. In this paper, we aim also to calculate the EP value of the RMS, as the most important evaluation criterion, if a certain command or a combination of commands is accepted. A study case is presented.

1. Introduction

Manufacturing control must be based on an integrated approach to:

- the two stages of the production process, namely the acceptance of the order and the execution of the order;
- all operations which should be carried out within a certain period of time, whether of an administrative, commercial or manufacturing nature;
- all the resources of the manufacturing system, even those consisting in the processing of materials or information.

The trade-off between cost and time should be achieved by using the specific rate of return - "earning power" (EP) as an evaluation criterion in all actions involving decision-making. In order to be able to calculate the value of EP, it is necessary to know the production costs, the production times, the resources involved and their operating costs as well as the value of the manufactured products. The specific rate of profit can be defined at the level of each technological operation, at the level of each manufactured product, at the level of each order and of course at the level of the manufacturing system. EP value is in fact the most important evaluation criterion in the analysis of optimal RMS management.

The purpose of reconfigurable manufacturing systems (RMS) is to make manufacturing orders for an assembly or a family of parts, in a certain number of copies, in order to maintain an optimal configuration of available resources according to the chosen criterion (minimum production time, maximum productivity, EP-Earning Power maximum), in conditions of system stability.

The manufacturing system is defined as the sum of all the resources that have been purchased to fulfill a certain class of orders received from customers. The basic cell of a manufacturing system is the resource. A resource can perform either manufacturing operations (processing, assembly, etc.), or administrative operations (such as monitoring, planning, or scheduling), or commercial operations (sourcing, for example).

From the resources that can perform manufacturing operations (machine tools, robots, buffers, parts or tool warehouses, AGVs, etc.) a virtual workshop will be assembled that will simulate the manufacture of the ordered product or products.

While simulating the operation of an RMS, a resource may be in one of the following states: Ready (R) - ready to work, Work (W) - in progress, Damaged (D) - faulty, Pause (P) - paused, Error correction (E) - in error correction or IntoRMS - integrable in RMS (I) - the resource is Ready but requires an administrative time to be entered in RMS.

Initially, all resources are considered, even if at the start of the manufacturing simulation they are not available, then select those available (which have the status R - Ready). The initial virtual workshop is made up depending on the technological operation to be performed and the correspondence between resource characterization fields.

Before any simulation and calculations, we must specify what the new three-dimensional model RPD3D represents.

The modeling of an RMS could have been done with the help of existing Petri nets, but the complexity and dynamics of manufacturing systems, mainly due to the reconfigurability feature, required the search for a more compact, multi-variable representation model as accurately as possible not only to be able to capture the normal operation of the manufacturing system but also to be able to capture and model the reconfiguration processes.

Thus, it was necessary to create a new class of Petri nets, called RPD3D (Three-dimensional Developed Petri Nets), the name showing both the offspring [1], but also the most important of the new defining features (transforming the model from a 2D to a 3D).

The idea was to introduce in the classic model of a third dimensional Petri net to be able to overlap several levels (layers) formed by 2D nets that interact with each other (receiving or giving commands to activate or deactivate the various modules which together simulate the operation of an RMS).

The entire RMS can be modularized by associating generic RPD3Ds to each resource and each technological operation, these differing only in the working parameters and status. These generic RPD3Ds can be predefined, are editable and are saved in the form of modules. The values of the working parameters that individualize the products, respectively their technological manufacturing process, are saved in the database associated with the manufacturing system.

Like any Petri RPD3D consists of two main types of elements: positions (represented by spheres) and transitions (represented by parallelepipeds), linked together by oriented segments of different weight (thickness).

A position can model operations - O-positions - (processing, transport, handling, assembly, etc.), fixed resources - R-positions - (machining centers, AGVs, robots), variable resources - V-positions - (pallets, parts, buffers) or various conditions required to carry out operations.

In the RPD3D model there are also intermediate positions, called I-positions that can simulate the operations of storage in buffers or warehouses, or to facilitate the maintenance of some behavioral properties of the processing systems.

The control of the subsystems of an RMS is modeled by C-positions.

The marking of the variable resources must be determined in such a way that the system can neither become blocked nor operate undercapacity.

We have transitions that represent the beginning or end of an operation - O-transitions and transitions that trigger subnets to handle errors - E-transitions.

The new model benefits from the advantages of storing and processing the data of its elements in a MySQL database, thus being able to easily store and process networks with hundreds of thousands or millions of elements that in turn have tens of features. Thus, the positions and transitions have a part metadata displayed graphically, the others being stored and circulated only through the RMS database [4].

Also as a novelty in the RPD3D model, the scale feature was introduced (the representation of the same element type at 0.5 of the normal size, 2x or 3x), A larger scale representation of an RPD3D element suggests a greater importance given to that element (robots, other important resources).

2. Identifying the activities and resources needed to process a product

The processing command comes from outside the reconfigurable manufacturing system whose operation we simulate and it contains the initial production specifications (initial data for RMS).

Obviously, the order data is stored and processed in the same MySQL database that serves the entire reconfigurable manufacturing system. The status of the order can have the following values: L-Launched, P-Paused, R-Resumed, F-Finish, J-Rejected, D- Deferred, N-New [4].

In the next step, for each technological operation, data about the resource that can perform the respective technological operation is introduced.

The last step is to create or load RPD3D modules [3] that model the operation of each resource or technological operation. The containing modules can be created and stored in the database in time due to the fact that each one of them correspond to a specific RPD3D. Therefore, these modules are predefined and the editing can be done only for the customization of some parametres.

The following table gives some examples of resources and technological operations modeled with RPD3D.

Table 1. Examples of resources and technological operations modeled with RPD3D.

RPD3D network related to a lathe-type resource on which a single technological operation is performed on a single piece	RPD3D network related to a lathe type resource on which 2 technological operations are performed dependent on each other, on the same part
RPD3D module for OT1S1P	**RPD3D module for OT2S2P**
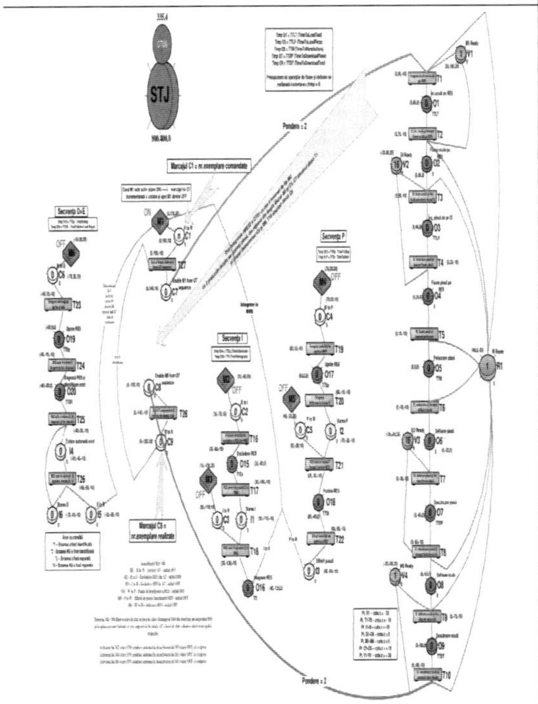	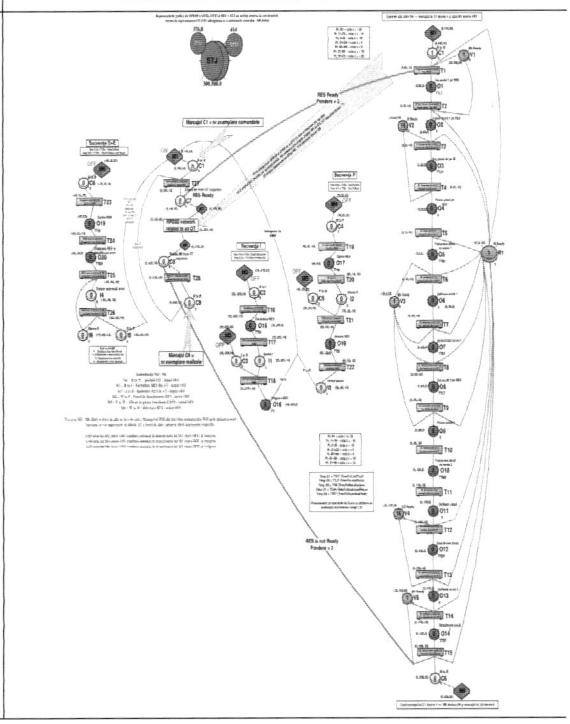

RPD3D network related to a self-governing vehicle type resource that can transport parts or tools	RPD3D network related to an industrial robot resource
RPD3D module for AGV	**RPD3D module for Industrial Robots (RI)**
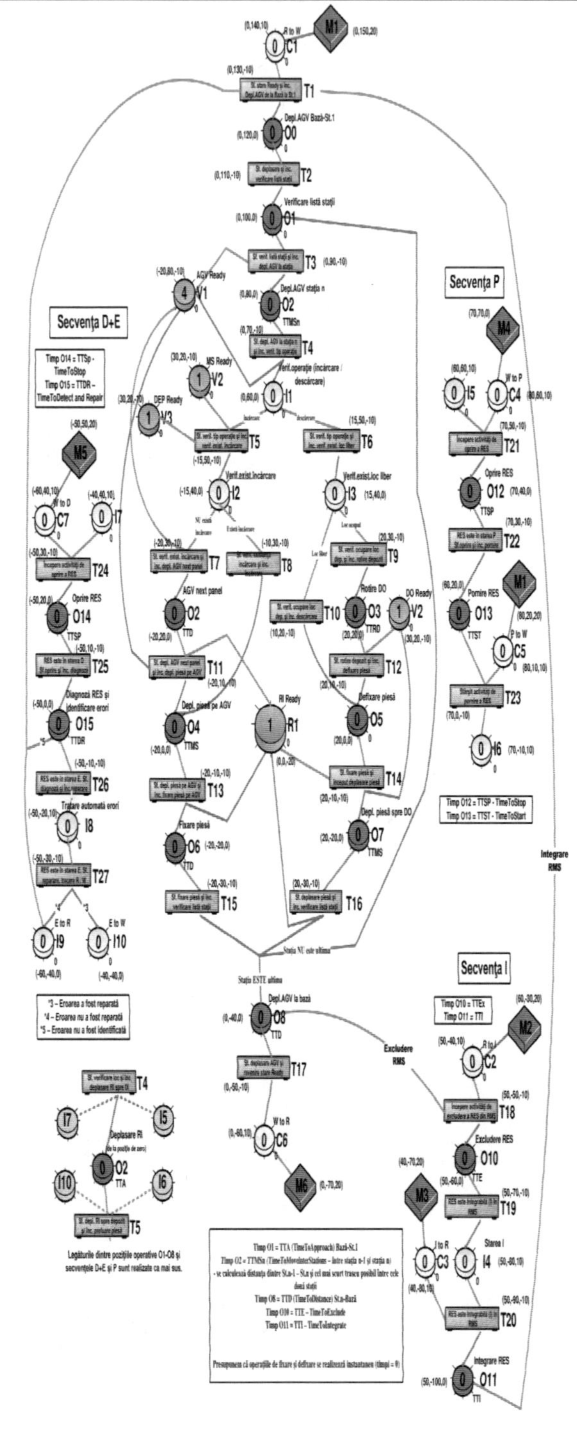 	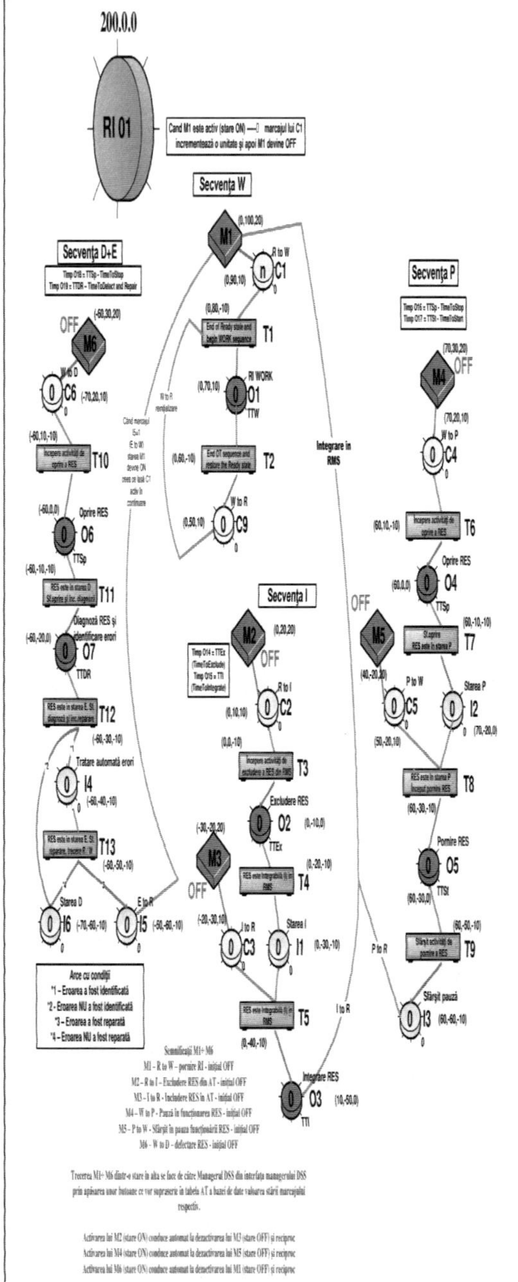

IOP Conf. Series: Materials Science and Engineering **968** (2020) 012005 doi:10.1088/1757-899X/968/1/012005

3. Case study

As specified in the introduction, the main purpose of this study is to present the method of choosing an optimal order in a reconfigurable manufacturing system. Depending on the evaluation of the indicator called the specific rate of return "earning power" – EP, the optimal order produced by RMS will be chosen.

In order to be able to calculate the value of EP, it is necessary to know the production costs, the production times, the resources involved, the operating costs and the value of the manufactured products.

3.1. The first command - CMD1 - Tubular axis

The first order - called CMD1 - Tubular shaft, must be completed by DD (DeadDate - DD - in TimeStamp format = 1443643032), consists of a single product - PRD1, from which it must be manufactured. a number of 10,000 copies, the time norm for a product unit being 6380 seconds.

The value of the EP (Earning Power) field is initially zero, and then, by simulation and calculation, the final value is determined.

The weight of the finished piece is 890 grams, so it can be handled very easily.

3.1.1. List of technological operations related to CMD1:

OT1 - Time for receiving and studying documentation, examining material, tools, devices and measuring devices: Nt = 4.80 min = 288 sec;
OT2 - The auxiliary time for catching the semi-finished product in universal - Nt = 0.6 min = 36 sec;
OT3 - Cutting with alternative saw - L = 190.8 mm - Tui = 1.10 min / pass - Nt = 6.5 min = 390 sec;
OT4 - Turning front turning - Tui = 0.88 min / pass - Nt = 6.28 min = 377 sec;
OT5 - Front turning finish - Tui = 1.5 min / pass - Nt = 6.9 min = 414 sec;
OT6 - Drills in full ф 16 × 143 mm - Tui = 7.56 min - Nt = 12.96 min = 778 sec;
OT7 - Drill widening ф 25.5 × 143 mm - Tui = 8.6 min - Nt = 14 min = 840 sec;
OT8 - External turning deflection ф 45.5 × 143 mm - Tui = 0.22 min / pass - Nt = 5.59 min = 335 sec;
OT9 - External finishing turning ф 45 × 143 mm - Tui = 0.71 min / passage - Nt = 6.08 min = 365 sec;
OT10 - Inner grinding turning ф 32 × 125.5 mm - Tui = 0.78 min / passage - Nt = 6.18 min = 371 sec;
OT11 - Turning of inner roughening ф 33.6 × 27 mm - Tui = 0.63 min / pass - Nt = 6.03 min = 362 sec;
OT12 - Inner grinding turning ф 36 × 15 mm - Tui = 0.63 min / pass - Nt = 6.03 min = 362 sec;
OT13 - Internal groove turning for the sealing gasket ф 39 × 1.5 mm - Tui = 0.63 min / pass - Nt = 6.03 min = 362 sec;
OT14 - Turning of interior finishing ф 35 × 12 mm - Tui = 0.84 min / passage - Nt = 6.24 min = 374 sec;
OT15 - Lathe cutting → L = 143 mm - Tui = 1.10 min / pass - Nt = 6.5 min = 390 sec;
OT16 - Internal finishing turning ф 27 × 17.5 mm - Tui = 0.8 min / pass - Nt = 6.2 min = 372 sec;
OT17 - Feather channel slackening - 256 sec;
OT18 - The auxiliary time for detaching the semi-finished product from the universal - Nt = 0.6 min = 36 sec;

There are 18 technological operations organized on 7 levels. According to the times associated with the technological operations and their ordering, there are several technological flows (figure 1.) that can be run in parallel in maximum 2833 seconds. The total time required to manufacture a product (with the passage of the semi-finished product through all the technological operations) is 6380 seconds.

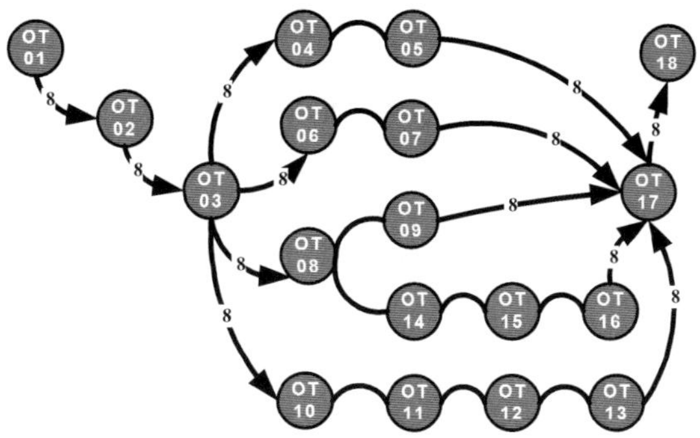

Figure 1. Dependence of the technological operations related to CMD1 command and the flows formed.

3.2. The second command - CMD2 - Flange for supporting a shaft

The second order - called CMD2, must be completed by DD (DeadDate - DD - in TimeStamp format = 1447989854), consists of a single product - PRD2 from which a number of 6000 copies, the time norm for a product unit being 7741 seconds.

As with CMD1, the value of the EP (Earning Power) field is initially zero.

3.2.1. List of technological operations related to CMD2:

OT19 - Cutting with alternative saw - L = 190.8 mm - Tui = 1.10 min / pass - Nt = 6.5 min = 390 sec
OT20 - Turn-side turning and widening - 5,013 min - 301 sec
OT21 - Turning off from the other side and widening of the turning - 9.64 min - 578 sec
OT22 - Finishing turn from one side and widening - 5.04 min - 302
OT23 - Turning finish from the other side and widening - 5.42 min - 325 sec
OT24 - Turning inside clearance D1 - 5.64 min - 338 sec
OT25 - Turning off D2 and turning T2 - 2.9 min - 174 sec
OT26 - Turning D3 and turning T4 - 2.84 min - 170 sec
OT27 - D4 turning and T3 turning - 2.5 min - 150 sec
OT28 - SP5 and SP6 milling milling - 2,917 min - 175 sec
OT29 - Milling milling SP1 and SP2 - 3.39 min - 203 sec
OT30 - Milling SP3 and SP4 - 3.15 min - 189 sec
OT31 - Drilling G1 - 22.8 min - 1372 sec
OT32 - Drilling G2 - 16.8 min - 1008 sec
OT33 - Drilling G3 - 5.5 min - 330 sec
OT34 - Internal rectification SRI 5 - 7.6 min - 456 sec
OT35 - Internal rectification SRI 1 - 7.6 min - 456 sec
OT36 - Internal rectification SRI 2 - 4.5 min - 270 sec
OT37 - External rectification SRE 1 - 1.6 min - 96 sec
OT38 - External rectification SRE 2 - 7.6 min - 456 sec

There are 20 technological operations organized on 7 levels in depth.

According to the times associated with the technological operations and their ordering, there are several technological flows (figure 2) that can be run in parallel in maximum 3377 seconds. The total time required to manufacture a product (with the passage of the semi-finished product through all the technological operations) is 7741 seconds.

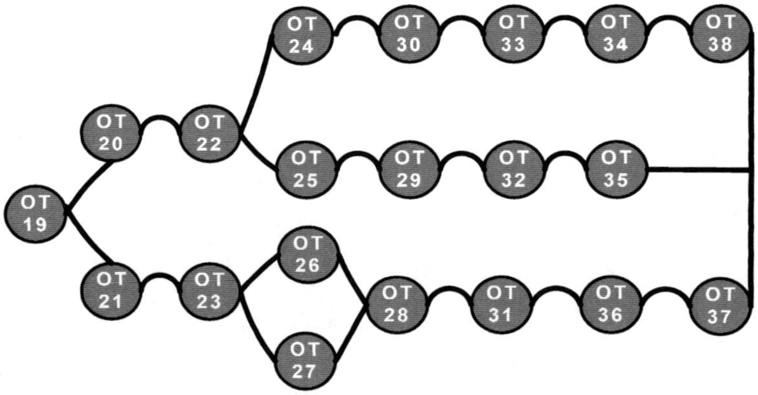

Figure 2. Dependence of the technological operations related to CMD2 command and the flows formed.

In order to calculate the EP_{system} values, some necessary data will have to be calculated first:

1. The average salary per economy is 2415 RON and the number of working hours per month is 160. Therefore, the employees were paid, on average, with 15,09375 RON per hour and for each minute with 0.2516 RON.

2. The salaries calculated above will have to be weighted with the value of the workshop management. The value can vary between 200% for the simple technological operations from the point of view of the labor force and 400% for the complicated ones.
3. As regards the cost of the products, they will be equal to the sum of the costs of the technological operations (working time x the average salary per economy) moderate with an addition between 10-20%;
4. The following formula will be used for the EP_{system} calculation:
   EP_{system} = (price of operations carried out [€] - costs of operations carried out [€]) / A_{sistem} x tperiod [€ x min], where A_{sistem} is the asset of the entire manufacturing system, and the tperiod is the period of use of this asset.

3.3. The association between the resources and the related technological operations CMD1 command and $EP_{system}1$ calculation

We will begin the calculations with the command CMD1 - 10000 pieces PRD1 in 6380 sec./product.

OT1 - Nt = 4.80 min => Cost OT1 = 4.8 * 0.2516 * 2 [RON] = 2.4155 RON
OT2 - Nt = 0.6 min => Cost OT2 = 0.6 * 0.2516 * 2 [RON] = 0.3019 RON
OT3 - Nt = 6.5 min => Cost OT3 = 6.5 * 0.2516 * 3 [RON] = 4.9062 RON
OT4 - Nt = 6.28 min => Cost OT4 = 6.28 * 0.2516 * 3 [RON] = 4.7401 RON
OT5 - Nt = 6.9 min => Cost OT5 = 6.9 * 0.2516 * 3.5 [RON] = 6.0761 RON
OT6 - Nt = 12.96 min => Cost OT6 = 12.69 * 0.2516 * 4 [RON] = 12.7712 RON
OT7 - Nt = 14 min => Cost OT7 = 14 * 0.2516 * 4 [RON] = 14.0896 RON
OT8 - Nt = 5.59 min => Cost OT8 = 5.59 * 0.2516 * 4 [RON] = 5.6258 RON
OT9 - Nt = 6.08 min => Cost OT9 = 6.08 * 0.2516 * 3 [RON] = 4.5892 RON
OT10 - Nt = 6.18 min => Cost OT10 = 6.18 * 0.2516 * 3 [RON] = 4.6647 RON
OT11 - Nt = 6.03 min => Cost OT11 = 6.03 * 0.2516 * 3 [RON] = 4.5514 RON
OT12 - Nt = 6.03 min => Cost OT12 = 6.03 * 0.2516 * 3 [RON] = 4.5514 RON
OT13 - Nt = 6.03 min => Cost OT13 = 6.03 * 0.2516 * 3 [RON] = 4.5514 RON
OT14 - Nt = 6.24 min => Cost OT14 = 6.24 * 0.2516 * 3 [RON] = 4.71 RON
OT15 - Nt = 6.5 min => Cost OT15 = 6.5 * 0.2516 * 3 [RON] = 4.9062 RON
OT16 - Nt = 6.2 min => Cost OT16 = 6.2 * 0.2516 * 3 [RON] = 4.6798 RON
OT17 - Nt = 4.27 min => Cost OT17 = 4.27 * 0.2516 * 2 [RON] = 2.1487 RON
OT18 - Nt = 0.6 min => Cost OT18 = 0.6 * 0.2516 * 2 [RON] = 0.3019 RON

Total cost of product PRD1 => 90,3139 RON, which means that for the entire order the cost of 10,000 pieces = 903139 RON = 200697,5 €.

PRD1 products can be used at a price = total cost * plus 10%, which means that for the entire order the price of 10,000 pieces = 1083766.8 RON = 220767.25 €

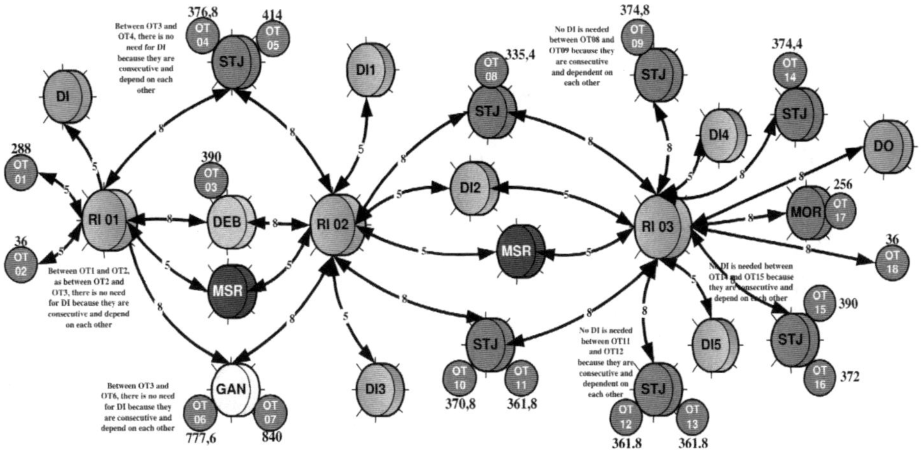

Figure 3. Virtual workshop with the technological resources and operations associated with the CMD1 command.

IOP Conf. Series: Materials Science and Engineering **968** (2020) 012005 doi:10.1088/1757-899X/968/1/012005

The resources that are not included in the Virtual Workshop make up the status I (Integrabila in the RMS) and do not participate in the CMD1 command. Therefore it doesn't influence the value of $EP_{system}1$. OT1 and OT2 did not associate with other resources;

Resource RI01 (figure 3) - industrial robot - passes the parts from OT1 and OT2 to the associated DEB (cutting) type OT3 resource, the parts input warehouse (DI), the tool rotary warehouse (MSR), to the associated GAN (drilling) resource OT6 and OT7 and to the STJ (turning) type associated with OT4 and OT5. RI01 is an RRR type robot and the working times associated with it are passed above.

Thus the resource RI01, which has the purchase value of € 15,000 excluding VAT, and uses a time of (5x2 + 5x2 + 5x2 + 5x2 + 8x2 + 8x2 + 8x2) = 88 seconds = 1.47 min, which means that the resource (asset) will participate in the calculation of the EP_{system} with the value of 15000 x 1.2 x 1.47 = 26460 [€ x min].

The DI resource (parts entry warehouse) realizes with the help of RI01 the supply of semi-manufactured resources. The DI is a rotating 16-seat warehouse (see position mark V2), powered by a linear conveyor and the working times associated with it are passed on the diagram above (5 sec. for each part).

Thus the DI resource, which has the purchase value of 2650 € excluding VAT, and is used for a time of (5x2) = 10 seconds = 0.17 min, which means that the asset will participate in the EP_{system} calculation with the value of (2650 x 1.2 x 0.17 = 540.6 [€ x min].

The resource MSR1 (rotary tool magazine) realizes with the help of robots RI01 and RI02 the supply of tools with the resources on which the parts are processed.

MSR1 is a 24-seater rotary warehouse, grouped on three levels, and the working times associated with it are passed on the diagram above (5 seconds for each tool).

Thus, the MSR1 resource, which has the purchase value of 3820 € excluding VAT, and uses a time of (5x2 + 5x2) = 20 seconds = 0.35 min, which means that the resource (asset) will participate in the calculation of the EP_{system} with the value 3820 x 1.2 x 0.35 = 1604.4 [€ x min].

OT3 is associated with the DEB (debit) type resource - RES31 - which has the purchase value of 297 € excluding VAT, and is used for a time of 6.5 min, which means that the asset will participate in the EPsystem calculation with the value 297 x 1.2 x 6.5 = 2316.6 [€ x min].

OT4 and OT5 are associated with the STJ type resource (turning) - RES1 - which has the acquisition value of 5150 € excluding VAT, and a time of (6.28 + 6.9) min is used, which means that the resource (asset) will participate in the calculation of the EP_{system} with the value of 5150 x 1.2 x 13.18 = 81452.4 [€ x min].

OT6 and OT7 are associated with the GAN (drilling) resource - RES15 - which has the purchase value of 6870 € excluding VAT, and a time of (12.96 + 14) min is used, which means that the asset (asset) will participate in the calculation of the EP_{system} with the value of 6870 x 1.2 x 26.96 = 222258.2 [€ x min].

OT8 is associated with the resource of type STJ (turning) - RES2 - which has the purchase value of 6560 € excluding VAT, and a time of 5.59 min is used, which means that the resource (asset) will participate in the calculation of the EP_{system} with the value 6560 x 1.2 x 5.9 = 46444.8 [€ x min].

OT9 is associated with the resource of type STJ (turning) - RES3 - which has the purchase value of 1720 € without VAT, and a time of 6.08 min is used, which means that the resource (asset) will participate in the calculation of the EP_{system} with the value 1720 x 1.2 x 6.08 = 12549.1 [€ x min].

OT10 and OT11 are associated with the STJ type resource (turning) - RES4 - which has the purchase value of 1820 € excluding VAT, and a time of (6.18 + 6.03) min is used, which means that the resource (asset) will participate in the calculation of the EP_{system} with the value of 1820 x 1.2 x 12.21 = 26666.6 [€ x min].

OT12 and OT13 are associated with the STJ type resource (turning) - RES5 - which has the purchase value of 3140 € excluding VAT, and a time of (6.03 + 6.03) min is used, which means that the resource (asset) will participate in the calculation of the EP_{system} with the value of 3140 x 1.2 x 12.06 = 45442 [€ x min].

OT14 is associated with the resource of type STJ (turning) - RES6 - which has the purchase value of 3640 € excluding VAT, and a time of 6.24 min is used, which means that the resource (asset) will participate in the calculation of the EP_{system} with the value 3640 x 1.2 x 6.24 = 27256.3 [€ x min].

OT15 and OT16 are associated with the STJ type resource (turning) - RES7 - which has the purchase value of 6780 € excluding VAT, and a time of (6.5 + 6.2) min is used, which means that the resource (asset) will participate in the calculation of the EP_{system} with the value of 6780 x 1.2 x 12.7 = 103327.2 [€ x min].

OT17 is associated with the MOR type resource (mortise) - RES22 - which has the purchase value of 7300 € excluding VAT, and is used for a time of 4.27 min, which means that the asset will participate in the calculation of the EP_{system} with the value 7300 x 1.2 x 4.27 = 37405.2 [€ x min].

OT18 has no associated resources;

The MSR2 resource (rotary tool magazine) realizes with the help of robots RI02 and RI03, the tooling of the resources on which the parts are processed. MSR2 is a 24-seater rotary warehouse, grouped on three levels, and the working times associated with it are passed on the diagram above (5 seconds for each tool).

Thus the MSR2 resource, which has the purchase value of 4550 € excluding VAT, is used for a time of (5x2 + 5x2) = 20 seconds = 0.35 min, which means that the resource (asset) will participate in the calculation of the EP_{system} with the value 4550 x 1.2 x 0.35 = 1911 [€ x min].

Resource Di1, Di3, Di4, Di5 (intermediate parts warehouses) realize by means of robots the supply of semi-manufactured resources. They are rotary deposits of 16 places, and the working times associated with them are passed on the diagram above (5 seconds for each piece).

Thus, the 4 resources of type DI, which have the purchase value of 2650 € without VAT, are used for a time of (5x2)=10 seconds = 0.17 min, which means that they will participate in the calculation of the EP_{system} with the value of 4 x (2650 x 1.2 x 0.17) = 2162.4 [€ x min].

The Di2 resource, which has the purchase value of 2650 € excluding VAT, is used for a time of (5x2 + 5x2) = 20 seconds = 0.35 min, which means it will participate in the EP_{system} calculation with the value of 2650 x 1.2 x 0.35 = 1113 [€ x min].

Thus the RI02 resource, which has the purchase value of 17,400 € excluding VAT, is used for a time of (5x5x2 + 8x5x2) = 130 seconds = 2.17 min, which means that the resource (asset) will participate in the calculation of the EP_{system} with the value 17400 x 1.2 x 2.17 = 45309.6 [€ x min].

RI03 is an RRR type robot and the working times associated with it are passed above.

Thus the RI03 resource, which has the purchase value of 16,350 € excluding VAT, is used for a time of (4x5x2 + 9x8x2) = 184 seconds = 3.07 min, which means that the resource (asset) will participate in the EP_{system} calculation with the value 16350 x 1.2 x 3.07 = 60233.4 [€ x min].

The DO resource (parts output warehouse) realizes with the help of RI03 the removal of the semi-finished products from the RMS. DO is a 16-seat rotary storage (see position mark V3), which supplies a linear conveyor and the working times associated with it are passed on the diagram above (8 seconds for each part).

Thus the DO resource, which has the purchase value of 2650 € excluding VAT, and is used for a time of (8x2) = 16 seconds = 0.27 min, which means that the resource (asset) will participate in the calculation of the EP_{system} with the value of 2650 x 1.2 x 0.27 = 858.6 [€ x min].

In total, the resources (assets) will participate in the calculation of the EPsystem with the value of 745311.4 [€ x min] / unit of product, which means that for the entire order of 10,000 units of product

$EP_{system}1$ = 20069.8 € / (745311.4 * 10000) = 2.69 e-6

The average asset used to carry out the CMD1 order was 7011€.

$EP_{system}1$ = 0.00000269 - means that the Virtual Workshop assembled to simulate the CMD1 order allowed to obtain a profit of € 0.0269 for every € invested in every working minute of the Workshop for 738.2 days (2.02 years).

3.4. The association between the resources and the related technological operations CMD2 command and $EP_{system}2$ calculation

CMD2 command involves the manufacture of 6000 pieces of PRD2 in 7741 sec./product.

OT19 - Nt = 6.5 min => Cost OT19 = 6.5 * 0.2516 * 3 [RON] = 4.9062 RON
OT20 - Nt = 5.013 min => Cost OT20 = 5.013 * 0.2516 * 3 [RON] = 3.7838 RON
OT21 - Nt = 9.64 min => Cost OT21 = 9.64 * 0.2516 * 3 [RON] = 7.2763 RON
OT22 - Nt = 5.04 min => Cost OT22 = 5.04 * 0.2516 * 3 [RON] = 3.8042 RON
OT23 - Nt = 5.42 min => Cost OT23 = 5.42 * 0.2516 * 3 [RON] = 4.0911 RON
OT24 - Nt = 5.64 min => Cost OT24 = 5.64 * 0.2516 * 3 [RON] = 4.2571 RON
OT25 - Nt = 2.9 min => Cost OT25 = 2.9 * 0.2516 * 4 [RON] = 2.9186 RON
OT26 - Nt = 2.84 min => Cost OT26 = 2.84 * 0.2516 * 4 [RON] = 2.8582 RON
OT27 - Nt = 2.5 min => Cost OT27 = 2.5 * 0.2516 * 4 [RON] = 2.516 RON
OT28 - Nt = 2.917 min => Cost OT28 = 2.917 * 0.2516 * 3 [RON] = 2.2018 RON
OT29 - Nt = 3.39 min => Cost OT29 = 3.39 * 0.2516 * 3 [RON] = 2.5588 RON
OT30 - Nt = 3.15 min => Cost OT30 = 3.15 * 0.2516 * 3 [RON] = 2.3776 RON
OT31 - Nt = 22.8 min => Cost OT31 = 22.8 * 0.2516 * 3 [RON] = 17.2094 RON
OT32 - Nt = 16.8 min => Cost OT32 = 16.8 * 0.2516 * 3 [RON] = 12.6806 RON
OT33 - Nt = 5.5 min => Cost OT33 = 5.5 * 0.2516 * 3 [RON] = 4.1514 RON
OT34 - Nt = 7.6 min => Cost OT34 = 7.6 * 0.2516 * 4 [RON] = 7.6486 RON
OT35 - Nt = 7.6 min => Cost OT35 = 7.6 * 0.2516 * 4 [RON] = 7.6486 RON
OT36 - Nt = 4.5 min => Cost OT36 = 4.5 * 0.2516 * 4 [RON] = 4.5288 RON
OT37 - Nt = 1.6 min => Cost OT37 = 1.6 * 0.2516 * 4 [RON] = 1.6102 RON
OT38 - Nt = 7.6 min => Cost OT38 = 7.6 * 0.2516 * 4 [RON] = 7.6486 RON

Total cost of product PRD2 => 106.6759 RON which means that for all CMD2 order the cost of 6000 pieces of product = 640055.4 RON = 142234.5 €

PRD2 type products can be used at a price = total cost * plus 10%, which means that for the entire order the price of 6000 pieces = 704061 RON = 156458 €

The resources that are not included in the Virtual Workshop (figure 4) have the status I (Integrable in the RMS) and do not participate in the CMD2 command. Therefore it doesn't influence the value of $EP_{system}2$.

In total, the resources (assets) will participate in the calculation of $EP_{system}2$ with the value of 2039947 [€ x min] / unit of product, which means that for the entire order of 6000 units of product $EP_{system}2$ = 14223.45 / (2039947 * 6000) = 1.162e-6

The average asset used to carry out the CMD2 order was € 15478.

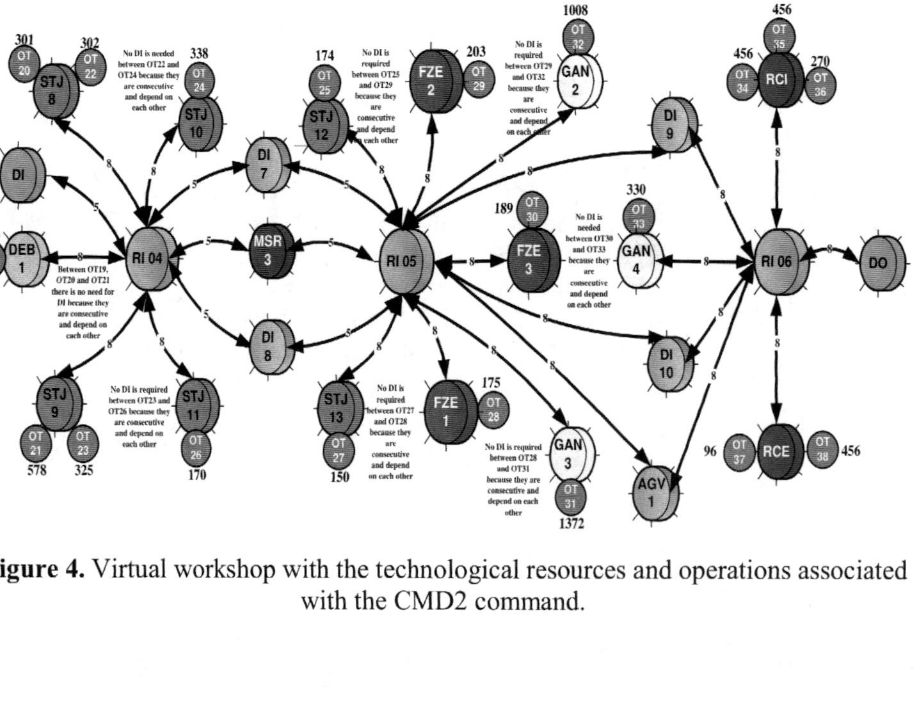

Figure 4. Virtual workshop with the technological resources and operations associated with the CMD2 command.

Table 2. The values of participation in the calculation of $EP_{system}2$ of the technological operations and the corresponding resources associated with the CMD2 command.

Nr. OT	Asset value [€]	Time of use the asset [min]	$EP_{sistem}2$ [€*min]
19	151	6,5	1177,8
20	15680	5	94080
21	18750	9,6	216000
22	15680	5	94080
23	18750	5,4	121500
24	12530	5,6	84201,6
25	16400	2,9	57072
26	25600	2,84	87244,8
27	5150	2,5	15450
28	18590	2,9	64693,2
29	17300	3,4	70584
30	27160	3,15	102664,8
31	7990	22,8	218606,4
32	15550	16,8	313488
33	16650	5,5	109890
34	18950	7,6	172824
35	18950	7,6	172824
36	18950	4,5	102330
37	18310	1,6	35155,2
38	18310	7,6	166987.2
DI	2650	0,17	540,6
DI7	2650	0,35	1081,2
DI8	2650	0,35	1081,2
DI9	2650	0,35	1081,2
DI10	2650	0,35	1081,2
DO	2650	0,27	829,4
RI 04	36450	2	87480
RI 05	64500	3,17	245358
RI 06	29640	1,87	66512,2
MSR3	1710	0,35	718,2
AGV1	6220	0,53	3955,9

$EP_{system}2 = 0.000001162$ - means that the Virtual Workshop assembled to carry out the CMD2 order allowed a profit of 0.01162 € for every € invested in every working minute of the AT for 537.6 days (1.47 years)

As an alternative to calculating the value of $EP_{system}2$, in order to compare the results obtained with $EP_{system}1$, the same resources as in the case of CMD1 control will be used to manufacture the CMD2 command:

IOP Conf. Series: Materials Science and Engineering **968** (2020) 012005 doi:10.1088/1757-899X/968/1/012005

Table 3. The participation values in the calculation of $EP_{system}2$ of the technological operations executed with the resources used in the realization of the CMD1 command.

Nr. OT	Asset value [€]	Time of use the asset [min]	$EP_{sistem}2$ [€*min]
19	151	6,5	1177.8
20	5150	5	30900
21	6560	9,6	75571.2
22	5150	5	30900
23	6560	5,4	42508.8
24	3140	5,6	21100.8
25	3640	2,9	10927.2
26	6780	2,84	23106.24
27	5150	2,5	15450
28	18590	2,9	64693.2
29	17300	3,4	70584
30	27160	3,15	102664.8
31	6870	22,8	185500.8
32	7990	16,8	161078.4
33	15550	5,5	102630
34	18950	7,6	172824
35	18950	7,6	172824
36	18950	4,5	102330
37	18310	1,6	35155,2
38	18310	7,6	166987.2
DI	2650	0,17	540,6
DI7	2650	0,35	1081,2
DI8	2650	0,35	1081,2
DI9	2650	0,35	1081,2
DI10	2650	0,35	1081,2
DO	2650	0,27	829,4
RI 04	36450	2	87480
RI 05	64500	3,17	245358
RI 06	29640	1,87	66512,2
MSR3	1710	0,35	718,2
AGV1	6220	0,53	3955,9

In total, the resources (assets) will participate in the calculation of $EP_{\mathbf{system}}2$ with the value of 1886596 [€ x min] / unit of product, which means that for the entire order of 6000 units of product $EP_{\mathbf{system}}2$ = 14223.45 / (1886596 * 6000) = 1.256 e-6.

The average asset used to carry out the CMD2 order was 12375.2 € (20.05% decrease compared to the version in which the more expensive resources are used).

$EP_{\mathbf{system}}2$ = 0.000001256 - (increasing by 7.48% compared to the variant in which more expensive resources are used).

4. Conclusions

With the help of the new RPD3D model - three-dimensional developed Petri nets we were able to evaluate a set of two orders given to the manufacturing system. The evaluation was based on terms of profitability of accepting one order or another in the factory and for optimal management system at the operational level.

The index by which the order evaluation was performed is the specific rate of profit - EP, an index that can be calculated at the level of a technological operation, at the level of a product from an order, at the level of an order or at the level of the entire virtual workshop.

In other words, the simulation of the manufacture of the two orders offers sets of data, on the basis of which the DSS manager has the possibility to choose one order or another, depending on the market requirements.

The simulation of the manufacture of the orders was carried out under comparable conditions, offering the possibility of using, as far as possible, the same resources. The resulting data is grouped in the following table:

Table 4. Comparative analysis of CMD1 and CMD2 commands.

Command	Profit [€]	Workshop value [€]	TTF [days]	Units produced [units]	Average Asset Value [€]	EP_{sistem} [$€_{profit}$ / $€_{investit}$ * min]
CMD1	20069,8	196308	738,2	10000	7011	2,69 e-6
CMD2	14223,5	383631	537,6	6000	12375,2	1,256 e-6

As it can be seen in the table, the first order is clearly the most advantageous in terms of manufacturing profitability. It is the one that should be chosen for production among all orders.

Also in order to provide data sets on which the DSS manager chooses one order or another, we simulated the situation in which the first order is accepted and the second order at the manufacturing site is in operation and accepted, in parallel with the first order. Obviously, the resources used for the second order were not used by the first command (more expensive and less used resources) and the result was that the value of $EP_{\mathbf{system}}2$ decreased by 7.48%.

Table 5. Comparative analysis between if only the CMD2 command is executed and if both CMD1 and CMD2 commands are executed.

Command	Profit [€]	Workshop value [€]	TTF [days]	Units produced [units]	Average Asset Value [€]	EP_{sistem} [$€_{profit}$ / $€_{investit}$ * min]
CMD2 only	14223,5	383631	537,6	6000	12375,2	1,256 e-6
CMD2 + CMD1	14223,5	479821	537,6	6000	15478	1,162 e-6

The conclusion of the simulation is that it is more cost-effective for the second order to be made alone and not at the same time as the first order.

5. References

[1] Marinescu V 2000 *Research on the flexible management of metalworking processes cold* PhD Thesis (Galaţi)
[2] Teodor F 2016 *Optimal management of reconfigurable manufacturing systems*, PhD Thesis (Galaţi)
[3] Teodor F, Marinescu V and Epureanu A 2016 Simulation of product manufacturing on RMS by virtual workshop assembling *The Annals of "Dunarea de Jos" University of Galaţi*
[4] Teodor F, Marinescu V and Epureanu A 2017 Applying relational database to simulate the reconfigurable manufacturing system *The Annals of "Dunarea de Jos" University of Galati*

[5] Teodor F, Marinescu V and Epureanu A 2013 Algorithm for compacting a three dimensional Petri Network that simulates the reconfigurable manufacturing system operation *The Bulletin of the Polytechnic Institute of Iași*
[6] Teodor F, Marinescu V and Epureanu A 2016 Optimal management of reconfigurable manufacturing systems modeling with Petri nets developed three-dimensional - RPD3D *The Innovative Manufacturing Engineering and Energy Conference (IManEE)* (Kallithea)
[7] Bi Z M, Lang S Y T, Shen W and Wang L 2008 Reconfigurable manufacturing systems: the state of the art *Int J Prod Res* 4 **46** 967–992
[8] Katz R 2007 Design principles of reconfigurable machines *Int J Adv Manuf Technol* **34** 430–439
[8] Koren Y and Ulsoy G 2002 Reconfigurable Manufacturing System Having a Method for Changing its Production Capacity *US patent 6,349,237*

Aknowledgement
This work was supported by the grant of the Romanian National Authority for Scientific Research and Innovation, CNCS – UEFISCDI, grant number PN-III-P1-1.2-PCCDI-2017-0446, within PNCDI III – 82/2018.

Numerical analysis of hydroforming process control using variable blankholder force

V Paunoiu, C Maier, I Iacob and V Marinescu
Dunarea de Jos University of Galati, Faculty of Engineering, 111 Domneasca Street, 800201, Galati, Romania

E-mail: catalina.maier@ugal.ro

Abstract. Hydroforming was developed to provide a cost effective means to produce relatively small quantities of drawn parts or parts with asymmetrical or irregular contours that are difficult to obtain by conventional stamping. This paper presents a study of hydroforming process with variable blankholder force in order to assure the parts quality. The main idea is to decompose the blankholder function of the elementary zone of the part contour corresponding to the linear and curvilinear zone. For each zone different blankholder force is applied in correlation with the hydrostatic pressure. A numerical analysis using finite element modelisation is performed considering different sets of blankholder forces and hydrostatic pressures, as process parameters. An optimum is determined in order to avoid parts defects (thickness reduction, wrinkles, fracture) for such types of parts.

1. Introduction

Sheet hydroforming is a process that consists of forming a blank into a certain shape by using a punch to mechanically press the blank into a pressure chamber filled with hydraulic fluid [1].

The process of sheet hydroforming offers several technological advantages for example higher deep drawing ratios, better geometrical accuracy and cost efficiency compared to conventional deep drawing processes. Additionally this technology is more suitable for deep drawing modern high strength steel or aluminium alloys which are utilized more and more in the automotive and aerospace industry [2].

Despite the advantages of this method, it has some limitations. Most research in robust process design focuses on problems with a single response but practical problems as sheet metal forming processes, often require the simultaneous improving of many quality characteristics (i.e. blank thinning and geometrical accuracy) and productivity considered as output parameters in the stage of process modeling.

Different researchers considers as input parameters the following process parameters:

- geometrical features within the tool design [3,4];
- fluid pressure path [5,6,7]; pressure path is the most important parameter for increasing the drawing ratio in HDD; finding the appropriate fluid pressure-punch stroke path while avoiding rupture and wrinkling instabilities is a critical and difficult step;
- time dependent blankholder force and the fluid pressure [8]
- blankholder force variation through the stroke and slide velocity [9];
- blankholder force variation through the stroke [10]

The studies performed for the optimization and control of the hydroforming process used both experimental and numerical methods. Finite element method is very common and the results obtained in this way can be exploited using ANOVA, neural network method, reduced order method [11,12] in order to obtain a process model with short time response or able to predict its results (quality characteristics of the piece and process productivity) [1].

During the deformation process the material flow into the die it is higher in the linear zones of the contour than in the corners zones. Considering the similarity with conventional deep drawing process using composed blankholder force [13-15] this paper propose a method to compensate the unequal material flow. This method consists in the use of a segmented blankholder with different value of the applied force corresponding to the different areas of the blank.

The blankholder forces or pressures values in a deep drawing or sheet hydroforming process determines the final properties and the quality of the formed part. If the loads applied on the blankholder are too low and the material flows too quickly, the blank will wrinkle, but if there are very high blankholder loads, the material flow may be slowed and that may cause the formed part to break.

For parts that require multiple different loads to be applied in different areas of the blankholder to improve their formability, there are methods like using a segmented blankholder or using a multi-point cushion system. These methods allow the possibility of using different loads on different locations of the blankholder for the areas of the blank where there needs to be more or less pressure (or force) than the other areas to prevent wrinkling, thinning or a possible rupture of the formed part [16].

Forming limit diagrams (FLD) are used to determine a material's formability to be able to predict if there are risks of necking (for values higher than the FLC) or breaking (for values higher than the FFL) of a part made from that material. The FLD are formed by using the maximum principal in-plane strains ε_1, named major strains and the minimum principal in-plane strains ε_2, named minor strains, that can be obtained either experimentally or numerically [23-26].

This paper intends to study the relation bewteen the different applied pressures on the different areas of a segmented blankholder and hydrostatic pressure toward the formability of a hydroformed square part by using the finite element software Abaqus. The formabilty is appreciated by considering the level of thickness, stresses, strain variation and fracture using the FLD.

2. FEM Model

A series of finite element analyses and experiments of a conventional deep drawing process for forming square parts have been performed and the results showed a good agreement between the simulations and the experiments [18].

The researches presented in this paper can be seen as a continuation of those tests and they consist in a numerical analysis of a hydroforming process considering a case study, presented in figure 1 [18]. The hydroforming process of a rectangular part obtained from an aluminum blank is modelled using Abaqus software. The hydroforming process model contains: the blank, the blankholder, the punch and the die that also acts as a pressure chamber (figure 3).

The main dimensions of the modeled assembly components are presented below. Similar assembly was used and analised by Hama [17] in their researches.

In figures 2 and 4 the following notations have been made:

Lp, Lbh, Lb, Ld, Lf – the lenghts of the punch, blankholder, blank, die chamber and formed part, respectively

Wp, Wbh, Wb, Wd, Wf – the widths of the punch, blankholder, blank, die chamber and formed part

hp, hd, hf– the heights of the punch, die chamber and formed part

tbh, tb – the thicknesses of the blankholder and blank

rp, rbh, rb, rd, rf – the radii of the punch, blankholder, blank, die chamber and formed part

The numerical values of all model components and formed part are presentd in the table 1.

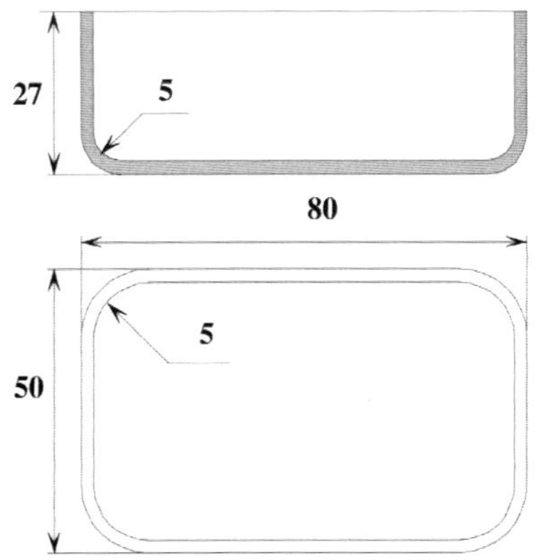

Figure 1. The final dimensions of the formed part [18].

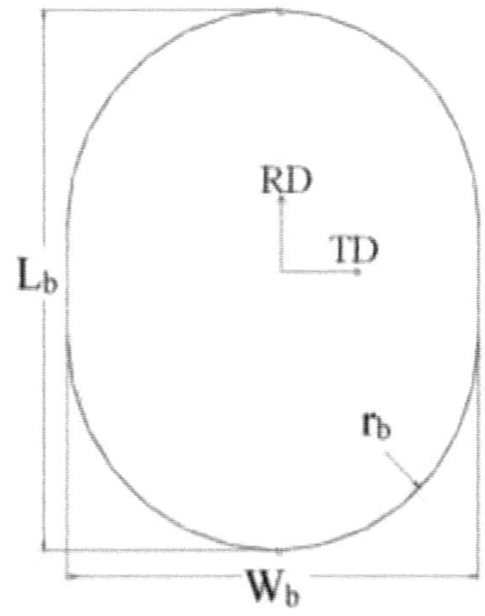

Figure 2. Dimensions of the blank.

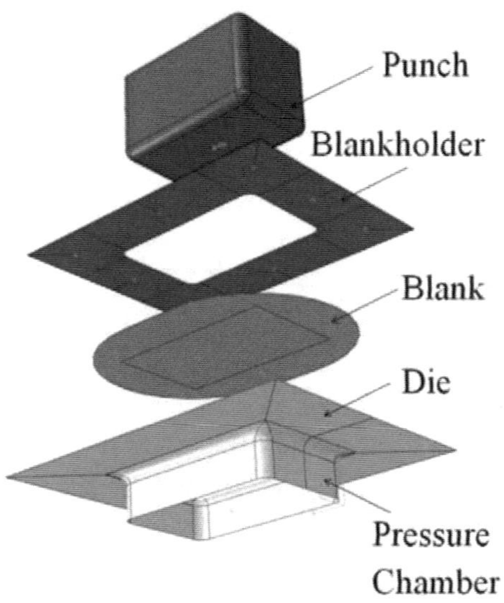

Figure 3. Assembly Components.

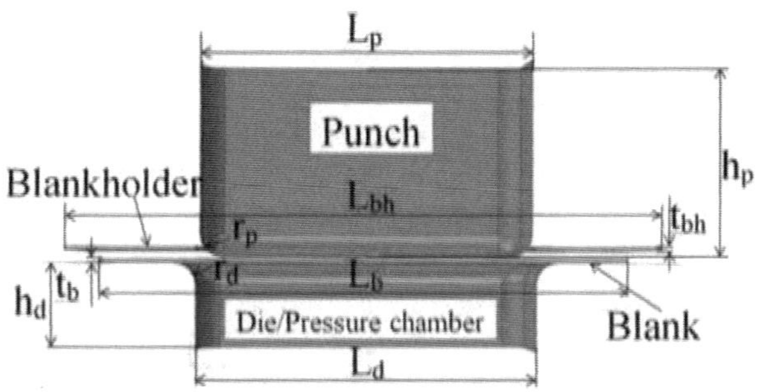

Figure 4. Dimensions of the sectioned assembly.

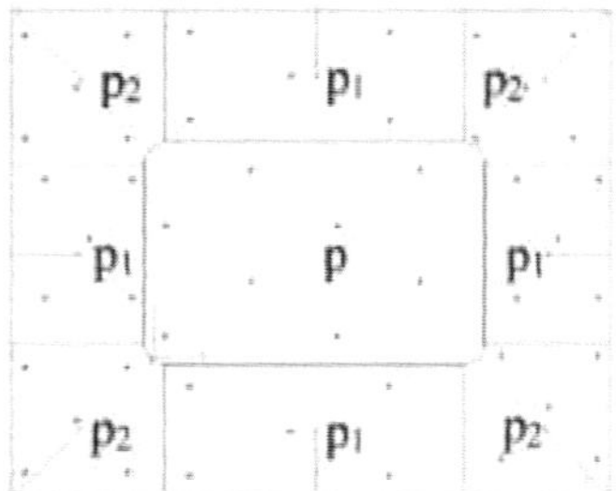

Figure 5. Values of pressures p_1, p_2 applied on the blankholder and of the hydrostatic pressure p.

Table 1. The dimensions of the modeled components.

Components	Length [mm]	Width [mm]	Height / Thickness [mm]	Radius [mm]
Punch	L_p=78	W_p=48	h_p=45	r_p=5
Blank holder	L_{bh}=140	W_{bh}=110	t_{bh}=1	r_{bh}=5
Blank	L_b=124	W_b=98	t_b=1	r_b=49
Die chamber	L_d=80	W_d=50	h_d=20	r_d=5
Formed Part	L_f=80	W_f=50	h_f=35	r_f=5

The components were 4odeled using shell elements (homogenous, thickness integration rule: Simpson, thickness integration points: 5) and the following properties:

- the blank was 4odeled as a 3D deformable shell with 1 mm thickness; the material assigned was aluminum and plastic properties were added (strength coefficient K=352.14, Young modulus E=61370 Mpa, yield strength σ_0=173 Mpa, strain hardening exponent n=0.163, tensile strength σ_r=262 Mpa, Poisson ratio 0.3). The yield criterion used was Barlat '89 and the hardening law was Swift [26];
- the blankholder was modeled as a 3D deformable shell with 2 mm in thickness and assigned material – steel;
- the punch and die were each modeled as a 3D discrete rigid body.

Loads (figure 5):

- the blankholder was divided in 8 areas corresponding to the linear and curvilinear zone of the part's contour; two different pressure values with a linear variation (starting from 0 to the applied value) were applied on these areas: pressure p_1 on the four linear zones of part's contour and pressure p_2 for each of the four corners (curvilinear zones of part's contour). The pressures values were taken from [17, 19].
- the hydrostatic pressure (p) of the hydroforming liquid, also with a linear variation, was defined without any extra initial pressure, without overcharging [17] and was applied on the bottom face of the blank. The pressure value was taken from [17, 19, 20].

The boundary conditions were:

- the die was considered fixed body (all degrees of freedom were zero);
- the punch had a translation on the z axis and a velocity was applied along this; all other degrees of freedom was blocked.
- The clearance between the die and the punch was 1t.

3. Numerical Simulation of Sheet Hydroforming

The finite element analysis of the hydroforming process of the rectangular part was carried considering as inputs the blankholder pressure in different areas and the hydrostatic pressure. Ten sets of values were prepared (table 2). The hydrostatic pressures were applied taken in account the avoidance of the leakage phenomenon. This phenomenon wasn't numerically studied in this paper, treating the problem in a classic way.

Table 2. Sets of input and output parameters.

Data set	Crt. No.	p [Mpa]	p_1 [Mpa]	p_2 [Mpa]	$t_{initial}$ [mm]	t_{min} [mm]	t_{max} [mm]
I.	1	4	7	7	1	0.891	1.450
	2	5	7	7	1	0.891	1.447
	3	6	7	7	1	0.891	1.453
II.	1	5	7	8	1	0.890	1.447
	2	6	7	8	1	0.891	1.449
III.	1	5	8	7	1	0.891	1.450
	2	6	8	7	1	0.891	1.450
IV.	1	5	9	7	1	0.892	1.444
	2	6	9	7	1	0.892	1.448
V.	1	25	26	30	1	0.874	1.585
VI.	1	25	30	26	1	0.873	1.545
VII.	1	30	31	35	1	0.874	1.753
VIII.	1	30	35	31	1	0.872	1.669
IX.	1	50	51	55	1	0.865	2.577
X.	1	50	55	51	1	0.866	2.378

For a global analysis of the quality assurance, the output parameter of the process was considered the thickness variation [21].

The first set has 3 different hydrostatic pressures and the blankholder pressure has the same value on the whole blankholder, in all 3 cases, acting as a rigid plate. The next 3 sets of data have only 2 diferent hydrostatic pressures each, while either the first or the second blankholder pressure is different than the first set.

The thickness values from table 2 can be observed in all the cases in the same locations: the minimum thickness value is in the bottom corner and the maximum value is in the top area of the formed part, at the contact zone between the linear and corner edges of the part (figure 8).

Because the differences of the first data sets results were small in terms of thickness variations, additional, higher values for the blankholders and hydrostatic pressures were also chosen, cases V-X from table 2.

For a detailed analysis of the process, two series of 20 points were picked by nodes, for each data set, in order to observe the evolution from the bottom to the top of some output parameters of the formed part. The first series of points were chosen from a line of the mesh in the TD of the formed part (figure 6.a). For the other series, the part was split diagonally, in order to obtain a reference line to help with picking points from the corner of the formed part (figure 6.b). The starting point was the same for both cases, in the center of the bottom of the formed part.

The output parameters that were observed in this case were the thickness of the formed part, the equivalent plastic strain, the Von Mises stress and the Cauchy stress tensor components σ_{11} and σ_{22} .

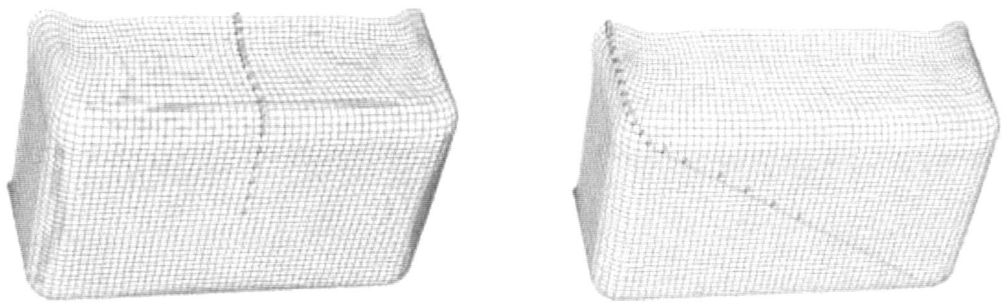

a) along the TD of the formed part b) in the corner of the formed part

Figure 6. Measured points.

4. Results and disscusion

Several studies considered that the optimum of the process regarding the quality of the part should be established considering the following aspects: *i)* the formability of the part (characterized by the capacity of the process to avoid the wrinkling due to excessive compression and tearing because of high local stress that cause thinning and failure of the material); *ii)* dimensional accuracy (springback caused by elastic recovery); *iii)* consistency (representing the minimization of dimensional variations due to the variation of different parameters as lubrication, material properties and thickness). The close connection between the control of the stress state in the material during the process and the quality of the obtained piece it is obvious. This is why, the critical zones of the part, from the point of view of stress state, were identified (figure 7). These zones are:

- Critical zone 1 represents the areas where the fracture of the material can appear due to: i) the equivalent von Mises stress exceeding tensile strength of the material, or ii) the biaxial tension stress state i.e. on the bottom piece area; both of these stress states determine a decrease in the value of the material thickness below the accepted value.
- Critical zone 2 represents the area where stress state is type tension+compression with tension dominance, determining wrinkles after the process, due to the springback;

- Critical zone 3 represents the area where the stress state is type tension+compression with compression dominance determining wrinkles during the process.

It can be observed that the critical zone are characterized by the stress state in the material, especially the value of the equivalent von Mises stress and Cauchy stresses tensor components, σ_{11} and σ_{22}. σ_{11} is the direct stress in the RD and σ_{22} is the direct stress in the TD [22].

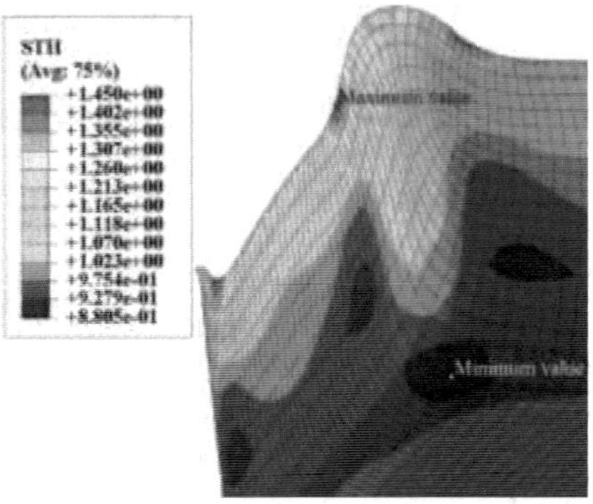

Figure 7. The minimum and maximum thickness values of a hydroformed part.

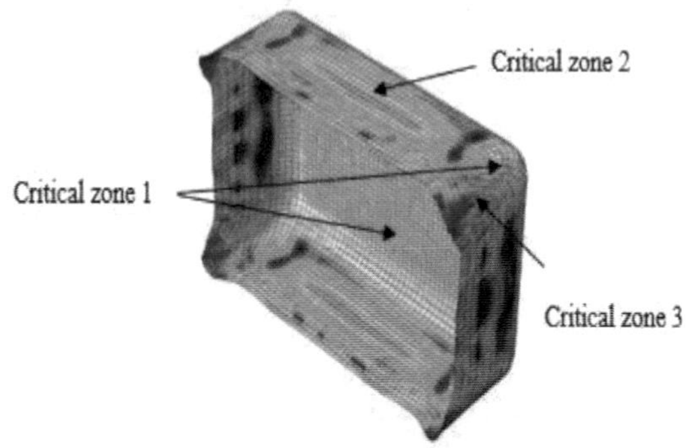

Figure 8. Critical zones of the hydroformed part.

After extracting the values of the thickness, the equivalent plastic strain, Von Misses stress and Cauchy stress tensor components σ_{11} and σ_{22} for the two series of 20 points, for each data set, the results were represented as graphs (figures 9 – 16). The graphs show the evolution of each considered output parameter for 3 zones: the bottom zone (points 1-5), the radius zone (points 6-10) and the wall zone of the formed part (points 11-20). Also in these graphs are represented the values of the first data set.

4.1. Thickness variation

The thickness of the formed part has small variations for the cases I – IV, both in the longer plane zone and in the corner sections of the part. The thickness along the TD has variations of 0.1 mm, plus or minus, between cases, meaning that the thickness results are very similar in this area for the first cases set. The cases V – X show more thinning than the first cases set, starting from the radius zone. In the wall zone there is a higher thickness in the cases IX and X, compared to the other cases.

The worst along the TD is the one from the data set X with a minimum value of 0.94 mm and a maximum value of 1.04 mm (figure 10). The cases II 1, II 2 and III 2 have the best thickness variations.

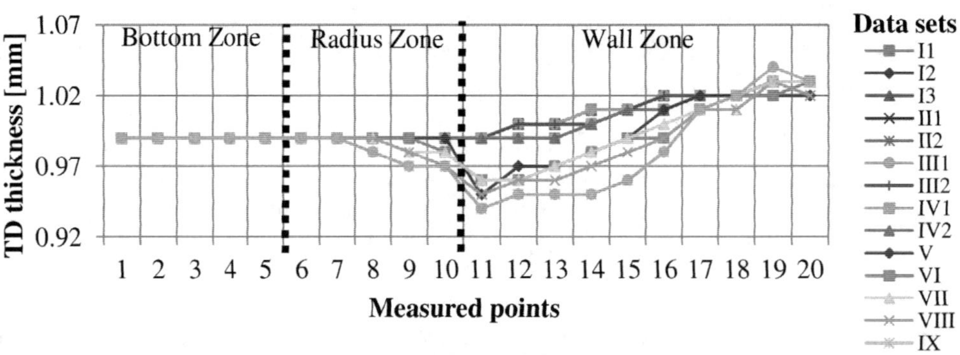

Figure 9. The thickness values from the points measured along the TD of the formed part.

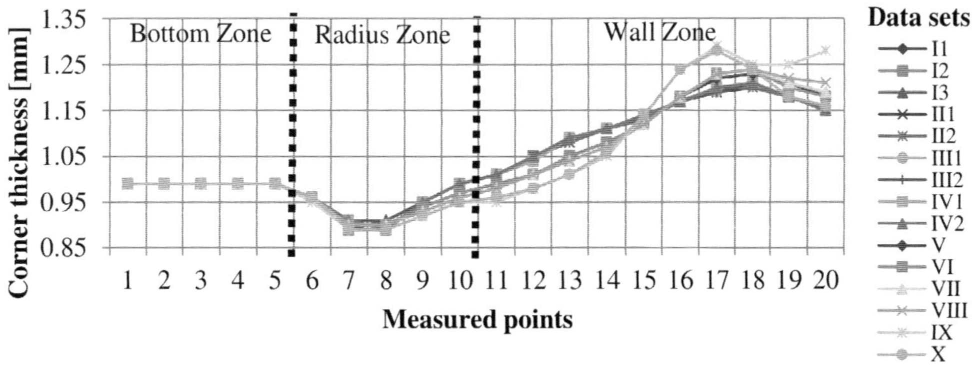

Figure 10. The thickness values from the points measured in the corner section of the formed part.

The best thickness situation from the points that were chosen for the corner section (figure 10), can be seen in the case III 2, that has a maximum thickness of 1.21 mm and on the opposite side is case IX with a maximum thickness of 1.29 mm, meaning that there is a bigger accumulation of material here than in the other cases. In all the cases there is a significant thinning at the bottom corner section with a value of 0.89 mm.

It can be concluded that the applied pressures have similar plots toward the thickness variations, both along the TD and corner zone. It is seen that the numerical simulations predict localized thinning around the punch radius, a limited, and uniform, straining at the bottom of the part, as well as the wall-thickening along and around the part edge, more pronounced in the corner zones [27, 28, 29]. The ears appears in these zones.

4.2. The equivalent plastic strain variation

Small variations can also be seen for the values of the equivalent plastic strain for the cases I – IV. In the along the TD of the formed part (figure 11), the maximum value is 0.37 for the case V. The lowest maximum value of 0.29 can be observed in the cases IV 1 and IV 2.

In the corner section, between most of the data sets there are only small variations, with the exception of the data sets IX and X, that have significant differences starting from the middle of the wall zone of the formed part. The biggest differences between all the cases can be seen at point 20, where the highest value is 0.73 in the case IX and the lowest value is 0.32 for the cases II – IV (figure 12).

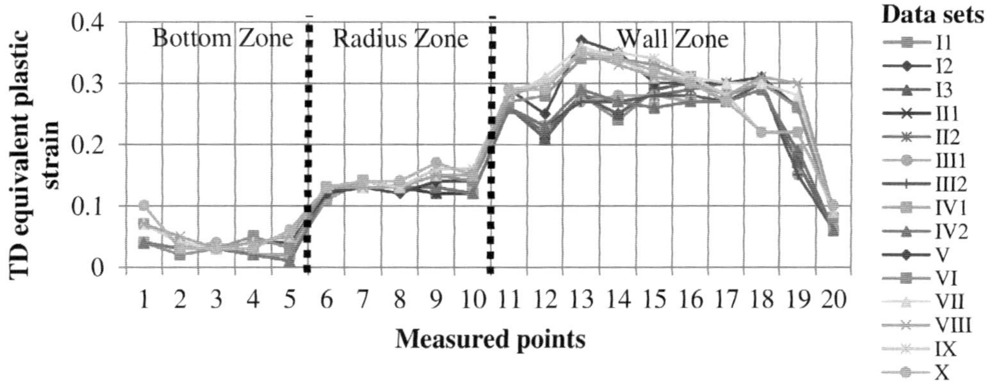

Figure 11. The equivalent plastic strain values from the points measured along the TD of the formed part.

It can be concluded that the applied pressures have similar plots toward the equivalent plastic strain variation, both along the TD and corner zone. It can be seen that the bottom of the parts develop limited amount of strain in comparison to the walls, which experience strains beyond the limit of uniform elongation in uniaxial tension [27, 28, 29].

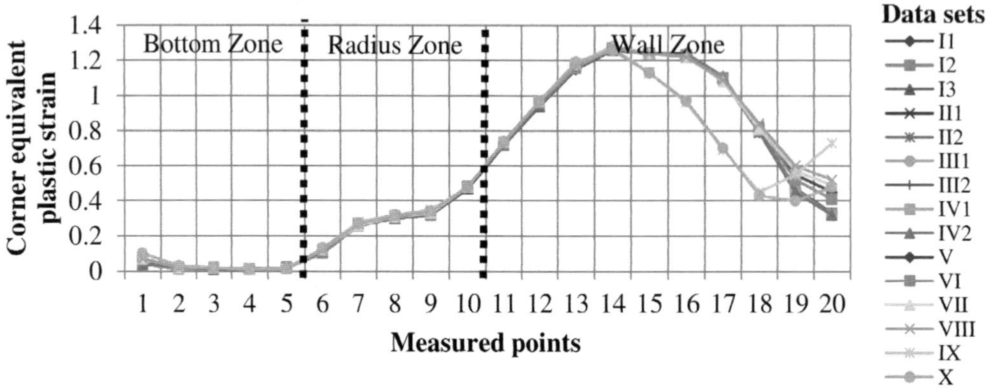

Figure 12. The equivalent plastic strain values from the points measured in the corner section of the formed part.

4.3. Von Mises stress variation

In the case of the Von Mises stress, the curves have bigger variations of values, but the shapes of these curves are similar between the cases I – IV. The cases IX and X have some significant differences compared to all the other cases.

The highest Von Mises stress value along the TD of the formed part was 448.29 MPa for the case IX at point 18, where the cases I – VIII have drops. The case I 1 has overall lower values than the other cases with a minimum of 80.13 MPa and a maximum of 396.98 MPa (figure 13).

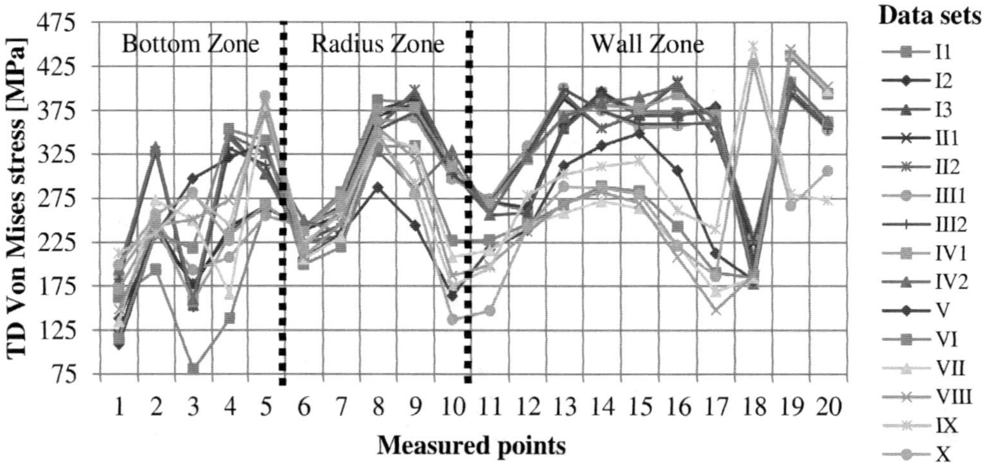

Figure 13. Von Mises stress values from the points measured along the TD of the formed part.

For the corner section of the formed part, the most noticeable differences ca be observed in the cases V – X, especially considering the fact that the cases IX and X have lower stress values in the radius zone and a few points from the wall zone. The highest Von Mises stress value is 531.49 MPa for the case IX, while the best case is II 2 with a maximum value of 507.85 MPa (figure 14).

The process simulation stopped before the hydroforming process was completed, in the cases IX and X. The finite element simulations with hydrostatic pressures of 50 MPa stopped before the full stroke of the punch, due to the large value of the equivalent von Mises stress. This situation is due to the wrinkles appearance in the top corner zones, blocking the flow of the material into the die. The fracture can appear in the critical zone 1. It can be concluded that the applied pressures have small influence toward the Von Mises stress variation, both along the TD and in corner zone.

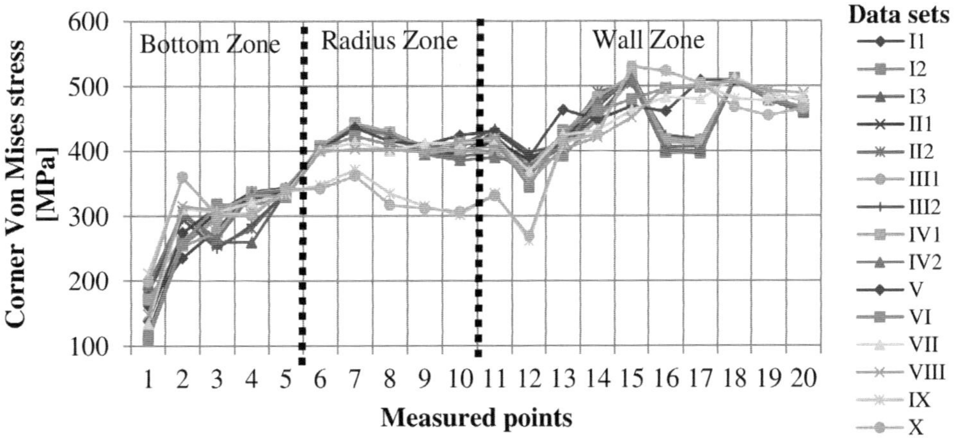

Figure 14. Von Mises stress values from the points measured in the corner section of the formed part.

4.4. Variation of σ_{11} and σ_{22} stresses
The graphs for the σ_{11} and σ_{22} stresses (figures 15-16) will be used to show what causes the high Von Mises values at the last few points of the corner section especially.

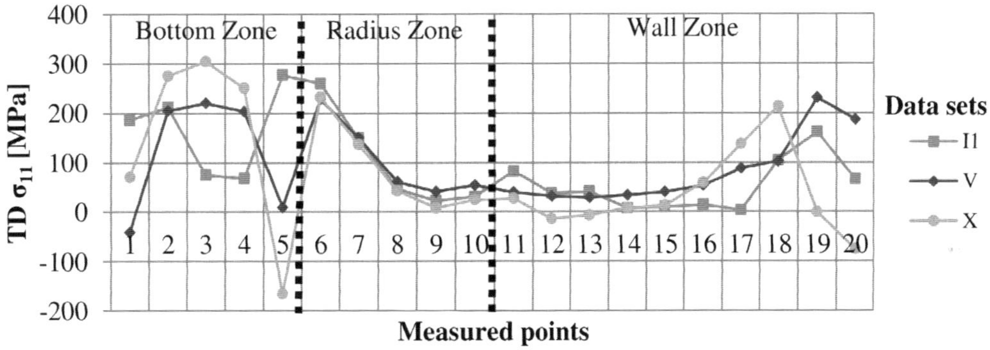

Figure 15. σ_{11} stress values from the points measured along the TD of the formed part.

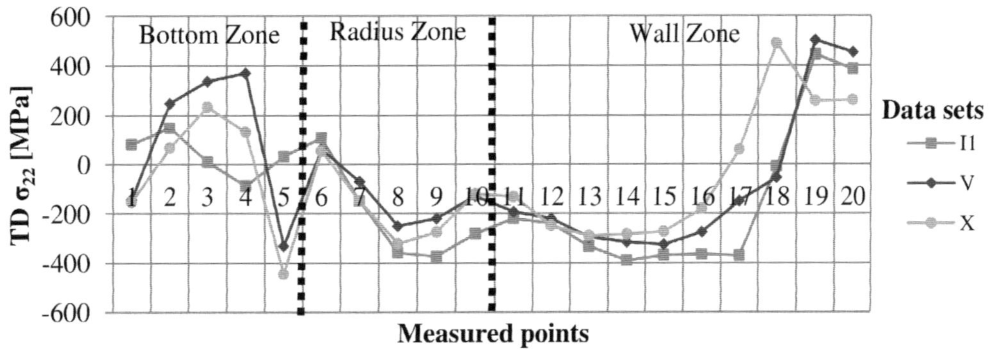

Figure 16. σ_{22} stress values from the points measured along the TD of the formed part.

For the TD of the formed part, the cases I - IV have similar σ_{11} (figure 15) and σ_{22} (figure 16) stress curves, but the cases V – X are very different in the bottom zone and the cases IX and X are also different in the wall zone. The minimum value is in the case X for both stresses, near the radius zone of the formed part: σ_{11} = -165.45 MPa and σ_{22} = -446.11 MPa. The maximum σ_{11} stress value is 304.32 MPa at the bottom of the part in the case X, and the maximum σ_{22} stress value is 490.23 MPa also in the case X, near the top of the part.

σ_{11} and σ_{22} stresses are also similar for all the I – IV cases for the corner section, except the cases IX and X. For σ_{11} stress, the maximum value is 495.63 MPa in the case X and the minimum value is -464,53 MPa in the case IX. σ_{22} graph has a maximum of 462.88 MPa in the case IX and a minimum of -467.9 MPa in the case X.

The fact that both graphs show at the top corner of the formed part, point 20, decreasing values compared to the previous points for most of the cases and even compression values for the cases V – X, means that the material flow is slowed down in that area and that is the reason that there are very high Von Mises stress values near the same area.

The ratio σ_{22}/σ_{11} shows a very high minimum value of -863.36 at point 16 in the case II 1 along the TD of the formed part, while in the corner section, in the same case – II 1, there is the highest maximum value of 113.44 at point 20. The lowest minimum value in the corner section is -51.46 at point 1 in the case VI. The positive values for the corner section are very small, the highest being 5.72 at point 20 in the case X.

4.5. FLD analysis

For the cases that were analyzed above, the major strain ε_1 and the minor strain ε_2 values were also extracted from the simulaions to be able to compare them with the FLC of the material [26]. In figure 17, for case I 1 and figure 18, for the case III 1, it can be seen that most points are in the tension/compression side [23], but there are a few points where the strain values of the formed part overlap the FLC values, meaning that there are some areas where there is a small risk of necking. The cases I – IV show similar results.

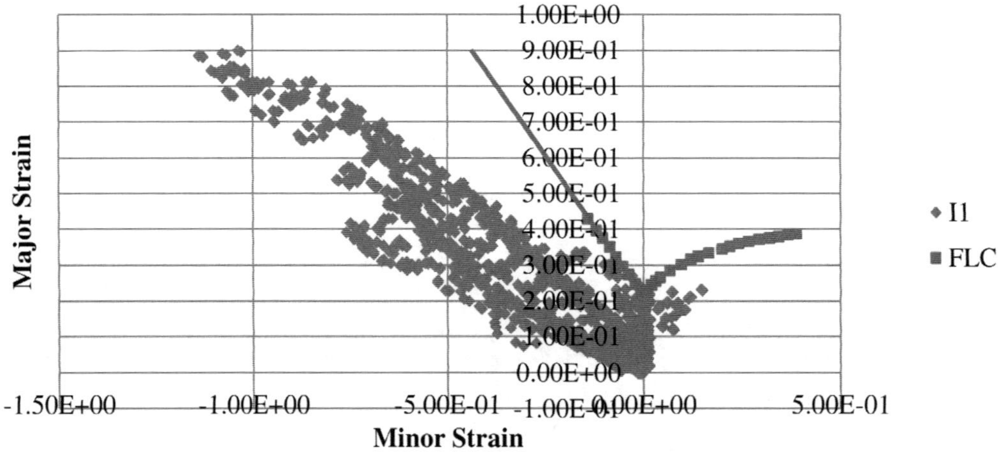

Figure 17. FLC comparison for the case I 1.

In the case VI there are points that are clearly over the FLC values, meaning that there is necking and a higher risk of fracture, as can be seen in figure 19.

IOP Conf. Series: Materials Science and Engineering **968** (2020) 012006 doi:10.1088/1757-899X/968/1/012006

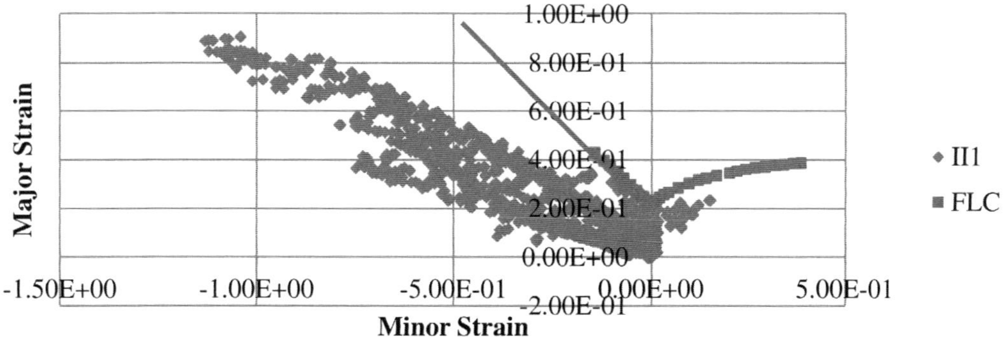

Figure 18. FLC comparison for the case III 1.

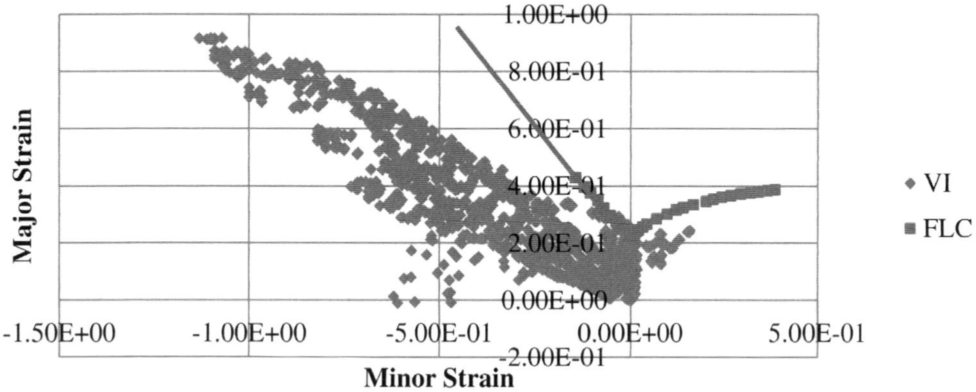

Figure 19. FLC comparison for the case VI.

5. Conclusions

The study analyse the influence of the hydroforming pressure p and p_1 and p_2 blankholder pressures applyed to different areas of the blank in order to analyses the quality of hydroforming part. The study revelead:

- Good results are obtained when the pressure p_1 applied by the blankholder on the linear areas of the blank is larger than the pressure p_2 applied by the blankholder on the curvilinear areas of the blank, for the same value of hydroforming pressure. The pressure p_1 lowers the material flow on the die in the linear areas so the material flow on the entire contour it becomes more uniform.
- From the data sets I – IV, the best overall results can be seen in the case III 2, that had the blankholder pressures p_1=8 MPa, p_2=7 MPa and the hydroforming pressure p=6 MPa and the worst results were in the case IV 2 with p_1=9 MPa, p_2=7 MPa, p=6 MPa.
- At higher pressures (data sets V – X), it was observed that some tests failed near the end. The failure occurred when the material stopped flowing because the applied pressures were too high.
- The best case between V- X was the case V that had the input parameters p_1=26 MPa, p_2=30 MPa, p=25 MPa and the two worst cases are: case IX with p_1=51 MPa, p_2=55 MPa, p=50 MPa and case X with p_1=55 MPa, p_2=51 MPa, p=50 MPa, that had the overall worst results.
- Comparing the case III 2 with the case V it results the following: the minimum value of the thickness is 0.891 in the case III 2 and 0.874 in the case V; the maximum value of Von Mises stress is 519.5 MPa in the case III 2 and 545.4 MPa in the case V; the equivalent plastic strain

has the maximum value of 1.27 in the case III 2 and 1.28 in the case V, showing the assurance of conditions for achieving a higher degree of deformation. Analysing these values, it results that the best case between allthat were studied is the case III 2.

- The quality of hydroformed part is well expressed by the stress (equivalent von Mises stress and components σ_{11} and σ_{22} of Cauchy stress tensor) evolution in the critical zones of the part, the minimum value of the thickness and the maximum value of the equivalent plastic strain.
- The applied pressures have similar plots toward the thickness, equivalent plastic strain and Von Mises variation, both along the TD and in the corner zone.
- Because in the top corners of the formed part there was an accumulation of material and the thickness had a very high value there compared to the rest of the part, those areas were trimmed at the end, similarly to what happens in reality. By removing the top corner areas the thickness distribution can be observed more accurately, as seen in figure 20 for the case I 1. The last four points from the 20 that were picked in the corner section previously are removed in this situation, the 16th point being the last one remaining after trimming. By trimming the part, most of the maximum values of stresses and thickness variation at the top areas are eliminated which leads to obtain a sound product.

The next step of the reasearch is to perform:

- numerical studies regarding the process optimization, considering ai inputs both design and process parameters and as outputs parameters the spingback values and degree of thinness;
- experimental tests to compare the obtained results with the ones from the finite element analysis;
- FEM simulations of the leakage fenomenon.

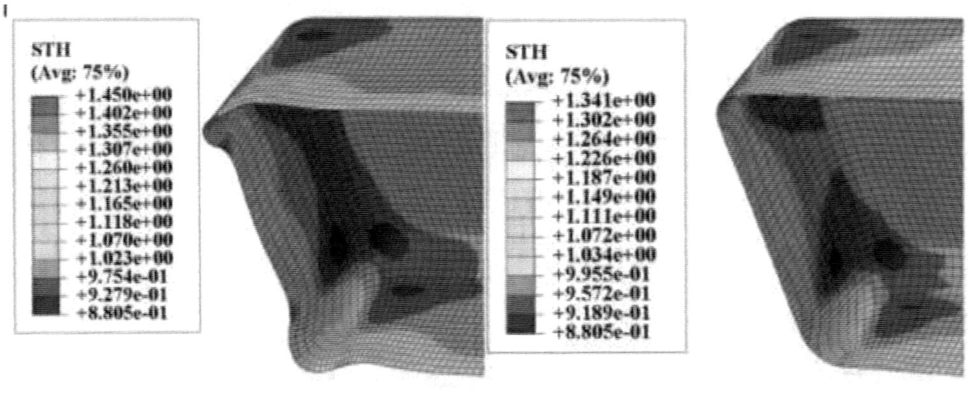

a) before trimming b) after trimming

Figure 20. The thickness variation of the formed part.

6. References

[1] Wang C, Meng B and Wan M 2017 Quick establishment of backpressure path in sheet hydroforming process 36th IDDRG Conference – Materials Modelling and Testing for Sheet Metal Forming *IOP Conf. Series: Journal of Physics: Conf. Series* **896** *012011*

[2] Maki T and Cheng J 2018 Sheet Hydroforming and Other New Potential Forming Technologies International Deep Drawing Research Group 37th Annual Conference *IOP Conf. Series: Mater Sci Eng* **418** *012117*

[3] Gather U, Hornberg W, Kleiner M, Klimmek C and Kuhnt S 2002 Parameter design for sheet metal hydroforming processes *Japan Proc. 7th ICTP* pp 1003

[4] Banabic D, Bunge H-J, Pöhlandt K, Tekkaya AE 2000, Formability of Metallic Materials: Plastic Anisotropy, Formability Testing, Forming Limits, *Springer*

[5] Shamsi-Sarband A, Zahedi SA, Bakhshi-Jouybari M, Hossinipour SJ and Banabic D 2012 Optimizitation of the pressure path in sheet metal hydroforming *Proc. of Romanian Academy Series* A **13**(4) 351

[6] Hama T, Ohkubo T, Kurisu K, Fujimoto H, Takuda H 2006 Formability of tube hydroforming under various loading paths, *Journal of Materials Processing Technology* **177**, pp. 676–679

[7] Paunoiu V, Teodor V, Susac F Researches regarding the hydroforming process of aluminum components 2015 *IOP Conf. Ser.: Mater. Sci. Eng.* **95** 012016

[8] Koller D and Ulbrich S 2011 Optimal control of hydroforming processes *Proc. Appl. Math. Mech.* **11** 795

[9] Kitayama S, Yokoyama M, Kawamoto K, Noda T, Miyasaka T and Echigo Y 2018 Practical approach of simultaneous optimization of variable blank *Int. J. Adv. Manuf. Tech* **98** 2693

[10] Colin Bell C, Corney J, Zuelli N and Savings D 2019 A state of the art review of hydroforming technology. *Int J Mater Form*

[11] Maier C, Epureanu Al and Marin F B 2013 Macromodelling of deep drawing process *Journal of Engineering and Architecture* 1/no.2 pp 1-9

[12] Maier C, Epureanu Al, Marinescu V, Paunoiu V, Afteni M, Marin FB 2010 Metal forming process control based on reduced order model *Proceedings of ModTech Int. Conf. - New face of TMCR* pp 363-366

[13] Paunoiu V and Nicoara D 2009 Numerical study in deep drawing with segmented blank-holder *Proceedings of ModTech Int. Conf. - New face of TMCR* pp 483

[14] Maier C, Paunoiu V, Marinescu V and Epureanu Al 2016 Method for Deep Drawing Process Control Using Segmented-Multiple Active Drawbeads *MATEC Web Conf.* **80** 14004

[15] Maier C, Paunoiu V and Marinescu V 2013 Developments in deep-drawing process control review *The Annals of Dunarea de Jos University of Galati, Fasc. V, Technologies in Machine Building* pp 25-30

[16] Altan T and Penter L 2010 Application of modern cushion systems to improve quality and productivity in sheet metal forming *Proceedings of the CIRP Conference Machine-Process Interactions,* Vancouver

[17] Hama T, Kitajima T, Nishimura Y, Fujimoto H and Takuda H 2012 Effect of outflow volume of pressure medium on fluid-lubrication effect during sheet hydroforming *Materials Transactions* **53**(5) pp 826-832

[18] Paunoiu V and Nicoara D 2002 An analysis of the blank shapes effect toward the quality of the square deep drawing parts, *The Annals of Dunarea De Jos University Of Galaţi*

[19] Modi B and Kumar DR 2013 Development of a hydroforming setup for deep drawing of square cups with variable blank holding force technique *The International Journal of Advanced Manufacturing Technology* **66**(5-8) pp 1159-1169

[20] Nakamura K. and Nakagawa T. 1987 Sheet metal forming with hydraulic counter pressure in Japan *CIRP Annals* **36**(1) pp 191-194

[21] Iacob I, Paunoiu V and Maier C 2019 FEM analysis of the hydroforming process with segmented blank holder (poster) *Romania SCDS-UDJG 2019*

[22] Abaqus 6.10-EF Analysis User's Manual 2010, from 130.149.89.49:2080/v6.10ef/books/usb/default.htms

[23] Banabic D, Bunge H-J, Pohlandt K and Tekkaya A E 2000 *Formability of Metallic Materials* (Berlin: Springer)

[24] Banabic D, Aretz H, Paraianu L and Jurco P 2005 Application of various FLD modelling approaches *Modelling and Simulation in Materials Science and Engineering***13**(5) pp 759.

[25] Silva MB, Martínez Donaire, A J, Centeno Báez G, Morales-Palma D, Vallellano Martín C and Martins PAF 2015 Recent approaches for the determination of forming limits by necking and fracture in sheet metal forming *Procedia Engineering* **132** pp 342-349.

[26] Banabic D, Lazarescu L and Comsa D S 2015 Predictive Performances of FLC Using Marciniak-Kuczynski Model and Modified Maximum Force Criterion *Key Engineering Materials* **651** pp 96-101

[27] Korkolis Y P, Kyriakides S, Giagmouris T and Lee L-H 2010 Constitutive modeling and rupture predictions of Al-6061-T6 tubes under biaxial loading paths, *J. Appl. Mech.* **77**(6) 064501

[28] Tian H, Brownell B, M Baral M and Korkolis YP 2017 Earing in cup-drawing of anisotropic Al-6022-T4 sheets *International Journal of Material Forming* **10** pp 329-34

[29] Ha J, Baral M and Korkolis YP 2018 Plastic anisotropy and ductile fracture of bake-hardened AA6013 aluminum sheet *International Journal of Solids and Structures* **155** 123

Aknowledgement

This work was supported by the grant of the Romanian National Authority for Scientific Research and Innovation, CNCS – UEFISCDI, grant number PN-III-P1-1.2-PCCDI-2017-0446, within PNCDI III – 82/2018.

IOP Conf. Series: Materials Science and Engineering **968** (2020) 012007 doi:10.1088/1757-899X/968/1/012007

Influence of some process input factors on the main dimensions of the grooves generated during the ball vibroburnishing

G Nagit, O Dodun, L Slatineanu, M Ripanu, A Mihalache and A Hrituc
"Gheorghe Asachi" Technical University of Iasi, Department of Machine Manufacturing Technology, Bd. D. Mangeron, 59 A, 700050 Iasi, Romania

E-mail: oanad@tcm.tuiasi.ro

Abstract. The ball vibroburnishing is a processing method based on the plastic deformation of the workpiece surface layer, as a result of a vibration movement achieved by the ball pressed with a known force on the workpiece surface. The surface obtained by ball vibroburnishing includes grooves with different directions and partially overlapped. To know better the influence exerted by the ball vibroburnishing conditions on the main dimensional characteristics of the grooves, an experimental research was designed and materialized. As the process input factors, the diameter of the ball, the force, and the workpiece rotation speed were used. The depth and width of the groves generated by the moving balls on the workpiece surface layer were measured. By mathematical processing of the experimental results, empirical mathematical models were determined. These models highlight the intensity of the influence exerted by the ball vibroburnishing process input factors on the main dimensions of the grooves.

1. Introduction

In the case of steel to which the thermal treatments could not be applied, to improve their functional properties, it is possible to affect surface roughness, and, on the other hand, the properties of the surface layer. With this aim in view, cutting processes and processes based on the plastic deformation of the workpiece surface layer could be used.

As finishing processes applied in the case of the metallic workpieces and when using the material removal from the workpiece, it can take into consideration the tools characterized by the existence of the metallic cutting edges (turning, milling, etc.), and tools based on the use of the abrasive particles (grinding, honing, lapping, superfinishing, etc.). The influence exerted by the process input factors specific to the cutting processes mentioned above was investigated and there is impressive knowledge on this aspect.

The surface roughness and the mechanical properties of the steels for which the thermal treatments could not be applied can be improved by processes based on the plastic deformation of the workpiece superficial layer. The burnishing and vibroburnishing are such processes applied to change the roughness and the hardness of the surface layer of the workpieces that possess flat or revolutions surfaces [1-8].

The burnishing is performed using balls or diamond tips having a certain rounding working surface and whose moving paths are thus established to entirely cover the surface to be processed. At the same time, the deforming tool is pressed on the workpiece surface to be processed.

In the case of *the vibroburnishing*, an additional vibration motion ensures a more complex movement of the deformation tool on the workpiece surface and groves characterized by a lower or higher area of the cross-section are generated on the workpiece surface. If these groves are of low height, it is expected to obtain low values of the surface roughness parameters that take into consideration the height of the surface asperities. It is also possible that the height and width of the grooves also exert influence on other functional properties of the surfaces obtained by vibroburnishing (friction coefficient, wear resistance, the capacity of retaining the oil, etc.).

There is relatively less information concerning the influence exerted by the surface deformation conditions on the dimensions of the grooves generated by vibroburnishing.

Thus, Schneider and Guzok investigated the influence exerted by the groove area on the value of the friction coefficient in the case of applying the vibroburnishing process to the parts of the refrigeration compressors [9]. They appreciated that, in this way, control of the bearing surface and capacity of retaining the oil and finally an increase of the parts working life are possible. Dmitriev et al. approached the problem of changing the feature of structure transformation as a consequence of applying a nano-burnishing process [10]. They presented inclusively aspects concerning the surface profile taking into consideration the model crystal and the proper working conditions at the level of micro space. Matalin and Svinitskaya performed an investigation concerning the submicroscopic microrelief generated by some different finishing processes, one of them being the vibroburnishing [11].

In the previously published works that were accessible to us, the problem of identifying non-linear empirical mathematical models that highlight the influence of working conditions on the channel sizes generated during ball vibroburnishing was not addressed. The objective of the research whose results are presented in this paper was to identify ways of highlighting the influence exerted by some input factors of the plastic deformation process on the width and depth of the groves generated by the vibroburnishing applied to test pieces that present cylindrical surfaces. The elasticity characteristics of the workpiece material can influence the geometry of the grooves generated by vibroburnishing as the phenomenon of elastic recovery manifests differently for different metallic materials.

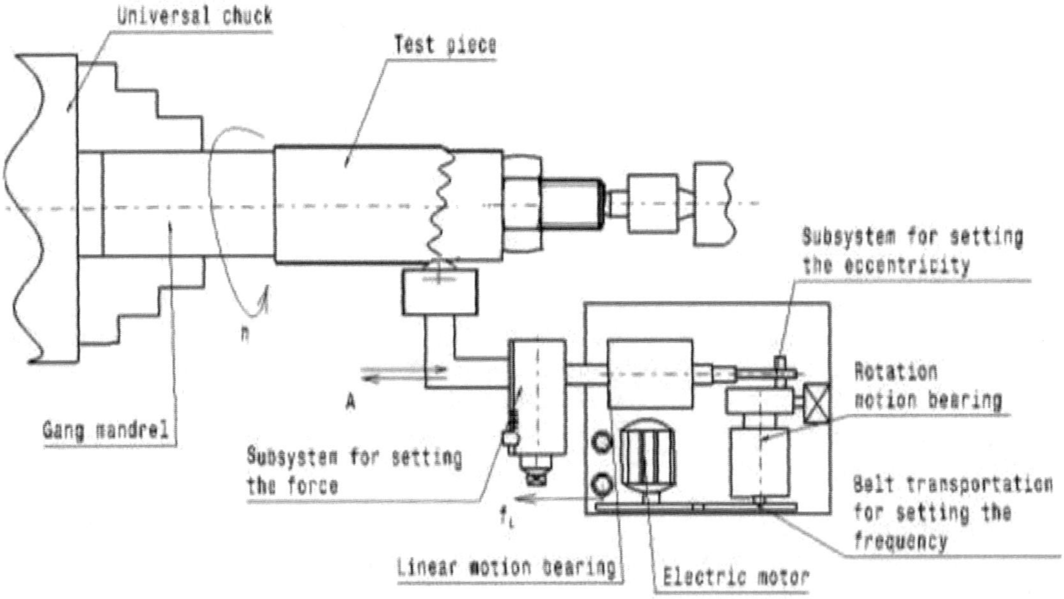

Figure 1. Schematic representation of the vibroburnishing process.

IOP Conf. Series: Materials Science and Engineering **968** (2020) 012007 doi:10.1088/1757-899X/968/1/012007

2. Vibroburnishing process

A processing scheme valid in the case of the vibroburnishing applied to cylindrical surfaces when using the lathe as machine tool and working devices adapted to this machine tool is presented in figure 1.

Table 1. Experimental conditions and results.

Line no. 1	Ball diameter, d_b, [mm]	Force F, [N]	Test piece diameter, d, [mm]	Groove depth, h, [μm]	Groove width, l_g, [μm]
Column no. 1	2	3	4	5	6
3	6.75	100	40	7	35.2
4	6.75	100	50	8.2	36.4
5	6.75	100	60	9	39.5
6	6.75	300	40	10.3	42.4
7	6.75	300	50	12.5	51.2
8	6.75	300	60	14.2	53.4
9	6.75	500	40	21	70.4
10	6.75	500	50	27	80.2
11	6.75	500	60	30	81.3
12	11.12	100	40	6	42.3
13	11.12	100	50	5.6	40.4
14	11.12	100	60	8	48.3
15	11.12	300	40	10	60.4
16	11.12	300	50	12	68.5
17	11.12	300	60	14.3	71.3
18	11.12	500	40	17	83.4
19	11.12	500	50	26	90.3
20	11.12	500	60	28	92.7
21	15.85	100	40	5.1	50.2
22	15.85	100	50	6.8	56.4
23	15.85	100	60	7.6	60.3
24	15.85	300	40	9	70.4
25	15.85	300	50	10.2	72.3
26	15.85	300	60	12.1	80.1
27	15.85	500	40	17	98.3
28	15.85	500	50	19	100
29	15.85	500	60	23	108

Input factors whose values have been kept constant: rotation speed n=250 rev/min, longitudinal feed f_l=1.5 mm/rev, frequency of vibration motion n_{ds}=350 double strokes/min, amplitude of vibratory motion $2A$=2 mm.

The ball vibroburnishing device was implemented on a medium-size universal lathe [12]. The device had a plate that facilitated its mounting instead of the common tool holder of the universal lathe. An electric motor transmits a rotational motion to the device shaft through a belt drive. The rotating shaft has a disc at one end where a bolt can be moved along a radial direction and fixed in the desired position. A certain value of the eccentricity of the bolt is thus ensured. A value corresponding to the double value of the eccentricity will be the amplitude of the vibration movement. A cylindrical joint allowed the bolt to drive a connecting rod. At the other end of the connecting rod, there is a spring subsystem for adjusting the size of the force exerted on the ball that will perform the vibroburnishing process. By placing the ball in contact with the cylindrical test piece surface and pressing it with known values using the arc, the value of the pressing force exerted by the ball on the sample surface was ensured. The device was moved in a longitudinal feed movement of preset value by the lathe apron. The bush type test piece was mounted on a mandrel supported at one end in the universal chuck and at the other end in a live center. The rotation of the mandrel on which the test piece was located was ensured from the main shaft of the universal lathe. Changing the wheels of the belt transmission, the change of the mandrel rotation speed was possible. As a deformation tool, a ball of known diameter and made of high hardness steel (bearing steel, 60 HRC) was used.

3. Design of the experimental research

The processing scheme presented in figure 1 was used to develop the experimental research. Cylindrical bushes with different diameters were used as test pieces. The test pieces were made of steel 1.060, whose chemical composition includes 0.6 % C, 0.6 % Si, 0.7 % Mn, 0.04 % P, 0.05 % S (185 Brinell hardness).

Before developing the proper experimental research, a calibration operation was applied to determine the values of the force that corresponds to the different levels of the spring compression.

As deformation process input factors, initially, the ball diameter d_b and the force F were considered. A detailed analysis of the processes developed at the zone of the work micro space showed that the quantity of the material moved as a result of the plastic deformation could be different for different diameters of the test piece and, for this reason, the test piece diameter d (the external diameter of the bush type test piece) was taken into consideration as a third process input factor. The groove depth h and width l_g were considered as the output parameters. The sizes h and l_g were defined as can be seen in the figure 2, *a*.

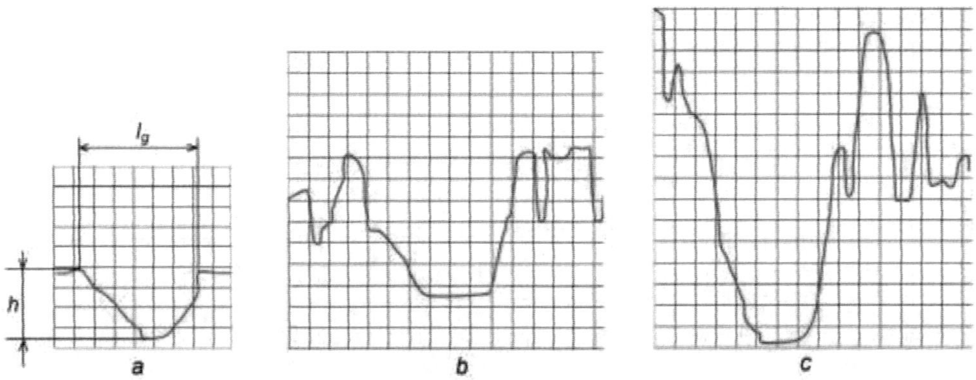

Figure 2. Shape and dimensions of the cross-section that correspond to the grooves performed using the vibroburnishing process and three different values for the deformation forces: *a* – 100 N, *b* – 300 N, *c* – 500 N (test piece material: steel 1.060; n=250 rev/min, f_l=1.5 mm/rev, f=250 double strokes/min=5.8 Hz, $2A$=2 mm, d_b=11.12 mm; vertical magnification of 2000, horizontal magnification of 20).

Appreciating that a maximum or a minimum of the process output parameters could correspond to certain values of the process input factors, experimental research at three levels of each process input factor was taken into consideration.

Aiming at developing a maximally completed experiment, the decision of developing a full factorial experiment with three process input factors at three levels was established. This means that 3^3=27 experiments had to be performed.

The values of the depth and height of the groove were measured using surface roughness tester type G4 (made in the former Soviet Union).

The values of the process input factors were mentioned in the columns nos. 2, 3 and 4 from table 1, while the values of the two process output parameters obtained as a consequence of the experimental research were mentioned in the columns nos. 5 and 6. The values of the input factors of the experimental research were established by taking into account the recommendations of other researchers and as a result of conducting preliminary research.

Three proper images that correspond to the grove cross areas obtained for three different values of the applied force *F* were presented in figure 2, *a*, *b*, and *c*.

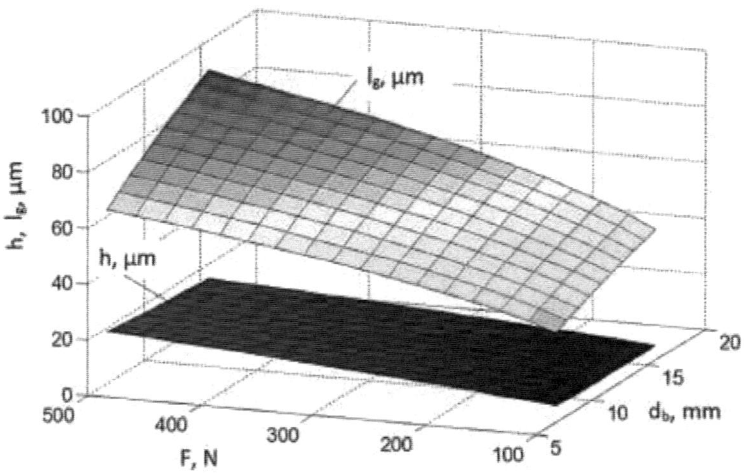

Figure 3. Effect of the ball diameter d_b and force *F* on the depth *h* and width l_g of the spiral groove generated by vibroburnishing of a test piece made of steel 1.060 (test piece diameter *d*=40 mm).

4. Experimental results and their analysis

Using the surface roughness tester G4 and a movement of the probe tip along the test piece axis, the profiles of all the 27 grooves were obtained. Based on the profilograms, the average depth and width of each groove were evaluated.

The experimental results were mathematically processed using specialized software based on the method of least squares [13]. In this way, the following empirical mathematical models of power type were determined:

$$h=0.0186d_b^{-0.258}F^{0.692}d^{0.841} \quad (1)$$

and

$$l_g=0.608d_b^{0.429}F^{0.405}d^{0.35} \quad (2)$$

Based on the empirical mathematical models (1) and (2), the diagrams from figures 3, 4 and 5 were developed. Taking into consideration the empirical mathematical relations (1) and (2) and the graphical representations from figures 3, 4 and 5, some remarks could be formulated.

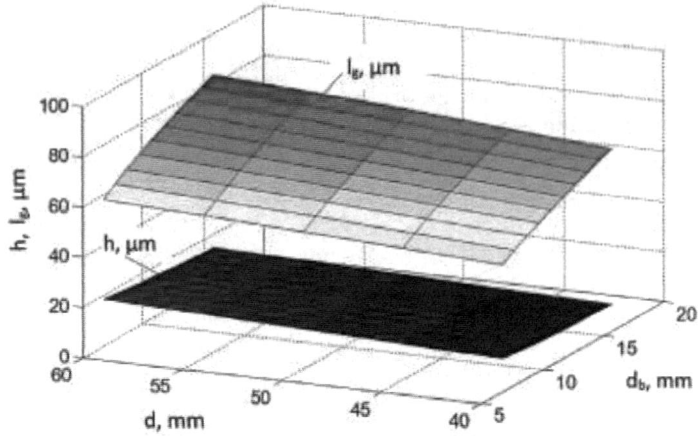

Figure 4. Effect of the ball diameter d_b and test piece diameter d on the depth h and width l_g of the spiral groove generated by vibroburnishing of a test piece made of steel 1.060 (force F=300 N).

Thus, in the case of the power type function that corresponds to the depth h of the groove (relation (1)), the increase of the ball diameter d_b determines a low decrease of the groove depth h, since the value of the exponent attached to this size is subunit and negative. At the same time, the increase of the force F and test piece diameter d leads to a low increase of the groove width l_g, since the exponents attached to the two sizes in the relation (1) are subunit and positive. Among the three process input factors, the test piece diameter d seems to exert the maximum influence, since to this factor the highest value is attached as an exponent.

An explanation of the decrease of the groove depth h when the ball diameter d_b increases could be based on the increase of the material resistance to be penetrated by the ball pressed on the test piece surface. The increase in the dimensions of the grove cross-area when the test piece diameter d increases also could be explained by a better flow of the material when the peripheral speed of the test piece increases, due to the increase in the test piece diameter d.

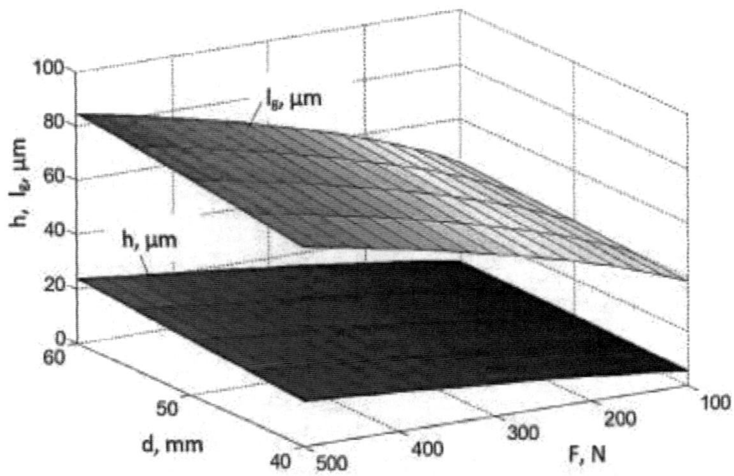

Figure 5. Effect of the force F and test piece diameter d on the depth h and width l_g of the spiral groove generated by vibroburnishing of a test piece made of steel 1.060 (ball diameter d_b=10 mm).

It was expected the increase of the groove dimensions when the force F exerted on the burnishing tool increases, due to the higher penetration of the burnishing tool in the test piece material.

In the case of the groove width l_g, the increase of each of the independent variables (ball diameter d_b, force F and test piece diameter d have, as a result, a low increase of the dependent variable, since all the exponents have positive subunit value. The highest influence is exerted by the ball diameter d_b, since the higher value of the exponent is attached to this independent variable in the relation (2). The influences of the force F and test piece diameter d on the depth h and width l_g of the groove have a similar character, as the two empirical mathematical relations show, but the influence is stronger in the case of the groove depth h, since the values of the exponents attached to the two sizes in the relation (1) are higher than those existing in the relation (2).

Experimental results can be taken into account in industrial practice, for example when it is necessary to ensure certain values of tribological characteristics for surfaces found in relative motion to each other.

5. Conclusions

The vibroburnishing is a process based on the plastic deformation of the workpiece surface layer using a ball pressed against the workpiece surface and relative motions able to ensure the covering of the entire workpiece surface to be processed. The grooves generated by burnishing ball on the workpiece surface during the vibroburnishing process could be characterized by their depth and width. The analysis of the experimental conditions in which the ball vibroburnishing process develops showed that some process input factors could exert influence on the two dimensions used to characterize the spiral groove generated by the vibroburnishing process. Experimental research aiming at determining the influence exerted by the ball diameter d_b, force F and test piece diameter d was designed and materialized. The conditions of a full factorial experiment with three independent variables at three experimental levels were taken into consideration. By mathematical processing of the experimental results, power type empirical mathematical models were determined. The analysis of the empirical mathematical models showed that only the increase of the ball diameter led to a decrease of the grove depth, while in the other cases, the increase of the independent variables has a result in an increase of the dependent variable. As main limitations of the technology investigated in this paper, we can mention the application only in the case of metallic materials characterized by a certain plasticity and the need to have adequate processing equipment. In the future, there is the intention to extend the experimental research, taking into consideration other steels used in the field of machine building.

6. References

[1] Odintsov VG 1981 *Final processing of details and diamond by burnishing vibroburnishing* (Moscow: Mashinostroenie)

[2] Nagîţ G., Slătineanu L, Dodun O, Rîpanu MI, Mihalache AM 2019 *J. Mater. Res. Technol.* **8** 4333-4346

[3] Slavov SD, Dimitrov D 2016 *Evraziiskii Soiuz Ucenyh –Tecnicheskii Nauki* **25** 11–22

[4] Georgiev DS, Kristev KA, Slavov SD 2002 *Methodology formeasurement the parameters full area and lubricant ability of surfaces manufactured by vibratory ball burnishing process Mashinostroitelna Tehnika i Tehnologii* (in Bulgarian) (Varna: Tehnicheskii Universitet)

[5] Patel MK, Patel MD 2012 *Int. J. of Res. in Engg. & Appl. Sci.* **2** 372-381

[6] García-Granada AA, Gomez-Gras G, Jerez-Mesa R, Travieso-Rodriguez JA, and Reyes G 2017 *Mater. Manuf. Process.* **32** 1279-1289

[7] Nikonova T, Zhetessova G, Yurchenko V 2018 Mathematical modeling of vibroburnishing of the hole of cylinder *J. of Appl. Engg Sci.* **16** 5-10

[8] Teimouri R, Amini S, Bami AB 2018 Evaluation of optimized surface properties and residual stress in ultrasonic assisted ball burnishing of AA6061-T6 *Meas.* **116** 129–139

[9] Shneider YG, Guzok YV 1973 *https://link.springer.com/content/pdf/10.1007%2FBF01155643.pdf* Accessed: 13.04.2020
[10] Dmitriev AY, Nikonov Y, Kuznetsov VP 2011 *Proceedings of XXXIX International Summer School–Conference APM Advanced Problems in Mechanics* (St. Petersburg) 114-120
[11] Matalin AA, Svinitskaya VG 1977 *Met. Sci. and Heat Treat.* **19** 1054–1057
[12] Nagîț G, Slătineanu L, Braha V 1996 *Device for vibrorolling of the external cylindrical surfaces* (in Romanian). Romanian Patent 118007
[13] Crețu G 1992 *The basis of experimental research. Laboratory guide* (in Romanian) (Iași: "Gheorghe Asachi" Technical University) pp 27–35

IOP Conf. Series: Materials Science and Engineering **968** (2020) 012008 doi:10.1088/1757-899X/968/1/012008

Influence of marine corrosion on the roughness of MAG welded joint surfaces

S N Florescu[1], M C Gheonea[2], D Mihailescu[2], V Teodor[2]
[1] National Institute of Marine Geology and Geo-Ecology (GeoEcoMar), 23-25 Dimitrie Onciul St., 024053 Bucharest, Romania
[2] "Dunarea de Jos" University of Galati, Faculty of Engineering, Domneasca Street, 47, RO-800008, Galati, Romania

E-mail: stefan.florescu@geoecomar.ro

Abstract. The study presents in the first part the behavior of the EH 36 steel with the thickness of 10 mm to MAG mechanized welding of butt-welded joints from the naval field, destructive and non-destructive testing for all welded samples. Furthermore, corrosion of the marine environment is analyzed by the electrochemical method. The study is completed with the corrosion influence on the roughness of the welded joints surfaces and conclusions.

1. Introduction

Steels for shipbuilding and offshore structures are low carbon steels, with good weldability, designed for high productivity processes of welded structures, which meet the conditions imposed by classification societies. In the early 1980s, in addition to normal-strength naval steels, the *High Strength Low Alloy* (HSLA) steels, which allow the production of narrower welded naval structures, were introduced to the same mechanical stresses. According to the rules of Lloyd's Register (LR), naval steels are classified as follows [1, 2]:

- with normal strength ($Rp_{0.2}$ min. 265 MPa): AH 27S, DH 27S, EH 27S, and PH 27S steels;
- high strength ($Rp_{0.2}$ min. 315 MPa): AH 32, DH 32, EH 32, and PH 32 steels ($Rp_{0.2}$ min. 315 MPa); AH 36, DH 36, EH 36, and PH 36 steels ($Rp_{0.2}$ min. 355 MPa); AH 40, DH 40, EH 40, and PH 40 steels ($Rp_{0.2}$ min. 390 MPa), and EH 47 steel ($Rp_{0.2}$ min. 460 MPa).

For the butt-welded joint of the steel plates from the naval field, in all welding positions, the welding process in fusible electrode gas environments is the most commonly used (MAG welding 135 and 136). Depending on the volume of work and the deadlines imposed, the decision will be made regarding the degree of mechanization necessary to ensure optimum welding productivity [3].

The experimental research presented in this study is structured in three directions: the welding behavior of the EH 36 naval steel with the thickness of 10 mm; the corrosion behavior by the electrochemical method of the base material and the welded joints; the influence of the marine corrosion on the roughness of the base material and welded joints surfaces. This research was carried out in three research units within "Dunarea de Jos" University of Galati, Faculty of Engineering, Center for Advanced Research in Welding (SUDAV), Competences Center: Interfaces-Tribocorrosion and Electrochemical Systems (CC-ITES), and the Center for Research Technological Engineering in Machine Construction (ITCM).

2. Experimental program

2.1. Welding materials and equipment

In the experimental program, the following were used [3]:

- eight EH 36 naval steel sheets with 500x150x10 mm dimensions;
- metal-cored wire E70C-6MH4 according to AWS A5.18 and rutile flux-cored wire E81T1-Ni1MJH4 according to AWS A5.29, with diameters of 1.2 mm;
- Corgon 18: 82% Ar + 18% CO_2 (M21) shielding gas mixture;
- flat ceramic backing with concave channel (channel width 9 mm and channel depth 1.3 mm);
- four stands for MAG-M mechanized welding in PA/1G, PC/2G, PF/3G, and PE/4G positions. The stands were equipped with the universal welding source Phoenix 405 Progress pulse MM TDM and the welding tractor K-BUG 5102.

2.2. Welding of samples

Within the experimental program, four butt-welding samples were welded in the positions involved in the construction and assembly of naval sections. The root layers were made with metal-cored wire E70C-6MH4 and RootArc transfer. The filling and final layers were made with rutile flux-cored wire E81T1-Ni1MJH4 and ForceArc transfer. The layers were deposited with pendulum and stationary on the edges (except for the final layer PC/2G welded sample made of three rows without pendulum). The samples were made on flat ceramic backing with concave channel.

The plates for the EH 36 naval steel samples, with dimensions of 500x150x10 mm, were processed with air plasma, according to Figure 1 (a - positions PA/1G, PF/3G, PE/4G; b - position PC/2G).

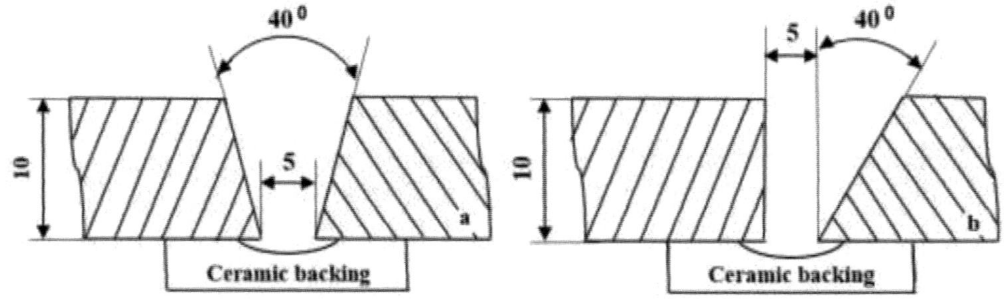

Figure 1. The preparation of the plates for welding [3].

During the mechanized welding some of the technological parameters were kept constant [3]: DC^+, reverse polarity; stationary time at the edges, t_{sm} = 0.2 ms; gas protection flow, Q_G = 18 l/min; gas flow time before starting the welding process, t_{bw} = 5 s; gas flow time after ending the welding process, t_{aw} = 5 s, and the sample-nozzle distance, h_{sn} = 15 mm [3]. Table 1 centralizes the welding process variables which were applied in the experimental program.

Table 1. Welding process variables applied in the experimental program [3].

Layer type	Welding position							
	Parameters of welding regimes*							
	v_e, [m/min]	I_w, [A]	U_a, [V]	v_t, [cm/min]	v_p, [cm/min]	L_p, [mm]	v_w, [cm/min]	E_l, [KJ/mm]
0	1	2	3	4	5	6	7	8
				PA				
Root	4.5	183	26.5	25	80	6	10	0.065
Final	11	302	31	30	80	8	13.7	0.091
				PF				

Root	2.7	103	13.8	17	60	6	8	0.024
Filling	5.4	158	18.9	27	60	8	14.3	0.028
Final	6	167	18,9	30	60	10	15.5	0.027
				PE				
Root	2.7	103	13.8	17	60	6	8.13	0.023
Filling I	5.4	158	18.9	27	60	8	14.3	0.028
Filling II	6	167	18.9	30	60	10	15.5	0.027
Final	6	167	18.9	27	60	12	16.7	0.025
				PC				
Root	5.9	170	18.9	60	-	-	60	0.071
Filling	8	220	25.3	60	-	-	60	0.012
	7	192	22.5	60	-	-	60	0.009
Final	6	167	18.9	60	-	-	60	0.007
	5.5	160	18.9	92	-	-	92	0.004

Note*: v_e- advance speed of the welding wire; I_w- welding current; U_a- arc voltage; v_t- speed of the welding tractor; v_p- pendulum speed; L_p- pendulum width; v_w- welding speed; E_l- linear energy.

Figure 2. Butt-welding of samples [3].

Figure 2 presents aspects during the welding of the four butt-welded samples: a - position PA; b - position PF; c - position PE, and d - position PC.

3. Non-destructive and destructive testing

3.1. Non-destructive testing

According to the rules of Lloyd's Register (LR), the butt-welded joints samples have been subjected to non-destructive, visual (100%), and radiographic (100%) testing. Following the non-destructive inspections, all four welded samples were of good quality (admitted), without surface and interior defects [1, 3].

3.2. Destructive testing

All four samples welded to each other had dimensions of 500x300 mm, and from each sample the following specimens were taken [1, 3]:

- one specimen for the macroscopic and microscopic examination and the Vickers HV1 microhardness test;
- four specimens for testing at the front transverse bending (two with the compressed root and two with the extended root), at an angle of 180 ^{0}, on a 40 mm diameter thorn;
- two flat specimens with the calibrated portion for cross-tensile testing;
- nine specimens for the bending test by shock on Charpy V-shaped test tubes, at a temperature of - 20 ^{0}C: three with a notch in the center of the weld; three with the notch on the fusion line, and three with the notch in the thermomechanically influenced area, at a distance of 2 mm from the fusion line.

The good results of the mechanical laboratory tests, and the macroscopic and microscopic analyzes, make possible the application of the welding technologies for the high strength EH 36 naval steel with the thickness of 10 mm in the convenient PA position and in the difficult positions (PF, PE, and PC).

4. Corrosion behavior by electrochemical method

The experimental program regarding the corrosion behavior of the base material and the welded joints in natural seawater was conducted with the electrochemical working equipment consisting of: potentiostat / galvanostat PGZ 100, the electrochemical cell, and the computer for data acquisition.

In the electrochemical cell, containing 200 ml of Black Sea water, the three electrodes were arranged, the working electrode (cathode), the auxiliary electrode (anode) Pt-Rd, and the reference electrode. The method of sampling the working electrodes from the specimens previously subjected to mechanical bending tests with the compressed root is shown in figure 3.

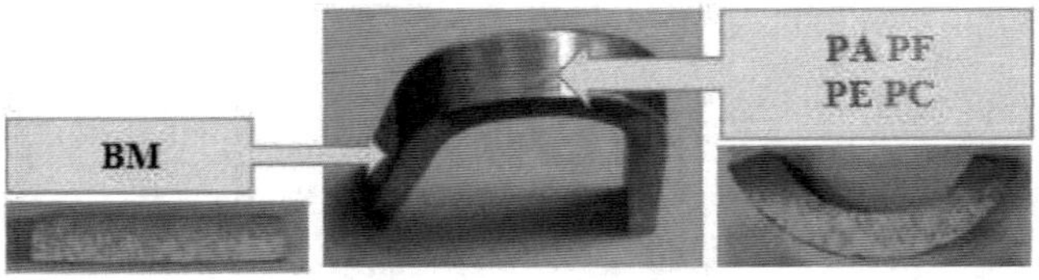

Figure 3. Sampling for the corrosion behavior analysis [4].

The resulting data were interpreted with the help of the VoltaMaster software 4. From the corrosion behavior of the base material and the welded joints in natural seawater, it can be observed that the corrosion speeds decrease with the exposure time, the lowest corrosion rate had the base material BM, and the corrosion speeds of the welded joints are different, depending on the welding regimes used [4].

5. The influence of marine corrosion on the surface's roughness

Experimental research on the influence of marine corrosion on the surface's roughness was performed using the SJ210 Mitutoiu roughness meter and SJ-Comunication-Tool analysis software.

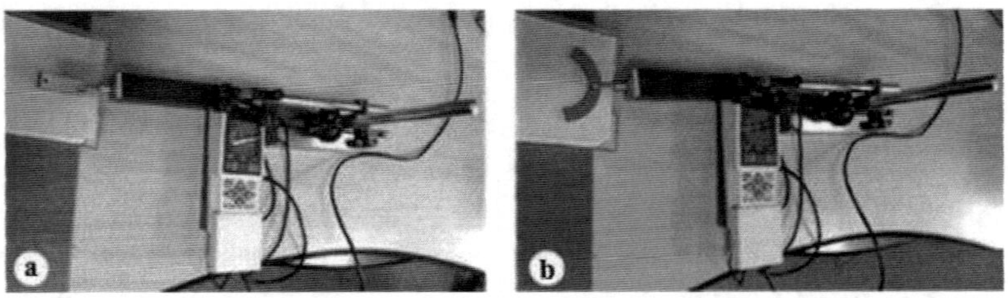

Figure 4. Determination of the depths of the asperities on the surfaces.

The determinations were made on a distance of 4 mm, with a speed of 0.5 μm/s on the surfaces of the five specimens, before and after conducting the behavior analyzes on corrosion (Figure 4: a - for the specimen taken from the base material BM, and b - for samples taken from butt-welded joints PA, PF, PE, and PC).

In the case of the base material, the depths of the asperities on the surfaces were determined in three areas, the base, the middle, and the upper part of the specimen taken on the thickness of the base material BM.

For the welded joints, the determinations were also made in three zones, corresponding to the root layer 1, the filling layer 2, and the final layer 3, in the joint between the welding components (PA, PF, PE, and PC samples).

In figures 5-7, the evolutions of the asperities' depths on the surfaces, before and after conducting the corrosion behavior analyzes, for a determination in the base material (BM) sample, and a determination related to the final layers are presented, in direct contact with the seawater (PA, PF, PE, and PC samples).

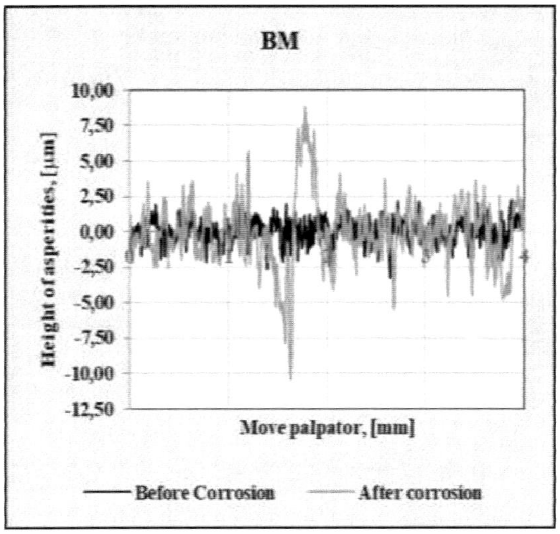

Figure 5. The evolutions of the depths of asperities for the BM sample.

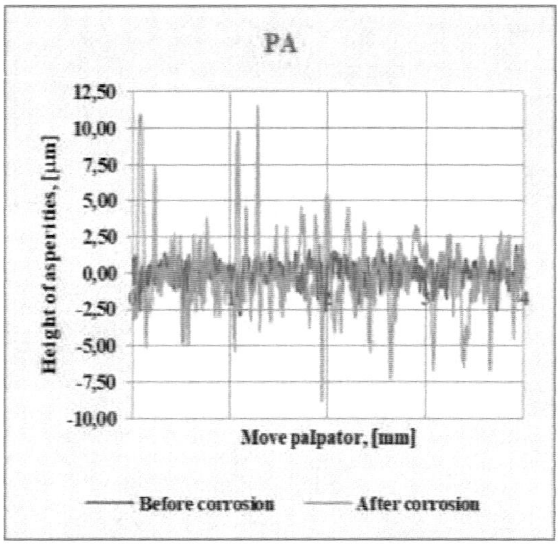

Figure 6. The evolutions of the depths of asperities for the PA sample.

With the SJ-Communication-Tool analysis software, the Ra roughness could also be determined. Figure 10 shows the average *Ra* values, before and after completing the corrosion behavior analyzes.

From figures 5...8, it is observed that both depths of the asperities on the surfaces, as well as the average roughnesses Ra of the surfaces, after the completion of the corrosion behavior analyzes by the electrochemical method, are higher, due to the corrosive action of the natural seawater [5, 6, 7].

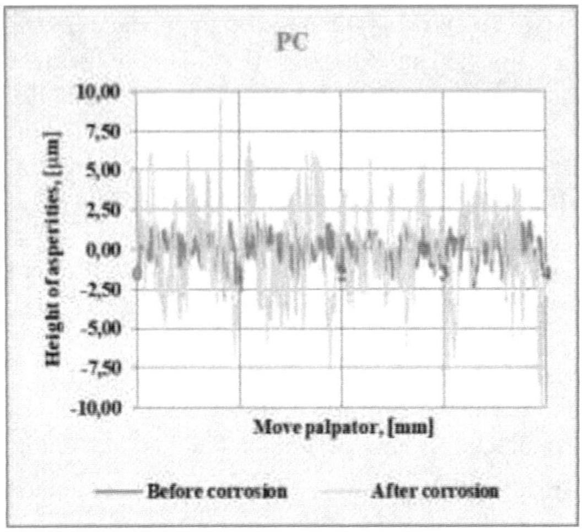

Figure 7. The evolutions of the depths of asperities for the PC sample

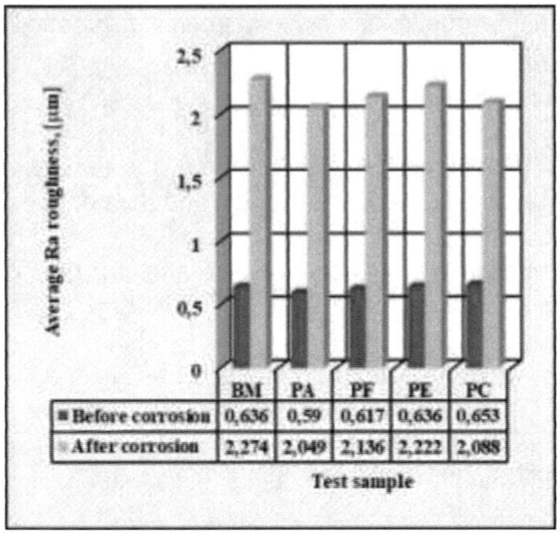

Figure 8. Average *Ra* roughness

6. Conclusions

The good results of the mechanical laboratory tests, and the macroscopic and microscopic analyzes, make possible the application of the welding technologies for the high strength EH 36 naval steel with the thickness of 10 mm in the convenient PA position and in the difficult positions (PF, PE, and PC).

After completing the behavioral analyzes on corrosion (the corrosion behavior of the base material and the welded joints in natural seawater) the roughness and the average roughness *Ra* of the test surfaces, were higher, due to the corrosive action of the seawater.

7. References

[1] *** *Rules for the Manufacture, Testing and Certification of Materials,* Lloyd's Register, July 2018.

[2] Suzuki S, Muraoka R, Obinata T, Endo S, Horita T and Omata K 2004 Steel products for shipbuilding *JFE Technical Report*

[3] Florescu S N, Mihailescu D and Gheonea M C 2019 Experimental research on the behavior of EH 36 naval steel at the MAG-M mechanized butt welding *The Annals of "Dunarea de Jos" University of Galati* pp 47-54

[4] Florescu S N, Mihailescu D, Gheonea M C and Mardare L 2018 Study of corrosion at MAG-M mechanized butt welding with metal-cored wire and rutile flux-cored wire from maritime ships in marine environment *Geo-Eco-Marina Journal* pp 127-131

[5] Hagen C M H, Hognestad A, Knudsen O and Sorby K 2019 The effect of surface roughness on corrosion resistance of machined and epoxy coated steel *Progress in Organic Coatings* **130** pp 17-23

[6] Pradhan D, Mahobia G S, Chattopadhyay K and Singh V 2018 Effect of surface roughness on corrosion behavior of the superalloy IN718 in simulated marine environment *Journal of Alloys and Compounds* **740** pp 250-263

[7] Sajid H U and Kiran R 2018 Influence of corrosion and surface roughness on wettability of ASTM A36 steels *Journal of Constructional Steel Research* **144** pp 310-326

Acknowledgements

The authors are grateful to Mr. Razvan Batrinu, General Manager of SC Ductil SA Buzau - Romania, for the support of the welding wires, and to Mr. Dr. EWE/IWE Adrian Campurean, General Manager of SC Omnisud Welding Research and Innovation SRL Bucharest - Romania, for the support with the welding equipment. Without their help, this experimental research could not have been achieved.

IOP Conf. Series: Materials Science and Engineering **968** (2020) 012009 doi:10.1088/1757-899X/968/1/012009

Influence of marine corrosion on the roughness of the dry hyperbaric underwater MAG welding joints

M C Gheonea[1], S N Florescu[2], D Mihailescu[1] and V Teodor[1]
[1] "Dunarea de Jos" University of Galati, Faculty of Engineering, Domneasca Street, 47, RO-800008, Galati, Romania
[2] National Institute of Marine Geology and Geo-Ecology (GeoEcoMar), 23-25 Dimitrie Onciul St., 024053 Bucharest, Romania

E-mail: marius.gheonea@ugal.ro

Abstract. The maintenance and repairing actions for underwater metallic structures from maritime and offshore field are usually achieved using the gas metal arc welding processes. The welding processes applied should take into consideration different underwater conditions, and experiments should be conducted for different pressures. In the first part of the study, the authors focused on the investigation of the EH 36 shipbuilding steel behaviour subjected to dry hyperbaric metal active gas (MAG) using semi-automatic welding devices. Hereinafter, the authors analyse, by using the electrochemical method, the behaviour of the welded joints to marine corrosion. In the last part of the investigation, the corrosion influence on the roughness of the specimen's surfaces, depending on the potentiodynamic polarization curves, is approached. The experiments revealed that increasing of pressure conditions leads to an increased corrosion resistance and moreover, to a decrease of the specimen surfaces roughness

1. Introduction

Given the increased interest in underwater interventions for the construction, maintenance, and repair of sunken metal constructions of particular importance (ships, underwater pipes, offshore drilling rigs, hydropower and hydrotechnical structures etc), underwater welding works is one of the basic solutions to be used. The subassemblies from maritime, offshore as well as in the chemical industry are subjected to rough conditions caused by corrosive environments and abrasive particles [1].

Executing underwater welded joints of high quality require specific welding technologies for underwater operations which need specialized, mechanized, and automated equipment which could, in some cases, facilitate the execution of works without the help of scuba divers. Underwater welding operations are carried out in a gaseous environment (protected from water), which is dry welding, or by direct contact with the water, which is wet welding [2], [3]. Beginning from the possibilities of obtaining the welds in accordance with the technical requirements for the field of shipbuilding, five working techniques are known: arc welding in water (wet welding); with local drying; dry hyperbaric in a mini-habitat; dry hyperbaric, and dry at atmospheric pressure [4], [5].

In case of underwater interventions, dry hyperbaric welding ensures a welded joint quality close to the joints made in atmospheric conditions. The created dry environment allows the preparation for the welding procedure, the actual welding of the joints, the non-destructive testing, as well as applying the anti-corrosion protection. Important attention should be paid to the preparation of the joint's surface, for applying the primer and the anticorrosive paint. In this study, the authors present an experimental

IOP Conf. Series: Materials Science and Engineering **968** (2020) 012009 doi:10.1088/1757-899X/968/1/012009

program that investigate the welding behaviour of EH 36 naval steel with 14 mm thickness. The purpose and novelty of the current investigation is to study the underwater hyperbaric welding technology using different pressures and to evaluate the corrosion behaviour by electrochemical method, and the influence of the marine corrosion on the surface's roughness of welded joints.

2. Materials and methods

2.1. Welding materials and equipment

Within the experimental program the following materials and equipment were used [2]: - six EH 36 naval steel sheets with the dimensions of 500x150x14 mm; cored wired with E70C-6MH4 eco-friendly metal powder and the diameter of 1.2 mm; M21 active shielding gas mixture (82% Ar + 18% CO_2); flat ceramic backing with concave channel (channel width 10 mm and channel depth 1.3 mm); dry hyperbaric underwater welding simulator equipped with Aristo Lud 320 universal welding power source and Railtrac FW 1000 welding tractor.

2.2. Welding of samples

Within the experimental program, a number of three butt welded samples were done, on the flat ceramic backing, at atmospheric pressure (P0) and at overpressures in the pressurized chamber of 2 bar (P2) and at 4 bar (P4) in the vertical ascending position. In table 1, the main welding parameters (I - amperage; U- voltage; v_t- speed of the welding tractor; w_{fs}- wire feed speed; L_p- oscillation amplitude; β- welding gun tilt angle; w_s- welding speed; Q- heat input) of the dry hyperbaric underwater welding regimes on the flat ceramic backing with E70C6MH4 eco-friendly metal powder cored wire are presented, according to the working pressure in the pressurized chamber [6]. In the case of the P2 and P4 welded samples, for removing the compressed air from the arc welding area, the M21 shielding gas mixture was administered at the overpressures of 7 bar and 9 bar.

Figure 1. Underwater and hyperbaric welding simulator [5].

The preparation for welding the samples was in V, with a narrow joint ($\alpha = 40^0$) and the opening of the joint of 6mm. In figure 1, the underwater and hyperbaric welding simulator is presented (a-simulator; b-welding preparation; c- welding samples) [5]. During the welding of the three samples, a part of the welding parameters were constant [2]: DC^+ reverse polarity; pendulum speed v_p = 5mm/s; stop time on the edges t_s = 0.5s; shielding gas flow Q_g= 18l/min; pre-gas time t_{preg} = 3s; post-gas time t_{posg} = 3s and the nozzle-layer distance deposited h_{d-s}= 15mm.

Table 1. Variable technological parameters of welding regimes [5].

No. of sample	Working pressure [bar]	Layer type	I_w [A] / L_p [mm]	U_a [V] / β [^{0}]	v_t, [cm/min] / w_s [mm/s]	w_{fs} [m/min] / Q_l [KJ/mm]
P0	0	root	100	25	10	3
			5	+10	0.7	2.9
		filling	100	25	20	3
			10	-10	1.7	1.2
		cover	100	25	25	3
			20	-10	2.0	1.0
P2	2	root	80	31	10	3
			5	+10	0,7	2.8
		filling	80	31	20	3
			10	-10	1.7	1.2
		cover	80	31	25	3
			20	-10	2.0	1.0
P4	4	root	80	32	10	3
			5	+10	0.7	2.9
		filling	80	32	20	3
			10	-10	1.7	1.2
		cover	80	32	25	3
			20	-10	2.0	1.0

3. Results and discussions

3.1. Non-destructive and destructive testing

According to the rules of the Lloyd's Register (LR), the three dry hyperbaric underwater butt welded joints were subjected to non-destructive testing-NDT (100% visual and radiographic) [5].

For the destructive testing (DT) of the three butt welded samples with the dimensions of 500x300x14 mm, the following test bars were sampled for the laboratory mechanical tests [8], [9]: 1 sample for the macroscopic and microscopic examinations and the Vickers HV1 microhardness test; 4 samples for the front side bending test, 2 samples for root and face bend test, on a bend former of 40 mm diameter and 180^0; 2 samples with the calibrated area for the tensile testing and 9 samples for Charpy V-notch impact testing, at -20^{0}C (3 samples with the V-notch in the centre of the weld-WM, 3 samples with the V-notch on the fusion line and 3 samples with the V-notch in heat affected zone - HAZ, at 2 mm from the fusion line-FL).

The results of DT and NDT (Table 2) revealed that the chosen process parameters are suitable for MAG dry hyperbaric underwater welding, using a combination of eco-friendly cored wire E70C6MH4, M21 gas mixture, and the high strength naval steel EH 36 of 14mm thickness.

Table 2. Results of NDT and DT according Lloyd's Register (LR).

Type of test	Extension/sample no.		Welded sample P0	Welded sample P2	Welded sample P4
Non-destructive testing					
Visual test	100%		Accordingly (no surface defects)		
Radiographic test with gamma radiation	100%		Accordingly (no interior defects)		
Destructive testing					
Transverse tensile test	2	$R_{p0,2}$ [MPa]	399.36 406.37	396.61 408.09	401.85 402.05
		R_m [MPa]	546.76 548.87	544.98 549.07	544.08 542.16
		Elongation [%]	33.05 33.35	33.15 33.17	33.03 33.07
Bending test	4 (bend former of 40 mm diameter, at 180^0)		Good (no defects)		
Charpy V-notch impact test at -20^0C	3- WM		41/49/43	42/40/48	40/42/46
	3- FL		82/87/87	88/85/82	88/73/84
	3- HAZ		109/111/106	102/110/105	99/92/101
Macroscopic examination (Nital 10%)	1		The three layers are clearly highlighted		
Microscopic examination (Nital 2%)	1	HAZ	Increased ferrito-perlitic granulation with the relative pressure increasing		
		WM	Acicular ferrite and modified perlite		
Vickers HV1 microhardness test	1 (3 investigation directions)		Increased Vickers HV1 microhardness in WM and HAZ with relative pressure increasing		

3.2. Corrosion behaviour by electrochemical method

The experimental program regarding the corrosion behaviour of the joints welded in seawater, with the characteristics presented in table 3, were carried out with the electrochemical work equipment, schematically presented in Figure 2.a, formed of the PGZ 100 potentiostat/galvanostat (Figure 2.b), the electrochemical cell, and the data acquisition system.

Table 3. Seawater characteristics.

Pressure, [db]	Sigma-Theta, [Kg/m^3]	Salinity, [PSU]	Conductivity, [S/m]
10.49	8.43	144.73	8.90
Oxygen, [mg/l]	**pH**	**ORP, [mV]**	**SeaTurbMtr, [FTU]**
5.28	10.00	15.30	2.22

The figure presents two images of the electrochemical cell before (Figure 2.c) and after (Figure 2.d) finalizing the corrosion analysis. In the electrochemical cell, containing 200ml natural seawater from the Black Sea, three electrodes were set, work (cathode) WE, auxiliary (anode) Pt-Rd AE, and reference RE.

The way of sampling the working electrodes, from the samples previously subjected to the mechanical testing, is shown in figure 3. The recording and interpreting the experimental research results was performed using the PGZ 100 potentiostat/galvanostat and the VoltaMaster 4 software.

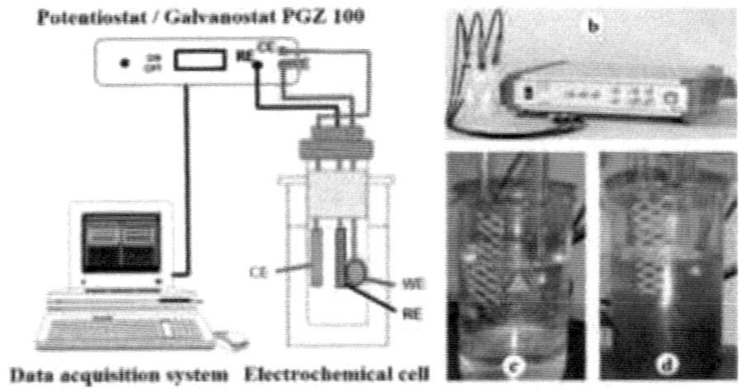

Figure 2. Electrochemical work equipment and the electrochemical cell [6]

Figure 3. Sampling for the corrosion behaviour test [6].

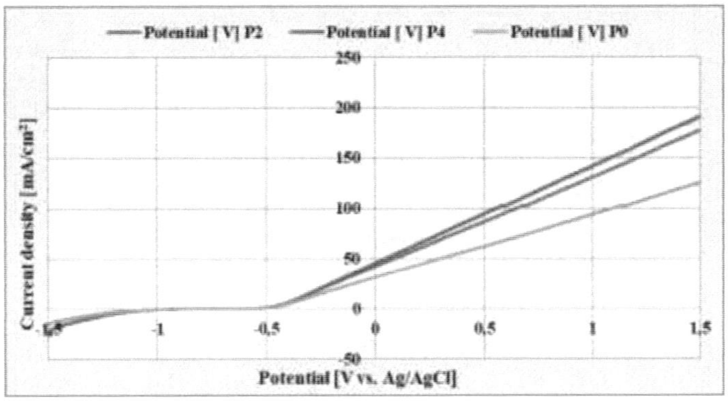

Figure 4. Potentiodynamic polarization curves.

Figure 4 presents the potentiodynamic curves when testing for corrosion in natural seawater of the test bars sampled from the welded joints.

3.3. The influence of marine corrosion on surface roughness

The experimental research regarding the influence of marine corrosion on the surface roughness of the three test bars was done by using the SJ210 Mitutoiu roughmeter and the SJ - Communication - Tool analysis software. The determinations were done on a 4 mm distance with a speed of 0,5 μm/s, on the three test bars' surfaces, before and after carrying out the corrosion behaviour analyses.

For the dry hyperbaric underwater welded joints at the three pressures, the determinations were carried out in the cover layer of the welded joints. Figure 5 presents the evolution of the roughness depths of the surfaces for the P0, atmospheric pressure, figure 6 for P2 (pressure - 2 bars) and figure 7 for P4 (pressure - 4 bars), before and after performing the corrosion behaviour analysis, related to the cover layers, basically in direct contact with seawater. The measurements were performed in three different areas of the samples on a 4mm length. It can be noticed that the profiles for height asperities is quite similar, the values varies between -8 μm and +8μm, but a decrease in the arithmetic average of the surfaces roughness (R_a) can be observed for P4 sample.

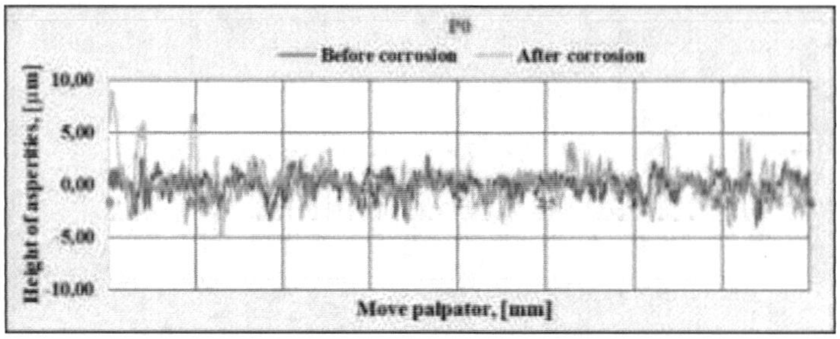

Figure 5. Evolutions of roughness depths for test bar P0.

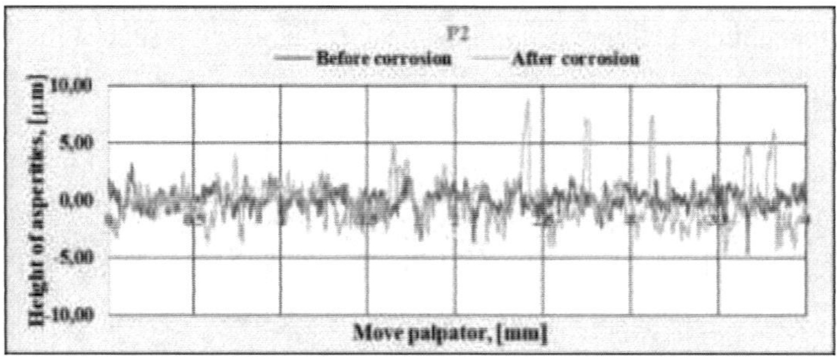

Figure 6. Evolutions of roughness depths for test bar P2.

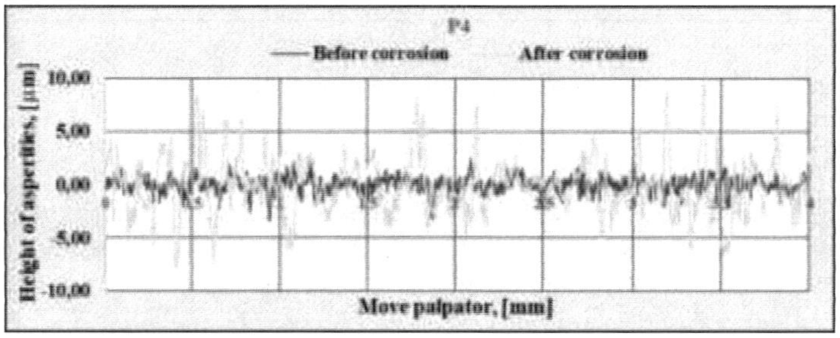

Figure 7. Evolutions of roughness depths for test bar P4.

The SJ-Communication-Tool analysis software was used for data acquisition and processing for the average roughness Ra. In the figure 8 it can be noticed the processed data for the average roughness Ra of the welded samples, before and after the corrosion behaviour analysis, for the case of P0, P2 and P4 pressures. It can be observed that by increasing pressure (P4 pressure condition) the average Ra roughness of the samples decreases by 30%.

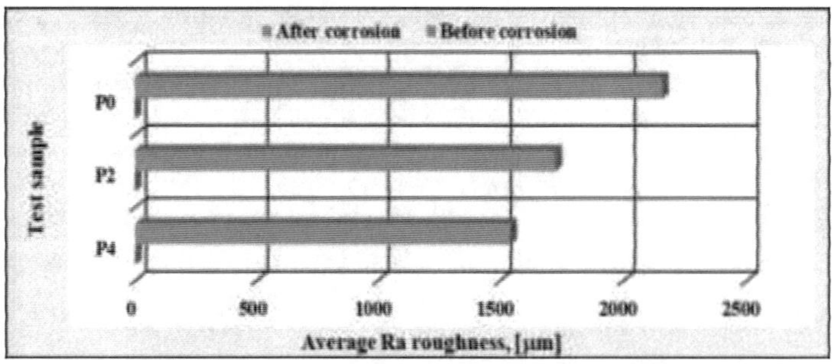

Figure 8. Average Ra roughness.

4. Conclusions

The analysis of the welded samples subjected to non-destructive and destructive testing from high strength naval steel EH 36 with 14mm thickness reveals that the MAG dry hyperbaric underwater welding technologies can be used for the maintenance and repairing of underwater metallic structures from maritime and offshore field. From the analysis of the potentiodynamic polarization curves of the three welded samples, it can be noticed that increasing the relative pressure inside the pressurized chamber, the corrosion resistance of the welded joints also increases. The roughness of the specimen surfaces, after the corrosion analysis, are higher, due to the corrosive action of the water from the Black Sea. The roughness values of the specimen decreased as the pressure inside the chamber increased.

5. References

[1] Heider B et al. 2020 Corrosion Resistance and Microstructure of Welded Duplex Stainless Steel Surface Layers on Gray Cast Iron *Journal of Thermal Spray Technologies* **29** pp 825–842

[2] Hu Y, Shi Y, Sun K, Shen X and Wang X 2018 Microstructure and mechanical properties of underwater hyperbaric FCA-welded duplex stainless steel joints *Journal of Materials Processing Technology* **261** pp 31-38

[3] Miklaszewski A, Jurczyk M U and Jurczyk M 2013 Microstructural Development of Ti–B Alloyed Layer for Hard Tissue Applications *Journal of Materials Science & Technology* 6 **29** pp 565-572

[4] Verma K and Garg H K 2012 Underwater welding-recent trends and future scope *International Journal on Emerging Technologies* 2 **3** pp 115-120

[5] Gheonea M C 2015 Research on dry hyperbaric underwater mechanized MAG welding with ecological cored wire *PhD Thesis "Dunarea de Jos" University of Galati*

[6] Carpenter K R, Monaghan B J and Norrish J 2009 Analysis of fume formation rate and fume particle composition for Gas Metal Arc Welding (GMAW) of plain carbon steel using different shielding gas compositions *ISI J.International* 3 **49** pp 416-420

[7] Gheonea M C, Mihailescu D, Florescu S N and Scutelnicu E 2018 Experimental assessment of gases and fumes developed during Gas Metal Arc Welding *IOP Conf. Series: Materials Science and Engineering* **40** (Constanta: IOP Publishing)

[8] *** 2018 Rules for the manufacture, testing and certification of materials *Lloyd's Register*

[9] Gheonea M C, Mihăilescu D and Florescu S N 2018 Research on dry hyperbaric underwater mechanized MAG-M welding with ecological cored wire *Book of Abstracts Conference on Material Science & Engineering* (Galati) p 49

Acknowledgements

The authors are grateful to Mr. Eng. Razvan Batrinu, general manager of SC Ductil SA Buzau - Romania, who kindly has provided the filler materials employed in this research.

IOP Conf. Series: Materials Science and Engineering **968** (2020) 012010 doi:10.1088/1757-899X/968/1/012010

Basic mechanical analysis of biodegradable materials

A D Mazurchevici[1], R I Popa[1], C Carausu[1], R Comaneci[2], S N Mazurchevici[1] and D Nedelcu[1,3]
[1] "Gheorghe Asachi" Technical University of Iasi, Department of Machine Manufacturing Technology, Blvd. Mangeron, No. 59A, 700050, Iasi, Romania
[2] "Gheorghe Asachi" Technical" University of Iasi, Department of Materials Science and Engineering, Str. Prof. Dr. Doc. D. Mangeron, No. 41, 700050, Iasi, Romania
[3]Academy of Romanian Scientists, Str. Ilfov, Nr. 3, Sector 5, Bucharest, Romania

E-mail: dnedelcu@tuiasi.ro

Abstract. The field of polymeric materials and manufacturing technologies is constantly evolving, offering the possibility to prototype 3D products in a responsible and ecological way, thus aiming to replace on a large scale the filaments of nonbiodegradable synthetic polymers (from fossil resources) with filaments of biodegradable materials, obtained from renewable resources. The paper supports the development mentioned above and follows to characterize biodegradable materials from the mechanical behavior point of view, tensile, bending, and impact tests. Also, the study reflects the influence of the technological parameters on the tensile test obtained results and also aims to optimize the obtained results. The studied materials were Extrudr Green-TEC Anthracite and Extrudr BDP Pearl which according to the obtained basic mechanical results can successfully replace conventional polymers such as Flexible, HIPS, PP and other ones.

1. Introduction

As biodegradable thermoplastic materials are increasingly entering on the filaments market used in 3D printing, the FDM method (Fused Deposition Modeling), knowing and understanding their functional characteristics is a very important aspect especially when it comes to the optimal selection of the material for specific applications of a certain field (engineering, medicine, food service, automotive etc).

Also, the study of the process parameters influence (printing speed, part orientation, layer thickness, printing temperature, infill degree, etc.) on the obtained results must be studied in order to optimize the 3D printing environment and at the same time the mechanical, thermal and structural characteristics of printed parts, [1, 2, 3].

To be considered filaments biodegradable, according to the FTC's Green Guide (Federal Trade Commission), [4] they must decompose in nature under the action of the microorganisms, into chemical elements that are found naturally in the environment (oxygen, carbon, etc.). It should be noted that these materials usually contain additives that allow them to decompose faster than conventional plastics, however, it is not recommended to be stored in nature to decompose but in well-established locations as the speed and type of biodegradation differ and depends on the constituents used in their production, [5].

The obvious advantage of the biodegradable polymeric materials is the possibility of degradation, but this is not the only one, other strengths are: the possibility of processing/manufacturing by using the same technologies as in the case of conventional polymers, reduced energy consumption due to the low printing temperatures, rigidity, and mechanical strength, pleasant aesthetic appearance, etc.

The results from the literature regarding the characterization of printed parts from biodegradable materials by the FDM technology look promising due to the relevant number of scientific articles published in this field recently, [5 - 12]. However, not all aspects have been examined and conclusive information can be obtained by understanding and improving the printing of biodegradable materials by the FDM method.

The novelty of the present manuscript consists of the fact that characterizes mechanically two biodegradable thermoplastic materials that until now were not study from any point of view. The objectives pursued by the authors are to determine the behavior of the Extrudr GreenTec Anthracite and Extrudr BDP Pearl materials during tensile, bending and impact tests. The tests were performed in order to make some recommendations regarding a possible replacement of conventional plastics with the studied materials.

2. Materials and methods

The prototyping of the samples necessary for the mechanical determinations was performed on a series of biodegradable materials selected according to their utility and the possibility of substituting nonbiodegradable synthetic plastics. Biodegradable materials in the form of filaments used in the study of this paper were *Extrudr Green-TEC Anthracite* and *Extrudr BDP Pearl*.

Extrudr Green-TEC Anthracite and Extrudr BDP Pearl filaments are produced by Extrudr company (Lauterach, Austria). Extrudr Green-TEC Anthracite is part of the BIO Performance products range, is made from renewable raw materials - contains PLA (based on corn starch, sugar cane or tapioca roots), copolyester and additives, [13]. Is biodegradable according to DIN EN ISO 14855, being created to be used in applications that require high tensile strength (excellent flexural strength), low deformation and high product surface quality. This material has no smell and is safe in the food field. Extrudr BDP Pearl is a material 100% from renewable resources, used mainly in applications where the aesthetic part is important (BIO Design) and with mechanical and thermal characteristics comparable to those of PLA. According to the manufacturer, its chemical composition is similar to that of the Extrudr Green-TEC Anthracite material, [14].

3D printing of specific tensile test samples was performed according to ISO 527, [15, 16], using Raise3D Pro2Plus printing equipment, available in the Laboratory of Fine Mechanics and Nanotechnology, Faculty of Mechanical Engineering and Industrial Management, "Gheorghe Asachi" Technical University of Iasi.

Prototyping of biodegradable materials was performed according to a complete factorial experimental plan of 2^3 type (8 experiments), in which the factors were the layer thickness (0.1mm, 0.2mm), the infilling speed (40mm / min, 80mm / min) and the sample orientation on the printing table (flat -  and on edge - ), using the ANOVA method, [17] and the Minitab 17 program. The samples are dumbbell shaped and the dimensions according to the standard mentioned above.

The tensile test samples were printed from Extrudr Green-TEC Anthracite and Extrudr BDP Pearl materials. The diameter of the nozzle was 0.4 mm. All samples were printed with three shell layers and a 100% filling degree. The temperatures of the extrusion nozzle and of the printing bed for the two printed materials were:

- For Extrudr Green-TEC Anthracite material, the nozzle temperature was 220°C and printing bed temperature 60°C;
- For Extrudr BDP Pearl material, nozzle temperature 180°C and printing bed temperature 55°C.

Uniaxial tensile tests were performed at the "Gheorghe Asachi" Technical University of Iasi, Faculty of Materials Science and Engineering, on the Instron 3382 universal test machine, with a constant transverse speed of 5mm/min according to ISO 527-3: 2003 (table 1), the distance between the jaws was 115mm, the data acquisition rate 10Hz and the determinations were performed at room temperature (23°C).

TableCurve3D v.4.0 application was used to *optimize the technological parameters* for 3D printing with filament from biodegradable materials.

In order to achieve a *three-point bending test*, three specimens were prepared (to demonstrate that the printing process is stable and reproducible) from each biodegradable material. A single test was performed on materials with a deformation greater than 5 mm. The used specimens were cut from those specific to tensile test and have been determined the distance between the sample placement supports, L = 60mm, the length of the tested sample 50mm, the thickness h = 4mm and its width b=10mm. A WTW 50 universal test machine was used to determine the bending strength, using a 1kN load and a 2mm/min load rate.

The *impact-specific samples* were printed according to the SR EN ISO 179 standard (table 1), with the following dimensions: length (L) = 80 ± 2 mm, width (l) = 10 ± 0.2 mm, thickness (h) = 4 ± 0.2 mm. The test parameters used to determine the impact resistance of the two biodegradable materials were: speed 2.9 m/s, hammer weight 1.189 kg, energy 5J.

Table 1. Samples specific to mechanical tests.

Material name	**Tensile-specific sample** *ISO 527-3: 2003*	**Impact-specific samples** *SR EN ISO 179*
Extrudr Green-TEC Anthracite	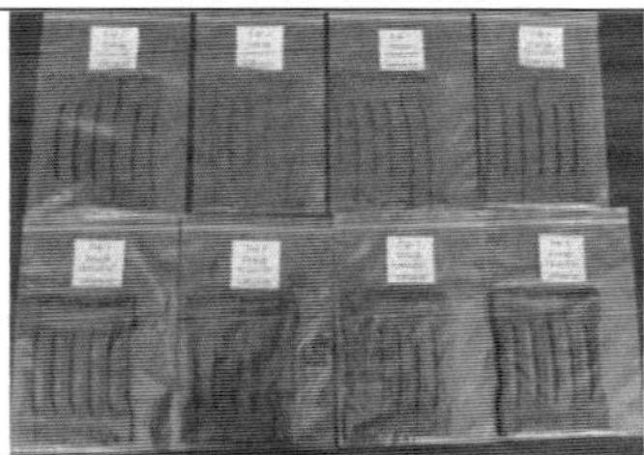	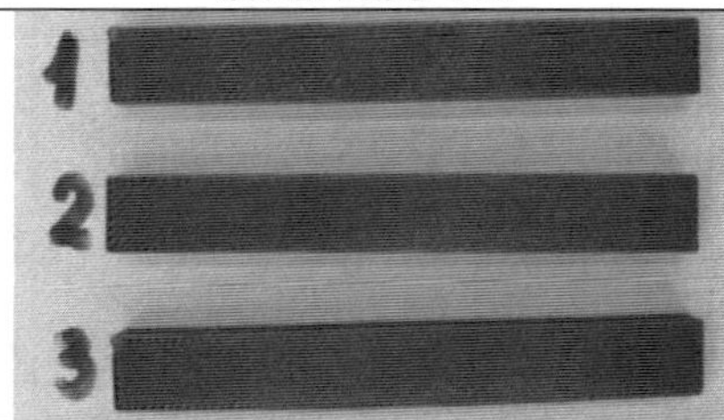
Extrudr BDP Pearl	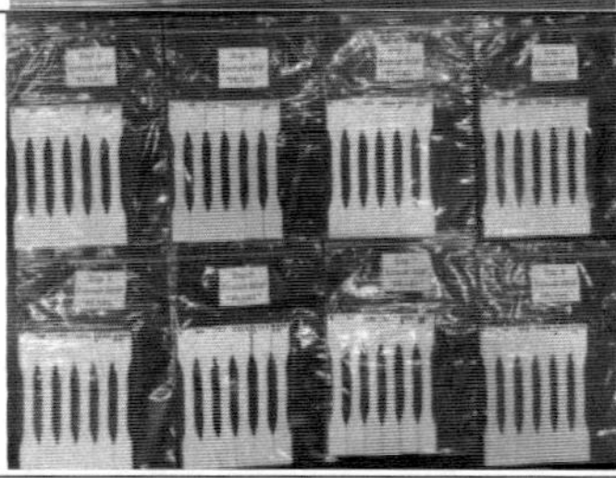	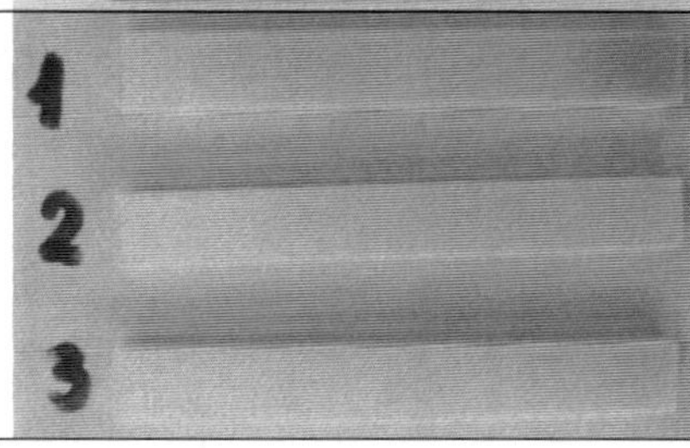

The temperature at which the tests were performed was 20°C. Three biodegradable samples of each material were printed for the impact test. The printing parameters set for obtaining samples were: printing temperature and printing bed temperature according to the statements made for the printed specimens for the tensile test; infill speed 60mm/s; the deposited layer thickness 0.2mm; filling type - grid; filling degree- 100%; flat orientation of the sample on the printing table; the temperature inside the printer was 23°C.

The test results are used to select a biodegradable material for a required load, to control the quality and to anticipate how this material will react under the action of different types of forces.

3. Results and discussions

3.1. Tensile strength

For each experiment, three samples were printed to highlight the process stability by calculating the mean and the dispersion of the values for each input parameters values set from the experimental plan.

Each test was generated responses regarding the tensile strength, σ [MPa], elongation, ε [%] and modulus of elasticity, E [MPa] of the analyzed biodegradable material, calculating for each one the mean value and dispersion. Using the ANOVA method, the influences of the input parameters on the tensile strength, on the elongation and on Young's modulus were determined, which allowed a hierarchy of the influence of these factors.

In table 2, the values of the mechanical characteristics obtained for the two studied materials are shown. The characteristic strength-strain curves for the two biodegradable materials Extrudr Green-TEC Anthracite (experiment 6) and Extrudr BDP Pearl (experiment 6), highlighted the relatively homogeneous behavior of the samples, figure 1(a) and figure 1(b), demonstrating the fact that the prototyping process, by the FDM method, was stable and reproducible, the mechanical properties of the materials varying within acceptable limits.

Table 2. Experimental results of uniaxial tensile tests of printed samples from Extrudr Green-TEC Anthracite and Extrudr BDP Pearl.

Exp. no.	Material	Input parameters l_t [mm]	i_s [mm/min]	s_o	σ_{max} [MPa]	ε [%]	ε_t [%]	E [MPa]
1	Extrudr Green-TEC Anthracite	0.1	40	flat	20.97±1.08	4.17±0.20	5.91±0.94	884.56±5.19
2		0.1	40	on edge	19.43±8.60	4.69±0.28	4.87±0.28	1201.87±16.35
3		0.1	80	flat	21.07±0.98	4.84±0.14	5.51±0.61	869.38±13.63
4		0.1	80	on edge	28.86±0.53	4.04±0.37	4.35±0.68	1169.71±26.12
5		0.2	40	flat	17.38±1.90	4.45±1.18	9.04±1.10	842.02±76.48
6		**0.2**	**40**	**on edge**	**30.26±1.89**	**4.50±0.59**	**6.00±0.68**	**1232.48±75.30**
7		0.2	80	flat	20.90±0.56	4.43±0.46	8.58±1.18	842.79±35.86
8		0.2	80	on edge	26.59±1.96	4.59±0.35	7.87±0.11	868.51±52.85
1	Extrudr BDP Pearl	0.1	40	flat	16.09±1.11	4.95±0.92	9.22±7.31	765.87±60.87
2		0.1	40	on edge	24.92±2.02	4.27±0.47	11.11±2.23	1062.65±65.13
3		0.1	80	flat	16.75±0.33	4.69±0.02	14.92±1.21	799.55±34.33
4		0.1	80	on edge	25.11±2.64	3.89±0.8	9.21±2.87	948.98±43.52
5		0.2	40	flat	18.32±2.53	5.35±0.35	15.28±0.89	835.87±94
6		**0.2**	**40**	**on edge**	**28.63±1.10**	**5.44±0.9**	**12.71±3.35**	**1036.86±44.83**
7		0.2	80	flat	19.40±0.49	4.58±0.08	15.96±0.18	858.61±2.15
8		0.2	80	on edge	27.91±107	5.08±0.01	13.21±3.21	1053.96±6.30

note: (l_t) - layer thickness; i_s - infill speed; s_o – sample orientation.

Analyzing the obtained results from the tensile tests it can be seen that the experiments with the best results, the highest values of tensile strength, σ_{max}, were recorded by the samples oriented "on edge" on the printing bed, Extrudr Green- TEC Anthracite - experiment number 6, (30.26 ± 1.89)MPa, Extruder BDP Pearl - experiment number 6, (28.63 ± 1.10)MPa.

Following the results dispersion for the three samples of each material studied, it can be concluded that the material Extrudr BDP Pearl figure 1(b) has the smallest dispersion, followed by the material, Extrudr Green-TEC Anthracite - experiment 6, which shows a higher dispersion of the results.

According to the obtained results regarding the strain at break (εmax) of the printed samples from the studied materials, it was found that the material that showed the most rigid behavior was the

Extrudr Green-TEC Anthracite biopolymer (4.84 ± 0.14)% having a value of the total strain at break of (5.51 ± 0.61)%. The Extrudr BDP Pearl material showed slightly lower elasticity than the Extrudr Green-TEC Anthracite material, the value of strain at break being (5.44 ± 0.9)%, and the total strain at break of (12.71 ± 3) 35)%.

The modulus of elasticity of the analyzed polymers is higher for the „on edge" oriented samples during the prototyping process. The highest mean value is presented by the material Extrudr Green-TEC Anthracite (1232.48 ± 75.30)MPa which reveals that this material is stiffer than the other material, Extrudr BDP Pearl (1062.65 ± 65.13)MPa which deform more because it has a smaller modulus of elasticity.

The behavior of the materials, Extrudr GreenTEC Anthracite, figure 1(a), and Extrudr BDP Pearl, figure 1(b), is approximately similar, their curves being characteristic for hard and resistant materials

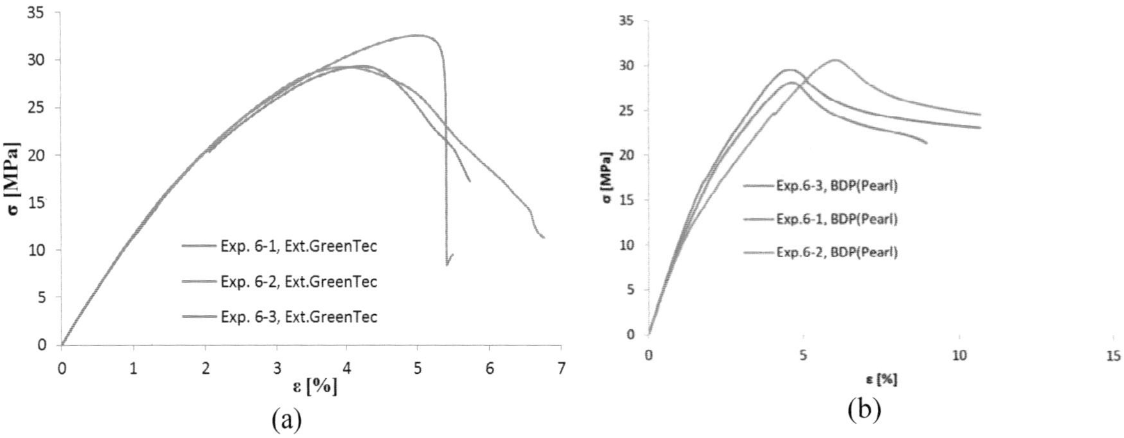

Figure 1. Strain-strength curve of materials: (a) Extrudr GreenTec Anthracite, experiment no. 6; (b) Extruder BDP Pearl, experiment no. 6.

The mechanical properties obtained for the studied materials are comparable to those of synthetic materials (Flexible - tensile strength 26-43MPa, HIPS (High Impact Polystyrene) - tensile strength 32MPa, PP (Polypropylene) - tensile strength 32MPa, Metal Filled - tensile strength 20-30MPa, etc.), [18] so that they can be used in various industrial fields and not only.

3.2. The influence of technological parameters on tensile behavior
The influence of the technological parameters on the mechanical behavior of the printed samples was made only for the Extrudr Green-TEC Anhtracite material because for it the best results in tensile and impact tests were obtained.

Table 3 presents the influence variance analysis results of the considered input parameters on the tensile strength, modulus of elasticity and strain of the Extrudr Green-TEC Anthracite material. The biggest influence on the tensile strength, but statistically insignificant ($p = 0.095 > 0.05$), is the sample orientation ($F = 4.76$). The infilling speed and the deposited layer thickness also do not have a statistically significant influence since $p > 0.05$. The same happens both in the case of the modulus of elasticity and in the case of strain, according to the obtained results no parameter has statistically significant influence ($p > 0.05$). For this reason, the influence of the parameters on the strain was no longer graphically represented.

Table 3. Results of the variance analysis on the tensile strength by the ANOVA method for the mechanical responses of the Extrudr Green-TEC Anthracite sample.

Parameter	Tensile strength		Modulus of elasticity		Strain	
	Fisher-Value	Probability of Fisher value, p-value	Fisher-Value	Probability of Fisher value, p-value	Fisher-Value	Probability of Fisher value, p-value
Sample orientation	4.76	0.095	44.27	0.003	0.06	0.821
Infill speed, [mm/min]	0.64	0.468	0.36	0.582	0.08	0.791
Layer thickness, [mm]	0.16	0.708	1.55	0.281	1.85	0.245

Figure 2 shows the main effects of the factors on the obtained mechanical responses for the Extrudr Green-TEC Anthracite material. For the tensile strength (mean value of the model 23.18MPa) the samples deposited „on edge" (variation +1 level) have a higher influence, with a higher infill speed (80mm / min) and at a higher deposition thickness (0.2mm).

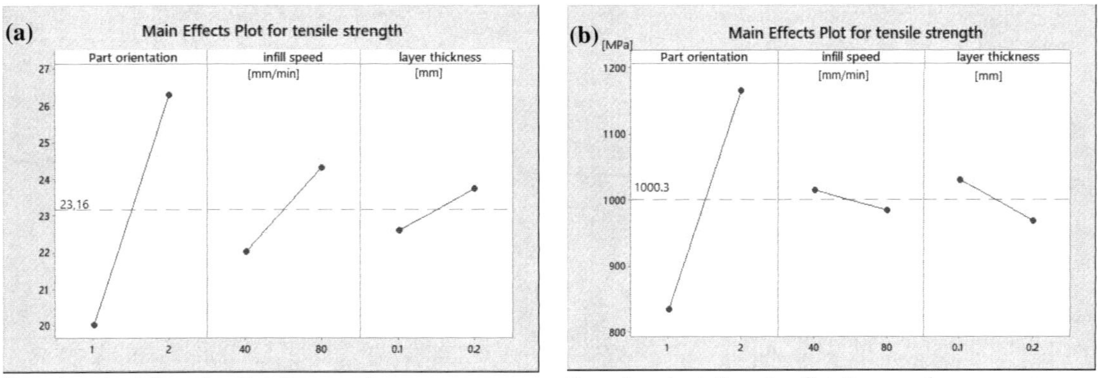

Sample orientation: 1 – flat; 2 – on edge   Sample orientation: 1 – flat; 2 – on edge

Figure 2. Influence of parameters on: (a) tensile strength; (b) modulus of elasticity.

The higher modulus of elasticity (model mean value 988.91MPa) was evidence by the „on edge" printed samples. A larger print layer thickness (0.2mm) and a higher infill speed (80mm / min) ensure the samples a higher modulus of elasticity.

3.3. Optimization of technological parameters in order to maximize mechanical responses

To optimize the technological parameters for 3D printing with filament from biodegradable materials, the application TableCurve3D v.4.0 was used and the same material was considered for the same reasons presented for the influence of technological parameters.

The optimization criteria are the mechanical responses maximization received from the tensile tests of the Extrudr Green-TEC Anthracite samples for tensile strength, modulus of elasticity and strain.

Quantitative technological parameters were taken into account to optimize the tensile strength: infill speed, s [mm/min], and the printing layer thickness, t[mm]. The printing direction was considered „on edge", for which, according to the analysis of the factors influence, the best results emerged. Table 4, the (a) graph shows the plan obtained by regression. The correlation coefficient is $R^2 = 0.383$ - being very small.

The equation that approximates the influence of the deposition layer thickness, t, and the infill speed, s, on the tensile strength, is presented in table 4, whose maximum in the experiments field is for t = 0.2 mm and s = 80mm/min. For these parameters and „on edge" printing was obtained σ = 29.865MPa. The deviation from the experimental data in this case is 12.32%.

For the modulus of elasticity quantitative factor, the considered printing direction was „on edge", because, according to the analysis of the factors influence, the best results emerged.

Table 4. Results regarding the optimization of the technological parameters in the case of the printed samples from Extrudr Green-TEC Anthracite.

Parameter	The regression equation	R^2	Max.	Erorr [%]	Optimum t [mm]	Optimum s [mm/min]
σ	$\sigma = 15.545 + 42.8 \cdot t + 0.072 \cdot s, [MPa]$	0.383	29.86MPa	12.32	0.2	80
E	$E = 1378{,}91 - 395{,}85 \cdot t - 2{,}559 \cdot s, [MPa]$	0.71	1235.5MPa	2.8	0.1	40
ε	$\varepsilon = 0.855 + 16.8 \cdot t + 0{,}0305 \cdot s, [\%]$	0.55	6.65%	12.32	0.2	80

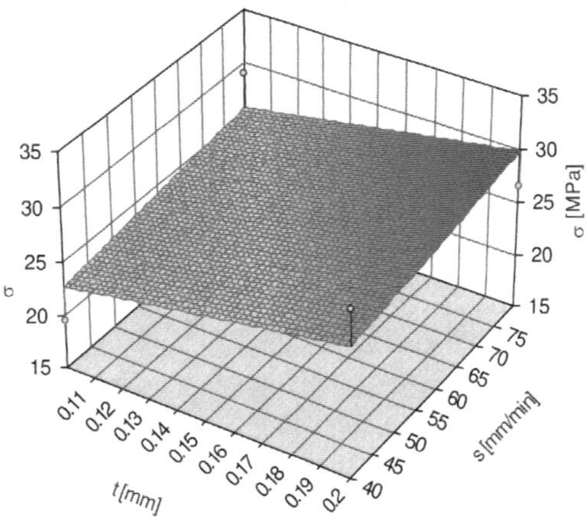

(a)

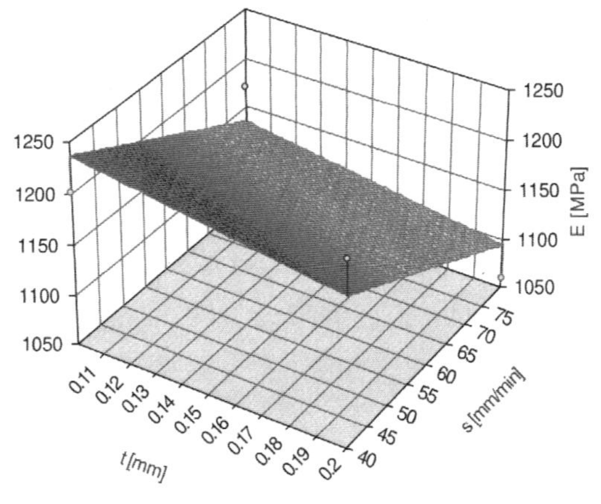

(b)

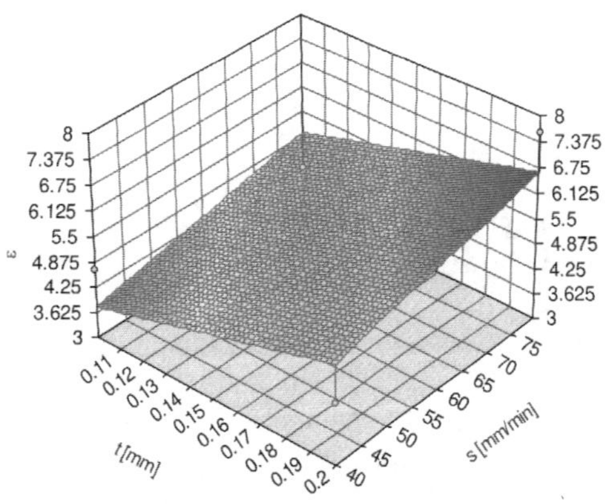

(c)

The regression equation for (a) tensile strength, (b) modulus of elasticity, (c) strain of Extrudr Green-TEC Anthracite samples when printed „on edge"

note: σ - tensile strength [MPa]; E - modulus of elasticity; ε - strain; R^2 - correlation coefficient;

In table 4, graph (b), is graphically represented the regression equation of the plane that provides information related to the infill speed and layer thickness influence on the modulus of elasticity in the case of „on edge" 3D printed samples made of Extrudr Green Anthracite material.

Also, the regression equation for the modulus of elasticity ($R^2 = 0.71$) is presented in table 4, whose maximum in the experiments field is for t = 0.1mm and s = 40mm/min. For these parameters and „on edge" printing, E = 1235.5MPa is obtained, with an error of 2.8% compared to the experimental data. The best results for elasticity came out when the printing direction was considered "on edge". The equation of the plane is for R2 = 0.552 (weak correlation) being presented in table 4, graph (c), whose maximum in the experiments field is for t = 0.2mm and s = 80mm/min. For these parameters and "on edge" printing is obtained ε = 6.655%. The error compared to the experimental data is 12.32%.

3.4. Bending strength

Before starting the bending test, certain conditions have been established according to the literature. Thus, it was considered that for materials with the maximum arrow in the centre of the sample (linear displacement in the direction of the application force) greater than 5 mm, the results obtained cannot be taken into account for this type of test.

Biodegradable materials, Extrudr Green-TEC Anthracite, figure 3 (a), Extrudr BDP Pearl, figure 3 (b), according to the characteristic bending test curves, are not suitable for this type of test as it does not yield easily under the action of a progressive load, because they are elastoplastic materials, with a maximum arrow greater than 5mm.

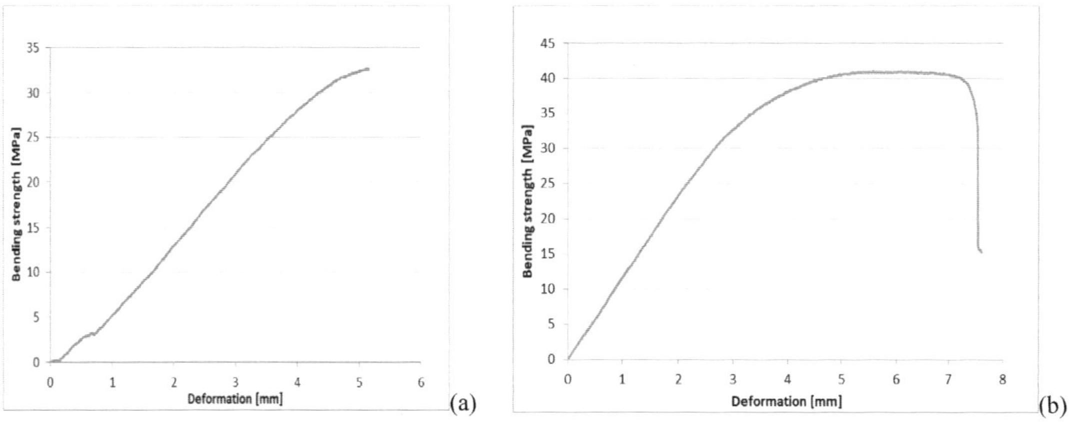

Figure 3. The bending strength curve depending on: (a) Extrudr Greentec, exp. no. 8; (b) BDP Pearl Extruder, exp. no. 8.

3.5. Impact strength

Determination of the impact strength, by the Charpy method, of the samples from Extrudr Green-TEC Anthracite and Extrudr BDP Pearl materials was realized in accordance with the SR EN ISO 179 standard, with the aim of observing their impact behavior. The obtained values during the impact test for Extrudr GreenTec and Extrudr BDP Pearl materials are shown in table 5.

The best value of impact strength was recorded by the Extrudr Green-TEC Anthracite material (8.87 ± 1.02)kJ/m^2, followed by the Extrudr BDP Pearl material (7.98 ± 0.26)kJ/m^2 . The obtained values are much lower than in the case of other printed samples made of biodegradable materials, for instance, compared to PLA (polylactic acid), which according to the literature is the most widely used biodegradable material in FDM printing and has an impact resistance of about 38kJ/m^2, [19]. The obtained results for the studied materials are up to five times lower than the material selected as reference (PLA). The major differences between the obtained results from the literature for PLA and the Extrudr Green-TEC Anthracite and Extrudr BDP Pearl biodegradable materials are cause by the different chemical compounds found in their structure. According to previous studies on the impact strength of biodegradable materials, the presence of renewable raw materials (lignin) and/or natural plant fibres (a substance that gives hardness) decreases the impact strength of polymers/composites, [20, 21]. The viscoelastic hardness is explained by the fact that the lignin molecule is a complex molecule which in its turn is made up of three other complex molecules and which, depending on an external deforming cause, tries to shorten the chemical bonds between them so that the lignin molecule to occupy the same volume.

Table 5. Mean values of the Charpy strength for printed samples from Extrudr GreenTec and Extrudr BDP Pearl.

Material	Sample	Sample dimensions [mm]		Impact strength (kJ/ m^2)	Mean values of impact strength (kJ/ m^2)
		length	width		
Extrudr Green-TEC Anthracite	1	36.4	100.0	8.15	8.87± 1.02*
	2	35.9	99.9	14.99	
	3	37.0	99.6	9.59	
Extrudr BDP Pearl	1	36.3	100.3	13.60	7.98± 0.26*
	2	36.1	100.4	7.79	
	3	36.3	100.1	8.16	

*mean value achieved between the impact strength of two samples, sample three being excluded from the study due to the very large difference in the value obtained for the impact strength (difference probably caused by possible temperature variations during printing)

Also, from the graph analysis presented in figure 4, the variations of the mean values of the impact strength but also can be observed the standard deviations calculated for each material. The mean value of the impact strength deviations is ± 0.73 kJ/m^2.

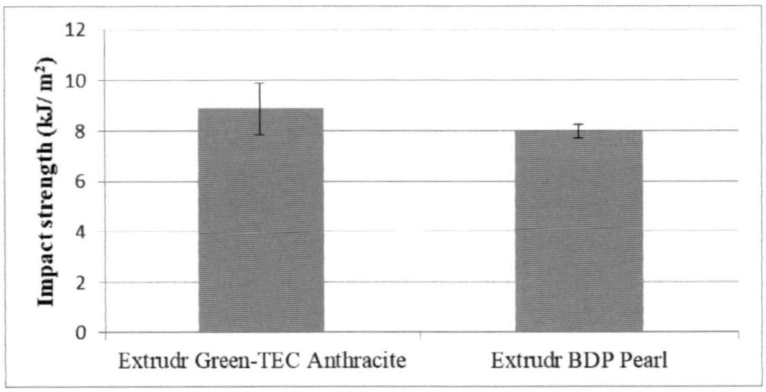

Figure 4. Impact strength diagram for the printed samples made of Extrudr Green-TEC Anthracite and Extrudr BDP Pearl.

Following the analysis of the obtained results, the printed samples from Extrudr Green-TEC Anthracite and Extrudr BDP Pearl materials can successfully substitut other plastics that have in their composition lignin and natural fibers but also polymers reinforced with glass microfibers or metal powders, [18, 22].

4. Conclusions

Samples obtained using FDM three-dimensional printing technology are significantly influenced by the process parameters that can be controlled directly or indirectly before the beginning of model printing. The selection of the optimal printing parameters is quite difficult if the aim is to obtain a sample with the highest possible mechanical characteristics, [7, 23].

The following conclusions can be drawn from the analysis of tensile tests:

- tensile strength is the mechanical property often used to evaluate the usefulness of the product manufactured by prototyping, and not only. Thus, the best values of tensile strength were obtained for

"on edge" oriented samples. The values of the tensile strength in the case of the "flat" orientation of the sample, varied/decreased compared to the other type of sample orientation by up to 40%;
- "flat" oriented samples have a more fragile behavior, offering lower strain results;
- the best mechanical behavior was highlighted by the biodegradable material Extrudr Green-TEC Anthracite, the material Extrudr BDP Pearl offering slightly lower results;
- also, at the moment of the mechanical determinations, the material interruptions must be taken into account, due to the layer's adhesion lack, which has a significant effect on the general resistance of the sample.

The mechanical properties obtained for the studied materials are comparable to those of synthetic materials such as Flexible - 26-43MPa tensile strength, HIPS - 32MPa tensile strength, PP - 32MPa tensile strength, Metal Filled - 20-30MPa tensile strength, but also with those of other plastics that have in their composition lignin and natural fibers, glass microfibers or metal powders (valid for impact resistance), [82, 94], so they can be used in various industrial fields and beyond.

5. References

[1] Redwood B, Schöffer F and Garret B 2017 *The 3D Printing Handbook Technologies, design and applications* 3D Hubs, Hardcover, November 14 p 13

[2] Pilla S 2011 Engineering Applications of Bioplastics and Biocomposites- An overview *Handbook of Bioplastics and Biocomposites Engineering Applications* (New Jersey: John Wiley & Sons) pp 1-14

[3] Rajpurohit S R and Dave H K 2018 Effect of process parameters on tensile strength of FDM printed PLA part *Rapid Prototyping Journal* **24**(8) 1317-1324

[4] https://www.ftc.gov/news-events/press-releases/2015/09/ftc-sends-warning-letters-about-green-certification-seals, Accessed on 02.03.2020

[5] Mazurchevici S-N, Mazurchevici A-D and Nedelcu D 2020 Dynamical mechanical and thermal analyses of biodegradable raw materials for additive manufacturing *Materials* **13** 1819

[6] Mazurchevici A D, Nedelcu D, Nitu E L, Racz S-G and Popa R 2020 Additive manufacturing of composite materials by FDM technology: A review *Indian Journal of Engineering and Materials Sciences* in press

[7] Mazurchevici A-D, Popa R I, Carausu C, Mazurchevici S-N and Nedelcu D 2020 Influence of layer thickness, infill rate and orientation on thermal and structural loading of FDM parts vol *Advances in Manufacturing Processes* (Singapore: Springer Nature Ltd) in press

[8] Chia H N and Wu B M 2015 Recent advances in 3D printing of biomaterials *Journal of Biological Engineering* **9**(4) https://link.springer.com/article/10.1186/s13036-015-0001-4

[9] Fountas N A, Benhadj-Djilali R, Stergiou C I and Vaxevanidis N M 2019 An integrated framework for optimizing sculptured surface CNC tool paths based on direct software object evaluation and viral intelligence *Journal of Intelligent Manufacturing* **30**(4) 1581-1599

[10] Fountas N A, Seretis G V, Manolakos D, Provatidis C and Vaxevanidis N M 2018 Multi-objective statistical analysis and optimisation in turning of aluminium matrix particulate composite using genetic algorithms *International Journal of Machining and Machinability of Materials* **20**(3) 236-251

[11] Cooke M N, Fisher J P, Dean D, Rimnac C and Mikos A G 2002 Use of stereolithography to manufacture critical-sized 3D biodegradable scaffolds for bone ingrowth *Journal of Biomedical Materials Research Part B: Applied Biomaterials* **64B**,(2) 65-69

[12] Fountasa N A, Kostazosb P, Pavlidisa H, Antonioua V, Manolakosb D E and Vaxevanidis N M 2020 Experimental investigation and statistical modelling for assessing the tensile properties of FDM fabricated parts *Procedia Structural Integrity* **26** 139-146

[13] Sin L T, Rahmat A R and Rahman W A W Abdul 2012 *Polylactic Acid: PLA Biopolymer Technology and Applications* 1st edition (Great Britain: Elsevier Publishing House) p 22

[14] https://www.extrudr.com/en, Accessed on 09.08.2019

[15] Tensile strength, final, ASTM D638, ISO 527, Accessed on 09.08.2019

[16] Harper C A 2016 *Handbook of Plastics Technologies – The Complete Guide to Properties and Performance* (United States of America: The McGraw-Hill Companies), p 1.6
[17] Batra P K and Jaggi S *Factorial Experiments* https://www.coursehero.com/file/9170119/5-Factorial-Expts/ Accessed on 09.08.2019
[18] https://www.simplify3d.com/support/materials-guide/properties-table/ Accessed on 09.08.2019
[19] Wangab L, Gramlichc W M, Gardner D J 2017 Improving the impact strength of Poly(lactic acid) (PLA) in fused layer modeling (FLM) *Polymer* **114** 242-248
[20] Kim H S, Yang H S and Kim H J 2005 Biodegradability and mechanical properties of agroflour–filled polybutylene succinate biocomposites *J Appl. Polym. Sci.* **97** 1513–21
[21] Jiang L, Chen F, Qian J, Huang J, Wolcott M, Liu L et all 2010 Reinforcing andtoughening effects of bamboo pulp fibre on poly (3-hydroxybutyrate-co-3-hydroxyvalerate) fibre composites *Ind. Eng. Chem. Res.* **49** 572–7
[22] https://www.mdpi.com/2073-4360/11/7/1094/htm, Accessed on 09.08.2019
[23] Rajpurohit S R and Dave H K 2019 Analysis of tensile strength of a fused filament fabricated PLA part using an open-source 3D printer *International Journal of Advanced Manufacturing Technology* **101**(5-8) 1525-1536

Acknowledgement: This work was supported by a grant of the Romanian Ministery of Research and Innovation, CCCDI – UEFISDI, project number PN-III-P1.2-PCCDI-0446/82PCCDI/2018, acronym TFI PMAIAA/ FAMCRIA, within PNCDI III.

IOP Conf. Series: Materials Science and Engineering **968** (2020) 012011 doi:10.1088/1757-899X/968/1/012011

Research on the optimization of stress relief holes applied in blanks used for body-in-white stamping parts with complex asymmetrical shapes

F Mihaila
Renault Technologies Romania, North Gate Business Center, Bd. Pipera Nr. 2/III, Oras Voluntari, Jud. Ilfov, 077190 Romania

E-mail: fmihaila@yahoo.com

Abstract. Increasing the degree of deformation when stamping body-in-white parts with complex asymmetrical shapes can be achieved by optimizing the discharge holes in the flange of the blank. For this purpose, the numerical simulation of the deep-drawing process with finite element is used. Considering the circular shape of the unloading hole, the expanded shape of this hole is determined using the FEM, taking into account the safety limit given by the final shape of the stamped part. To check that the deformation that occurs at the edge of the expanded hole not cause cracks in the part the area with deformations whose values exceed the permissible limit is delimited. The arc that stretches this area along the contour of the expanded hole is divided into a convenient number of equal segments. The points at the end of each segment are then translated on the edge of unloading hole by measuring the distances between these points on the expanded blank hole. Using a mathematical modelling of the allocated points on the contour of the unloading orifice, the corrected shape of this orifice is determined, which is perforated in the flat blank.

1. Introduction

In body parts manufacturing stress relief holes applied in blanks are used for deep drawn parts with complex shapes to allow increasing of deformation degree. Often without the use of such holes the parts are infeasible and has cracks. Examples of parts that need presence of such holes are body side outer, door inner, trunk lid inner and other symmetrically or different parts that have been connected in the same die.

If there are two parts in the same blank the size of the deformed hole in the space between parts influences blank dimension and material flow ratio. Always we try to optimize this flow ratio who is one important parameter of process performance.

Figure 1 shows the case of stamping two symmetrical pieces, arranged face to face, having the longitudinal plane of the die as a plane of symmetry.

The parts are represented in blue, with the filled surface of the holes in red. It can also see in beige the surface of the die on which the following elements are highlighted: the drawbead (green surface), the red line representing the die entry line and the blue line corresponding to the boundary of the part.

The surface between the two parts is scrap and is used to place a hole to eliminate internal stresses.

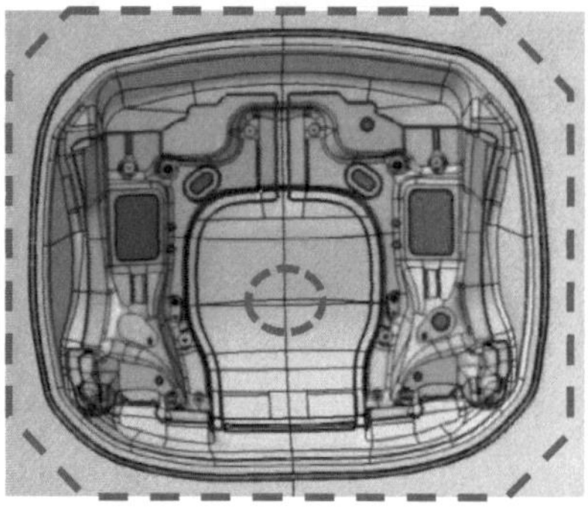

Figure 1. Parts arrangement in the die.

The blank shape is shown by dark blue dashed lines.

To avoid any crack propagation on the part we must predict the risk of cracks on deformed hole boundary. FEM simulation could be used for this [10]. Also using FEM simulation, the position, dimension and shape of the holes could be evaluated.

One of the most used FEM simulation program in sheet metal forming is Autoform [11]. The robustness of the results obtained with Autoform allows to use the simulated geometry directly to the die machining and then the validation is done at try-out where the results obtained theoretically through simulation are checked. For this reason, are very important to start the try-out with one blank who has a corrected and tested hole.

The originality of this work consists in finding a method to apply some corrections on the shape of the initial hole in order to eliminate the risk of cracks on the boundary of the expanded boundary hole after deformation that would propagate on the part causing to reject it. The purpose is to identify some rules to establish position, dimension and shape of the hole for reducing the number of simulation iterations which require a lot of time. Autoform and an original program for optimize the hole geometry are used.

2. Degree of deformation analysis

In deep drawing of complex body parts its difficult to define a global degree of deformation, because this differs from region to region. So, this parameter is defined in sections with high degree of deformations.

For the case presented in figure 1, the degree of deformation of the blank in two sections, longitudinal and transversal, was analyzed after completely deformation of the blank.

The initial dimensions of the blank, shown in yellow dotted line in figure 2, are:

$L_{1\ ini}$ = 1370 mm the cross section (red curve)

$L_{2\ ini}$ = 1195 mm the longitudinal section (blue curve).

At the end of the stamping stroke, highlighted in figure 3, the deformed blank has the following dimensions measured in the two sections:

L_{1fin} = 1825 mm the cross section (red curve)

$L_{2\ fin}$ = 1468 mm the longitudinal section (blue curve).

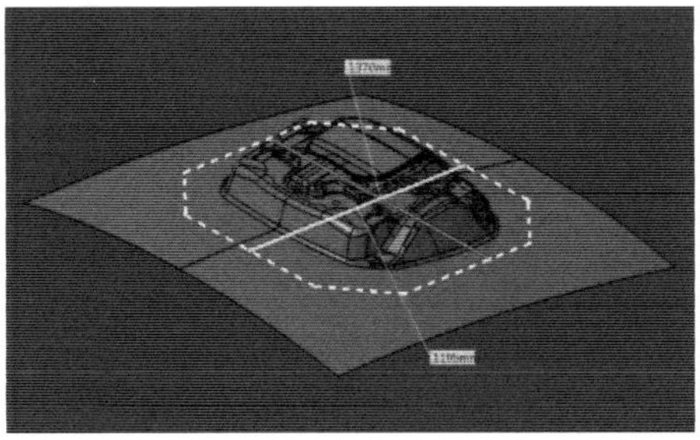

Figure 2. Dimensions of the blank.

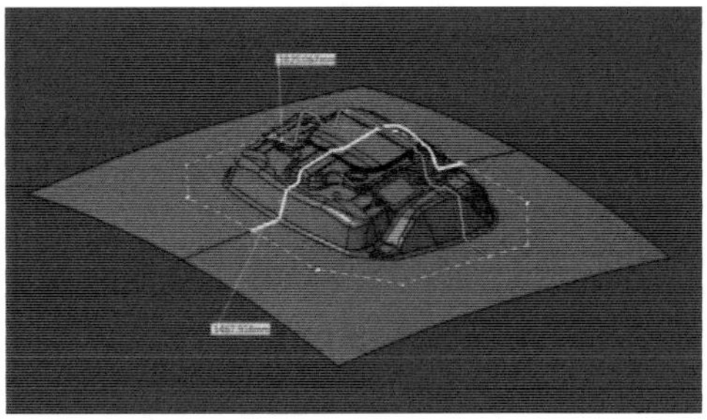

Figure 3. The lengths of the sections of deformed part.

The degree of deformation for the complete stroke is:

$$\varepsilon_{tranv} = \frac{1825-1370}{1370} = 33.21\ \% \tag{1}$$

$$\varepsilon_{long} = \frac{1468-1195}{1195} = 22.84\% \tag{2}$$

These values are used only to identify the localization of the biggest theoretical deformation and to choose first hole position for following iterations.

For each material there is a file charged in AUTOFORM simulation program who give the information like hardening curve, anisotropy values, yields surface model and forming limit curve [9]. These values is one average of values from different suppliers to assure that simulation results are the same like in reality.

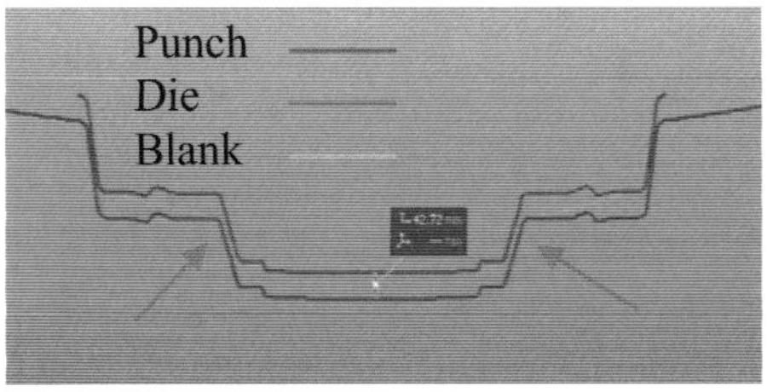

Figure 4. Timing of break.

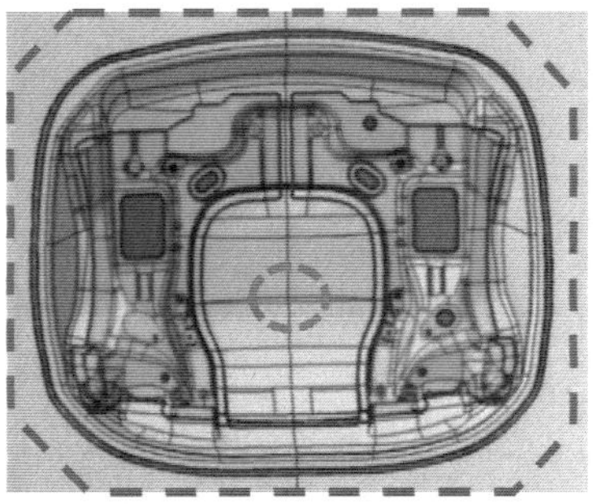

Figure 5. Additional stress relief hole.

The maximum allowed limit for the material is 29% according to [8] and is available for pure traction ($\varepsilon_2 = 0$) and it's correspond of intersection between forming limit curve and y-coordinate axis. To have robust results is used one safety limit of 20% below FLC curve.

From the analysis of the deformation behavior of the blank, it was found that the first breaking primer indicated with red arrows appears 42.7 mm before the end of the stroke (figure 4).

It was noticed that the initialization of the rupture is in the area where we have the counter-drawing on the piece. For this reason, it is necessary to add a technological stress relief hole between the two parts in the scrap area (figure 5). In the initial case we have draw-in only for external boundary of the blank and we need to reduce internal tensions between those die radii.

By adding a hole internal tension are removed and hole boundary are free to expand and finally reduce tensions in areas where are cracks.

3. Deformation state analysis around the hole

A simulation was performed with a blank containing a 200 mm diameter hole. The presence of an area with a risk of rupture identified by the yellow color on the boundary of the deformed hole was observed and the position of these finite elements was determined on the FLD diagram (figure 6).

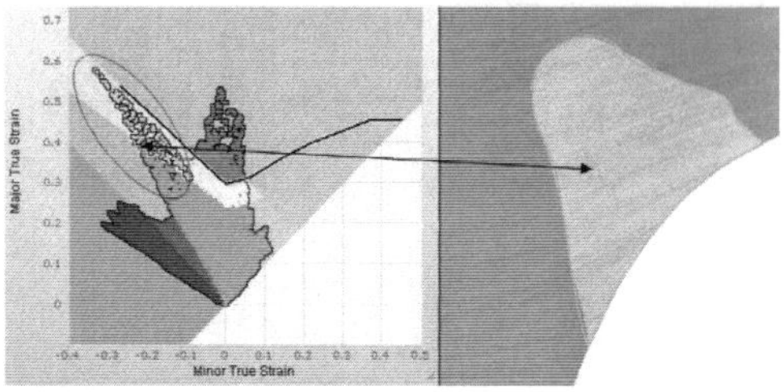

Figure 6. The position of the elements on the diagram.

The network of colored triangular finite elements according to the corresponding area of the FLD and the main directions of deformation which are indicated by red for expansion or blue arrows for compression are shown in figure 7 and 8.

After selecting a finite element on the part, its position on the deformation limit diagram is highlighted by a dot and the values of the degree of deformation in the three directions are indicated in figure 8.

For example, the following values were highlighted for the specified item:

$$\epsilon_1 = \quad 0.529$$
$$\epsilon_2 = -0.297$$
$$\epsilon_3 = -0.232.$$

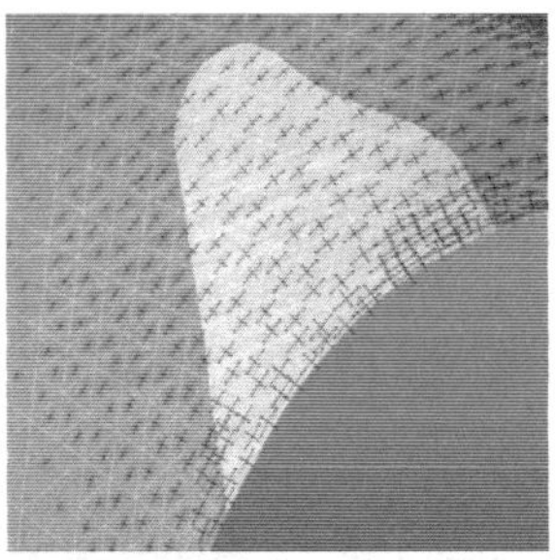

Figure 7. Highlighting the directions of deformation.

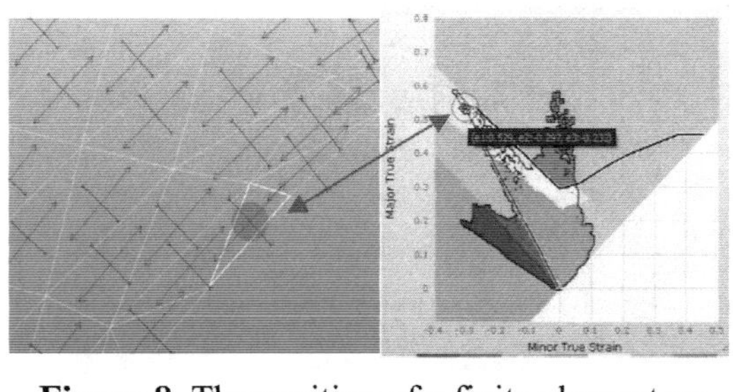

Figure 8. The position of a finite element on the FLD diagram.

It was observed that the elements on the contour of the hole are deformed in the plane in two directions. Tensile stress is present on the tangential direction to the contour, and compression stress is present on the perpendicular direction (in the radial direction).

This position of elements on FLD diagram shown increasing of deformation on traction, over the limit of pure expansion, because elements are subject of compression on the other two directions.

Finite elements have been numbered on contour of the hole as shown in figure 10. In figure 9 are presented deformations in graphical form associated with the table.

IOP Conf. Series: Materials Science and Engineering **968** (2020) 012011 doi:10.1088/1757-899X/968/1/012011

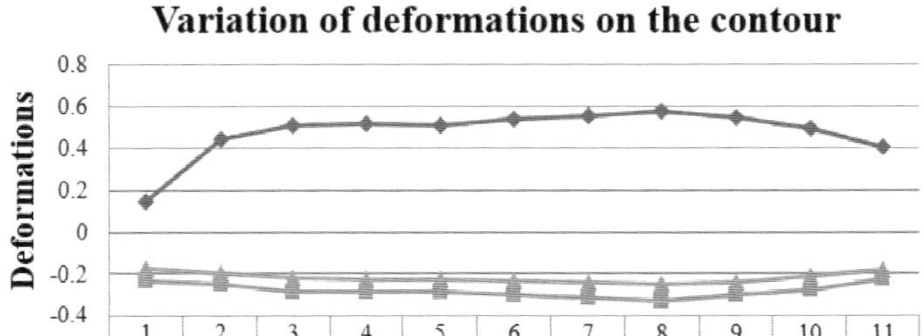

	1	2	3	4	5	6	7	8	9	10	11
ε1	0.143	0.444	0.507	0.516	0.511	0.538	0.558	0.579	0.549	0.494	0.407
ε2	-0.233	-0.249	-0.285	-0.289	-0.283	-0.303	-0.318	-0.333	-0.304	-0.281	-0.228
ε3	-0.175	-0.195	-0.222	-0.227	-0.228	-0.235	-0.24	-0.247	-0.24	-0.213	-0.179

Figure 9. Graph of the variation of deformations on hole boundary.

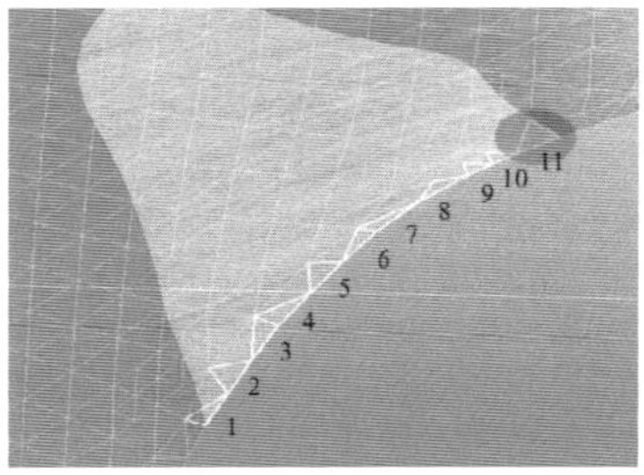

Figure 10. Items numbering.

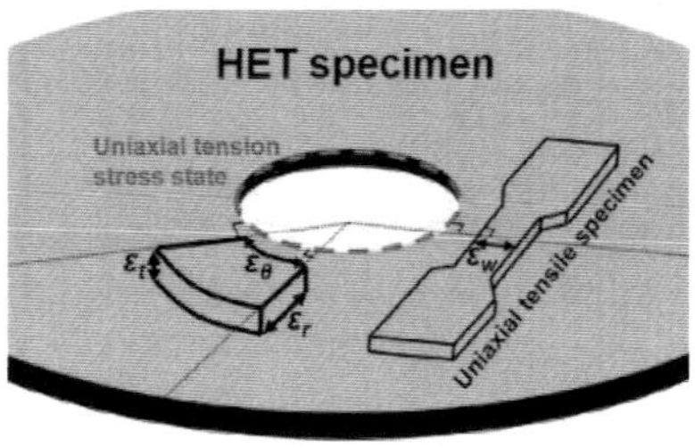

Figure 11. Components of strains on the periphery of the hole [3].

In paper [3] have shown three types of deformations on a blank which include a hole subject to expansion: deformations in the circumferential direction on the periphery of the hole ϵ_θ, in the radial direction ϵ_r and in the thickness direction of the material ϵ_t, figure 11.

It is considered that the deformation of the radial expansion that occurs in the case of holes deformation may be comparable with that of the tensile test specimen case.

In the case of perforated holes, the degree of expansion is lower because during processing, plastic deformations are generated along the edge of the hole, which are accompanied by local hardening, microcracks and burrs.However, it is preferable to use this process because it is less expensive and has a higher productivity than the other methods.

The degree of expansion of the hole, λ, according to ISO 16630-2009 is determined by the relation:

$$\lambda=\frac{D_f-D_o}{D_0}\cdot 100\ [\%] \tag{3}$$

where : D_f is the final diameter of the hole; D_0 - the initial diameter of the hole.

The paper [1] highlighted the criterion of maximum edge tension for the analysis of rupture on the contour of the hole. Maximum extension of the edge of the opening (ε_{max}) is equivalent to the degree of expansion of the hole.

$$\varepsilon_{max}=\lambda\ [\%] \tag{4}$$

The maximum deformations are calculated with the relation:

$$\varepsilon_{max}=\ln\ \left(1+\frac{\lambda}{100}\right) \tag{5}$$

This criterion can be used to estimate the occurrence of cracks on the edge of the hole, is compatible with the formability simulation using CAE applications and may have limited use in the analysis of deformed grid pattern marked on parts. In this paper is considered that the state of stress at the edge of a circular hole is approximately uniaxial tensile.

Starting from the equation of the degree of expansion of a hole we can write:

$$\lambda=\frac{D_f\text{-}D_0}{D_0}\cdot 100=\frac{\pi(D_f\text{-}D_0)}{\pi D_0}\cdot 100=\frac{L_f\text{-}L_0}{L_0} \tag{6}$$

where L_f is the length of the diameter circle D_f and L_0 is the length of the diameter circle D_0.

It was considered that for the irregular holes the two initial and deformed lengths of the hole can be used, respectively. The deformed contour was divided into segments according to the deformations resulting from the simulation, called L_{fi}, where i corresponds to the number of segments. By determining the position of the points on the initial contour, the corresponding L_{0i} segment is determined.

Equation (6) becomes:

$$\lambda=\frac{L_f\text{-}L_0}{L_0}=\frac{\sum_{i=1}^{n} L_{fi}\text{ - }\sum_{i=1}^{n} L_{0i}}{\sum_{i=1}^{n} L_{0i}} \tag{7}$$

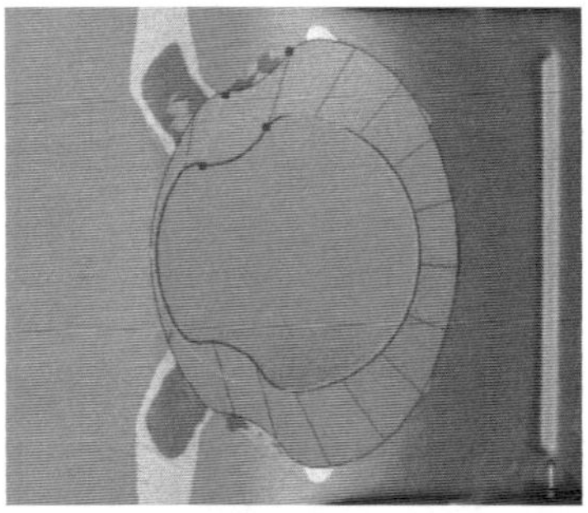

Figure 12. The result of the correction by reducing the radius.

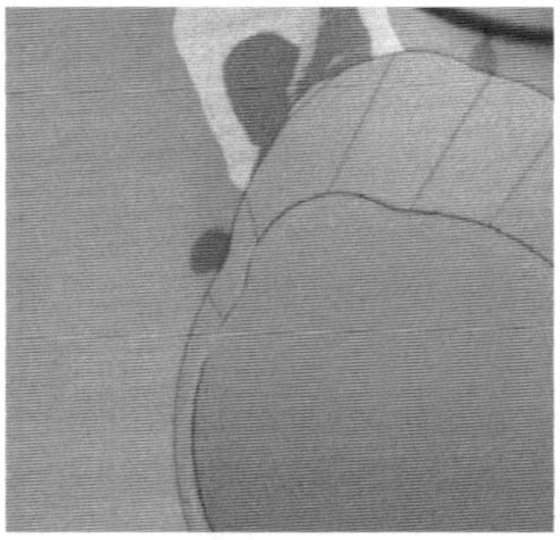

Figure 13. The result of correction by increasing the radius.

Basically, in order to eliminate the thinning on the edge of the hole, the degree of deformation of the respective segment must be equal to the maximum tensile deformation. For this, the length of the segment subjected to expansion was modified by increasing it while maintaining the position of the points that delimit the segment. Tests have shown that the shape leading to the reduction of the deformation of the hole is obtained by changing the segment to the inside of the hole according to figure 12.

4. Analysis of the trajectories of the points on the contour of the hole during the deformation

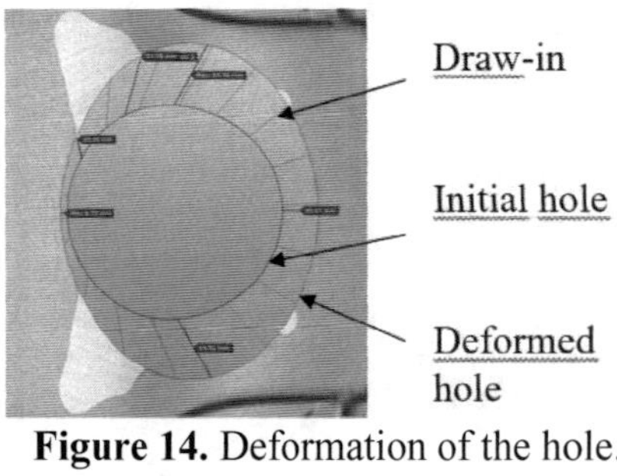

Figure 14. Deformation of the hole.

IOP Conf. Series: Materials Science and Engineering **968** (2020) 012011 doi:10.1088/1757-899X/968/1/012011

The Autoform software dynamically shows the change in the shape and position of the punch made in the semi-finished product on the entire drawing stroke. At the end of the drawing stroke, the direction of the material flow is indicated by lines and these values can be measured. The values of the minimum hole expansion of 6.73 mm and a maximum of 64.96 mm are indicated.

The values can be measured both in the direction of material flow and perpendicular to the contour of the initial blank (figure 14).

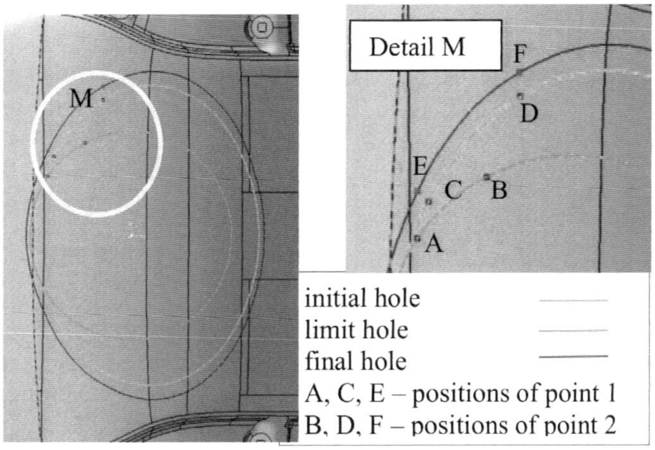

Figure 15. Displacement of points on the boundary hole.

Two points were chosen on the expanded hole delimiting the area in which the thin limit was reached at the final stroke (marked with E and F). The punch is then retracted to the stroke 24.42 mm before the bottom dead center when the first finite element on the contour of the hole reaches the limit of thinning. We note the position of the points chosen at this stroke with C and D. Then the punch is withdrawn at the beginning of the stamping operation when the blank is undeformed. I noted these points with A and B.

5. Working method for hole correction

5.1 Graphic determination of the profile

1. After the first simulation, the areas where the finite elements on the contour of the hole exceeded the safety limit were identified

2. Two points on the contour delimiting this area were chosen (figure 15) and the length of the deformed curve was identified. The deformed curve and the 2 points in (.igs) format were exported and overlapped on the punch surface.

3. The punch was moved until the first element that reaches the safety limit appeared to identify the position of the two points on the deformed hole. These elements were also exported and designed on the surface of the punch.

4. The curve of the deformed semi-finished hole was designed under the action of its own weight as well as the two points.

Figure 16 shows the curves mentioned above:

- the deformed contour is indicated by the color blue;
- the contour deformed when the first element reaches the safety limit with green;
- the contour deformed under its own weight by the magenta color.

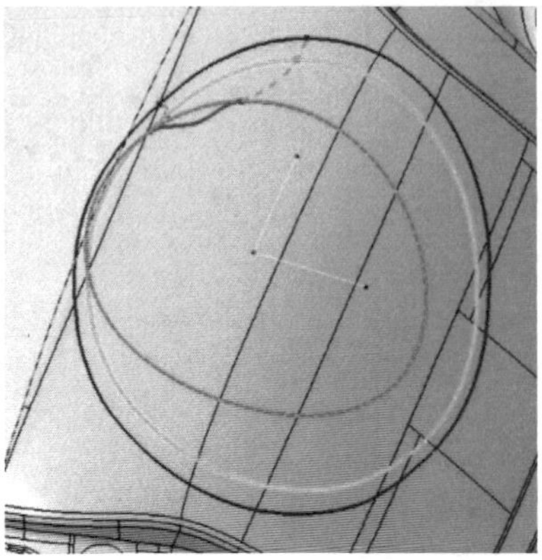

Figure 16. Variation of the hole during deformation.

5. The system of axes was chosen with origin in the center of the circle corresponding to the hole in the blank, all the coordinates of the points on the three curves being measured in relation to it.

6. Three curve lengths measured on the punch surface shown in figure 17 were identified:

- the first corresponds to the initial orifice deformed under its own weight AB;
- the second corresponds to boundary of the hole when the first element reaches the safety limit CD;
- the third, EF, which corresponds to the contour of the hole at the end of the stroke.

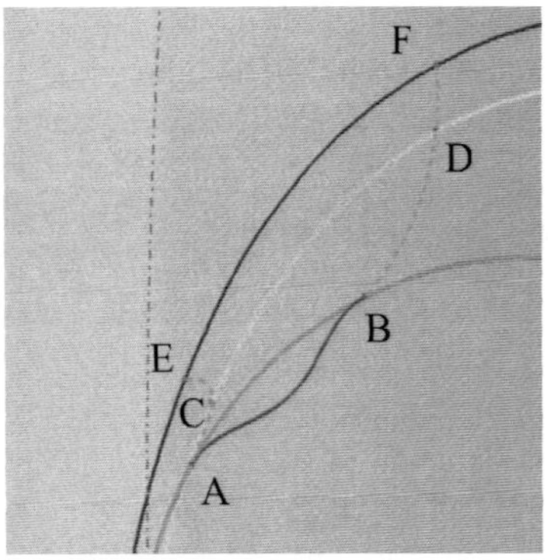

Figure 17. The trajectory of the analysis points during the stamping stroke.

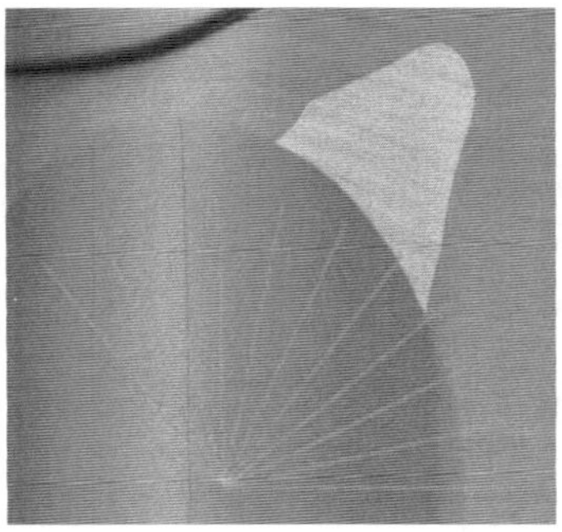

Figure 18. Critical area and directions of analysis.

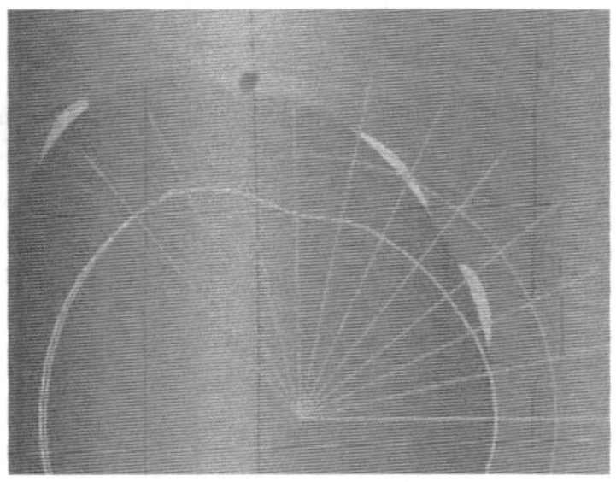

Figure 19. Simulation result after hole correction.

7. The length of the CD corresponds to the maximum deformed segment without breaking.

8. Until the end of the operation we have another stroke of 24.42 mm, so additional deformation,

9. To allow further deformation without breaking, the length of the segment AB must be longer.

10. The curve was drawn, brown color, which was then connected by tangents to the initial profile. The length of the sinusoid, L_{sin}, was calculated with the relation:

$$L_{sin} = L_{AB} + L_{EF} - L_{CD} \tag{8}$$

where: L_{AB} is the length of the arc of the circle measured on the inner hole; L_{EF} is the length of the deformed arc between the points EF; L_{CD} is the length of the deformed arc when the safety limit is reached between the CD points.

For the dimensional analysis of the circular hole, ten equidistant radial directions (blue lines in figure 18) were analyzed at 13 degrees between them.Thus we have analyzed four sectors of a circle in the central area which corresponds to the critical area, and every three sectors in the two lateral zones.

After several tests, a profile of the corrected hole which eliminate the thinning was obtained in (figure 19) which the radial directions for measuring the points and the intial hole having the color blue, respectively the corrected hole having the color green are indicated.

5.2 Determining of corection radius

The length of the arc of the circle between the 2 points A and B is calculated with the relation:

$$L_{CD} = \frac{\pi \cdot R \cdot \alpha}{180} \tag{9}$$

where: R is circle radius;α - the arc angle in degrees.

In figure 20 is presented determination of radius which corected the blank hole. Where R is initial hole and corected radius is R_1 circle. In order not to reach the breaking limit, it is necessary to compensate with K = 29% - the elongation limit of the material.

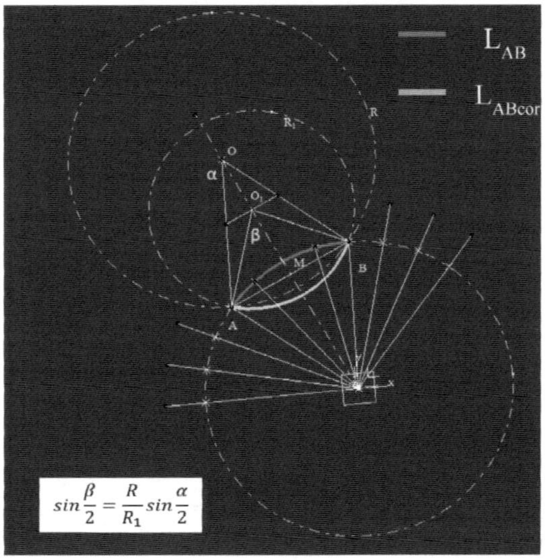

Figure 20. Determination of the correction radius.

$$\varepsilon = \frac{\frac{\pi \cdot R \cdot \alpha}{180} - \frac{\pi \cdot R_1 . \alpha}{180}}{\frac{\pi \cdot R_1 . \alpha}{180}} = \frac{R - R_1}{R_1} = k \tag{10}$$

$$R_1(1 + k) = R \tag{11}$$

The radius of the new arc has the value:

$$R_1 = \frac{R}{1{,}29} \tag{12}$$

The R_1=77-ray circle is used for correction, which is connected to the initial hole with radii of different values. Those results are presented in table 1.

IOP Conf. Series: Materials Science and Engineering **968** (2020) 012011 doi:10.1088/1757-899X/968/1/012011

Table 1. The obtained results.

Test result	Hole geometry	Description
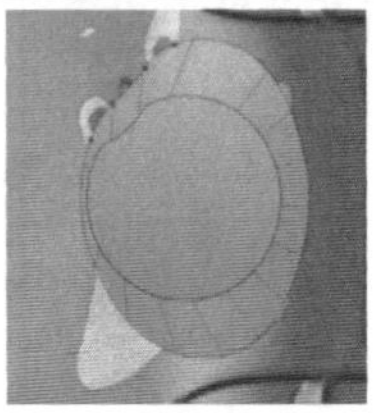	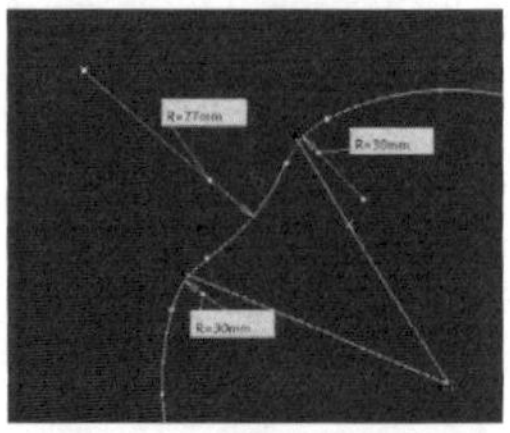	We use R30 to fillet
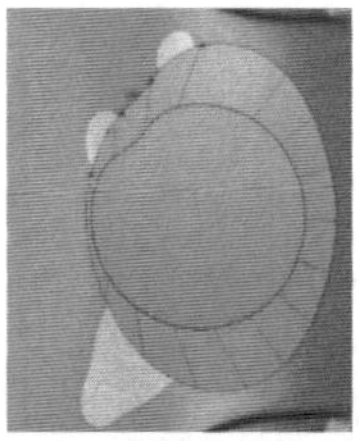	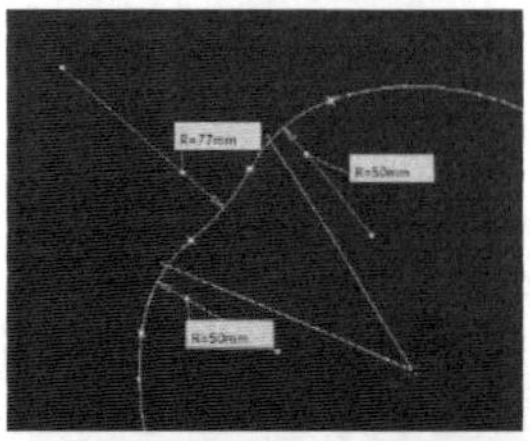	We use R50 to fillet
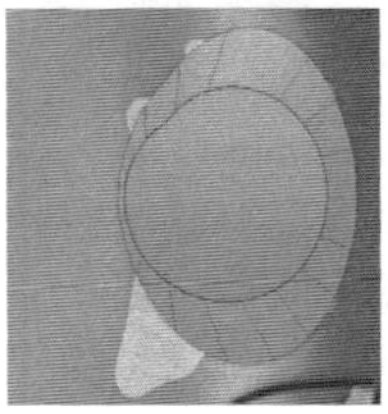	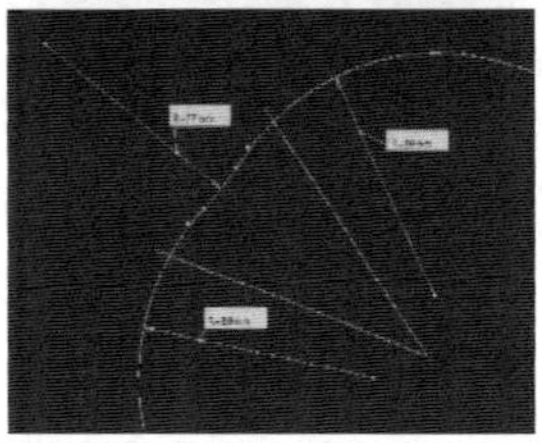	We use R80 to fillet

5.3 Determining the profile equation

To determine the function that will generate the profile of the hole, we worked in Cartesian coordinates with the origin in the center of the hole and longitudinal plane of the die being the axis Ox. The number of the point at the intersection of the direction and whose coordinates are written in table 2 was noted.

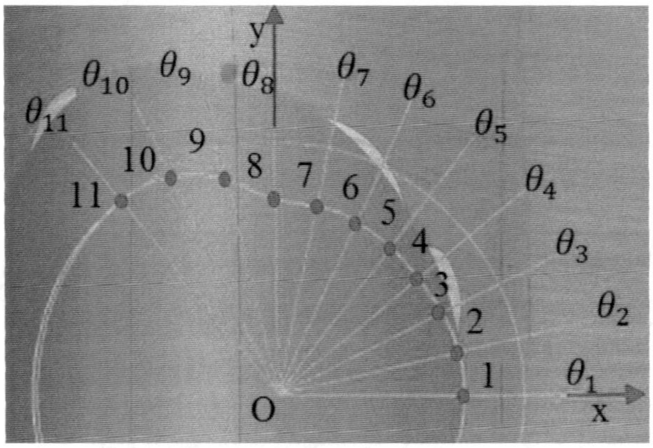

Figure 21. Determining the coordinates of the points on the corrected hole.

Table 2. Punched profile coordinates.

No. crt.	xi	yi	Θi
1	72.096	0	0
2	69.887	16.135	13
3	63.489	30.966	26
4	54.263	43.941	39
5	43.4	55.549	52
6	30.355	65.097	65
7	15.079	70.939	78
8	1.284	73.588	91
9	-20.23	81.136	114
10	-41.844	82.123	127

Relationships are used to calculate the profile equation:

$$Y = f(\mathrm{x}) \tag{13}$$

$$Y = b_0 + b_1 \cdot x + b_{11} \cdot x^2 \tag{14}$$

where b_0 , b_1 si b_{11} are the coefficients to be determined.
To determine them, the following system of equations is used:

$$B = (X^T \cdot X)^{-1} \cdot X^T \cdot Y \tag{15}$$

$$B = \begin{bmatrix} b_0 \\ b_1 \\ b_{11} \end{bmatrix} \; Y = \begin{pmatrix} y1 \\ y2 \\ y3 \\ y4 \\ y5 \\ y6 \\ y7 \\ y8 \\ y9 \\ y10 \end{pmatrix} \; X = \begin{pmatrix} 1 & x_1 & x_1^2 \\ 1 & x_2 & x_2^2 \\ 1 & x_3 & x_3^2 \\ 1 & x_4 & x_4^2 \\ 1 & x_5 & x_5^2 \\ 1 & x_6 & x_6^2 \\ 1 & x_7 & x_7^2 \\ 1 & x_8 & x_8^2 \\ 1 & x_9 & x_9^2 \\ 1 & x_{10} & x_{10}^2 \end{pmatrix} \tag{16}$$

After performing the calculations, the following coefficients were determined:

$b_0 = 73{,}89189;$
$b_1 = -0{,}28686;$
$b_{11} = -6{,}95388.$

The equation that determines the profile of the hole is:

$$Y = f(x) \rightarrow \quad Y = 73{,}892 - 0{,}287 \cdot x - 6{,}954 \cdot x^2 \tag{17}$$

6. Conclusions

Stress relief holes influence the deformation process and are important in terms of shape, size and position. Cracks that appear on the contour of the holes in the blank must be removed because they represent a risk of propagation on the part causing rejects. These cracks may be different from one batch to another due to the anisotropy of blank material. A balance must be made between the costs associated with the assimilation of such holes in the dies.

By this method of correction decreased the depth and width of the area of risk of propagation from edge of the hole.

There was a tendency for two areas to appear on the side of the corrected area depending on the connection radius between the correction radius and the initial radius. The risk of rupture is inversely proportional to the size of the connection radius.

A mathematical relationship must be determined to determine the connection radii in a future work. Also the results must be validated experimental.

7. References

[1] Hance B 2017 Practical Application of the Hole Expansion Test *SAE Int. J. Engines* **10**
[2] Paul S K 2019 Effect of Punch Geometry on Hole Expansion Ratio *Proc I. Mech. E.* B 1-6
[3] Kim J H, Kwon Y J, Lee T, Lee K A, Kim H S and Lee C S 2018 Prediction of Hole Expansion Ratio for Various Steel Sheets Based on Uniaxial Tensile Properties *Met. Mater. Int* 1 **24** pp187-194
[4] Yoon J I, Jung J, Lee H H, Kim G S and Kim H D 2016 Factors Governing Hole Expansion Ratio of Steel Sheets with Smooth Sheared Edge *Met. Mater. Int* 6 **22** pp 1009-1014
[5] Paul S K 2019 The Efect of Deformation Gradient on Necking and Failure in Hole Expantion Test *Manufacturing Letters* **21** pp 50-55
[6] *** 2009 Hole expanding test *ISO 16630:2009*
[7] Yoshida H, Yoshida T, Sato K,Takahashi Y, Matsuno T and Nitta J 2013 Evaluation and Improving Methods of Stretch Flageability *Nippon Steel Technical Report No. 103*
[8] Păunoiu V, Nicoară D, Cantera Lopez A and Arroyo Higuera P 2005 Numerical simulation of forming limit curves using reduced scale samples *The Annals of Dunarea de Jos University of Galati*
[9] Hama T, Takamura M, Makinouchi A, Teodosiu Cristian and Takuda Hirohiko 2008 Effect of Tool-Modeling Accuracy on Square-Cup Deep-Drawing Simulation *Materials Transactions* **49** pp 278-283
[10] Banu M, Hama T, Alves L 2008 Numerical prediction of the stress fields in a bending-unbending stage using STAMP3D and DD3IMP *Steel Research International* 1 **79** pp 186-193
[11] *** AUTOFORM *Software manual*

IOP Conf. Series: Materials Science and Engineering **968** (2020) 012012 doi:10.1088/1757-899X/968/1/012012

Experimental investigation and regression modelling to improve machinability in CNC turning of CALMAX® tool steel rods

N M Vaxevanidis[1], N A Fountas[1], I Papantoniou[2] and D E Manolakos[2]
[1] Laboratory of Manufacturing Processes & Machine Tools (LMProMaT), Department of Mechanical Engineering Educators, School of Pedagogical and Technological Education (ASPETE), Amarousion GR 151 22, Greece
[2] School of Mechanical Engineering, National Technical University of Athens (NTUA), Zografou, GR 157 80, Greece

E-mail: vaxev@aspete.gr

Abstract. This paper studies the effect of cutting conditions (spindle speed - rpm; feed rate - mm/rev and depth of cut - mm) on main cutting force and surface roughness during CNC turning of the commercial tool steel CALMAX® by Uddeholm (Sweden). Experiments have been designed using the Central composite design (CCD) approach. The experimental design involved twenty base runs with eight cube points, four center points in the cube, six axial points and two center points in axial direction. Statistical analysis to examine the effect of cutting conditions on the responses of main cutting force and surface roughness included ANOVA under the scope of generating a full quadratic model for predicting the responses. Finally, a feed-forward back-propagation neural network was applied to predict the responses of cutting force and surface roughness. It was found that regression models corresponding to the responses as well as the neural network developed can efficiently explain much of the variation in terms of main cutting force and surface roughness and thereby they may be implemented to practical applications either for predicting actual machinability parameter values or for setting up objective functions to be evaluated by intelligent algorithms for process optimization.

1. Intoduction

Machining processes of industrial products may affect directly production costs at almost 15 to 20% and is estimated that reach down to over 80% of their value. Thus, it is essential to study the process parameters, especially when it comes to difficult-to-machine materials. Machining productivity and surface quality may be significantly improved when appropriate selection of cutting tools and machining parameters is made.

The success of material removal operations lies heavily on several important technological aspects such as lubrication; [1-4], machining strategy selection; [5], cutting tools selection [6] and machining parameter selection [7]. Under this prism, noticeable contributions have examined the possibilities of modelling and optimizing discrete quality objectives by controlling parameters yielding strong influence on the objectives. It turns out that selecting proper cutting conditions, suitable cutting fluids as well as taking into consideration the properties of cutting tools, work materials and system stability, are of paramount importance. It is claimed in [8,9] that even small changes in any of the aspects reported, may affect the results in terms of the produced part surface. A variety of research works have been proposed so far regarding modelling and optimization of discrete manufacturing criteria reflecting quality characteristics. Such quality characteristics are surface roughness, production cost, machining time, material removal rate, cutting forces, etc.

Musfirah et al. in [10] presented results concerning their work on tool wear and surface integrity during dry and cryogenic high-speed machining of Inconel 718. Cryogenic and dry cutting experiments were conducted with the utilization of a CNC milling machine capable of achieving a maximum speed of 8000 rpm. Devillez et al. [11] focused on the effect of dry machining on surface integrity of Inconel 718 alloy. Wet and dry turning tests were performed at various cutting speeds, with semi-finishing conditions using a coated carbide tool. For each cutting test, cutting force was measured, machined surface was observed, and residual stress profiles were determined. Xavior and Adithan in [12] investigated the effect of cutting fluids on tool wear and surface roughness during turning of AISI 304 with carbide tool. They also tried to identify the influence of coconut oil in reducing tool wear and surface roughness during turning. Lalwani et al. [13] examined the influence of cutting speed, feed rate and depth of cut on cutting force components; and surface roughness in finish hard turning of MDN250 steel which is equivalent to 18Ni (250) maraging steel, using coated ceramic tool. The machining experiments were conducted based on response surface methodology (RSM) and sequential approach using face centered central composite design. The results show that cutting forces and surface roughness do not vary much with experimental cutting speed in the range of 55–93 m/min. To predict optimal parameter sets satisfying productivity and quality demands, several artificial intelligence (AI) techniques for modelling and optimization have been employed so far. Most popular AI techniques involve the usage of genetic and evolutionary algorithms (GAs-EAs) and artificial neural networks (ANNs) [14-16], being supported by analyses of results obtained from experimental design.

2. Experimental

2.1. Design of experiments

In this work experiments have been conducted to investigate the effect of spindle speed, *n* (*rpm*); feed rate, *f* (*mm/rev*) and depth of cut, *a* (*mm*), on main cutting force, *Fz* (*Nt*) and surface roughness, *Ra* (*μm*). A rotatable central composite design (CCD) consisting of twenty experimental runs was established according to the number of cutting parameters and their experimental levels (see Table 1).

Table 1. Cutting parameters and corresponding experimental levels.

Central composite design of experiments					
Parameter	Symbol	Level			Unit
		Low (-1)	Center (0)	High (1)	
Spindle speed	n	1500	1750	2000	rpm
Feed rate	f	0.050	0.125	0.200	mm/rev
Depth of cut	a	0.500	1.000	1.500	mm

The central composite design sampling method is widely used in response surface applications. By selecting *corner*, *axial*, and *center* points, it is an ideal solution for fitting a second-order response surface model [17]. However, as it requires a relatively large number of sample points, the CCD method should only be chosen in a later stage of the RSM application when the total number of important variables is reduced to an acceptable figure. In this work the three cutting parameters produce only few experiments. The rotatable central composite design is the most widely used experimental design for modelling a second-order response surface. A design is called rotatable when the variance of the predicted response at any point depends only on the distance of the point from the center point of design. The rotatable design provides the uniformity of prediction error and it is achieved by proper choice of α. In rotatable designs, all points at the same radial distance "r" from the center point have the same magnitude of prediction error. For a given number of variables, the α required to achieve rotatability is computed as $\alpha = n_f^{\,1/4}$, where n_f is the number of points in the 2^k factorial design. A rotatable CCD consists of 2^k fractional factorial points, augmented by two k axial points [($\pm\alpha$, 0,...0),(0, $\pm\alpha$,...0), (0, 0,... $\pm\alpha$)] and n_c center points (0, 0, 0, 0...,0). Here also, the center points vary from three to six. With proper choice of n_c the CCD can be made orthogonal or it can be made uniform precision design. It means that the variance of response at origin is equal to the variance of response at a unit distance from the origin. Considering uniform precision for three-factor experimentation, eight (2^3) factorial points, six axial points and six center runs, a total of 20 runs are considered whilst the value of "α" is $(8)^{1/4} = 1.682$.

2.2. *Materials and equipment*

The material under investigation was the "CALMAX®" tool steel by Uddeholm (Sweden) with the chemical composition given in Table 2. The steel has been used in its soft-annealed condition, having a hardness of 220 HV.

Table 2. General properties of CALMAX® tool steel of Uddeholm-Sweden.

Chemical composition	
Element	Content amount (%)
C	0.60
Si	0.35
Mn	0.80
Cr	4.50
Mo	0.50
V	0.20
Properties	
Modulus of elasticity (at 20°C)	194000 MPa
Hardness (Rockwell C)	56 HRC

Machining experiments were conducted using two Ø30x300mm pre-machined bars comprising ten discrete zones per bar for facilitating chip removal among the cuts. Machining experiments were conducted using a *HAAS® TL-1* CNC turning center equipped with a typical HAAS CNC control unit. The cutting insert used was a *SECO® TNMG 160404-MF2 TP200* with the suitable insert holder. Figure 1 illustrates the two CALMAX bars; one before preparation and one after preparation. Figure 2 depicts the *HAAS TL-1* CNC turning center used for conducting the experiments.

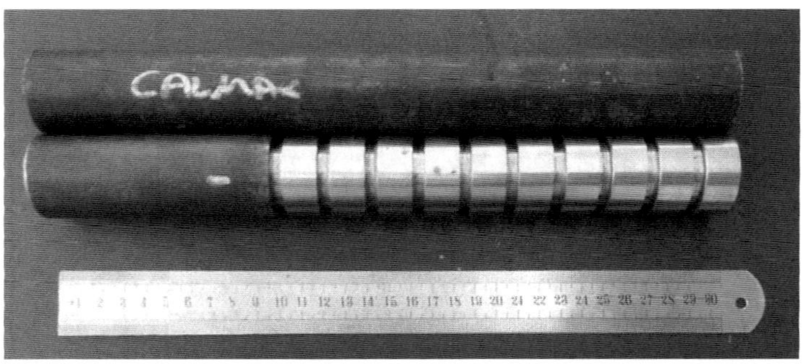

Figure 1. Ø30x300mm CALMAX® bars for CNC turning experiments.

Figure 2. The HAAS® TL-1 CNC turning center used for conducting the experiments.

The CNC turning center was integrated with a computer-assisted apparatus based on a *Labview®* module capable of taking performing online measurements for the three cutting force components, *Fx, Fy* and *Fz* (*Nt*). A typical and commercially available three-component dynamometer of *Kistler®* was selected to measure the main cutting force. Figure 3 depicts the trend for online measurements for the main cutting force *Fz*, obtained during the experiments.

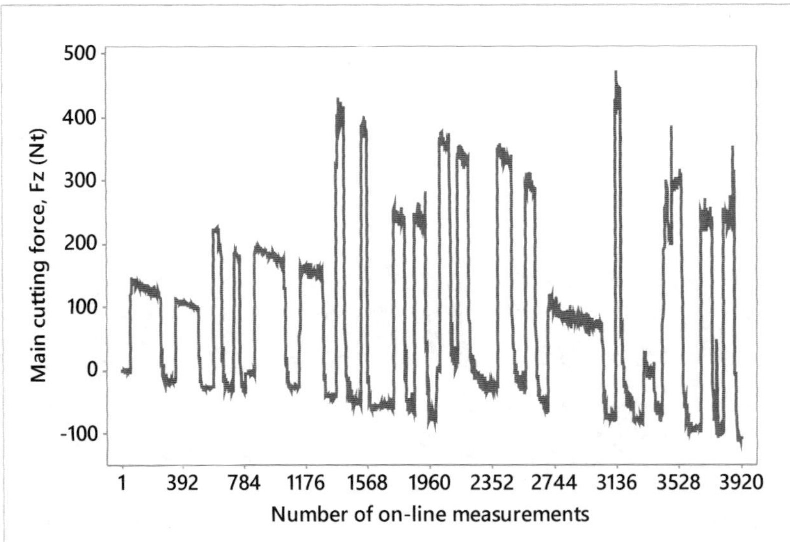

Figure 3. Trend for online measurements for main cutting force, Fz (Nt).

These measurements provided the raw data for further processing on the context of computing the final output for cutting forces. From the trend depicted in figure 3, the average value from the regions where large cutting forces exhibited, was computed to formulate the first response. Mean arithmetic surface roughness *Ra (μm)* was treated as the second response. Surface roughness was measured using a *TESA® Rugorsurf® 10G* portable roughness tester along with its respective software, according to ISO-4287 standard. Figure 4 depicts the results obtained from measuring the 6th cutting zone of the first tool steel bar. For each cutting zone three roughness measurements were taken by rotating the workpiece at an angle of 120°. As a final output for surface roughness corresponding to a cutting zone, the average value from the three measurements was computed.

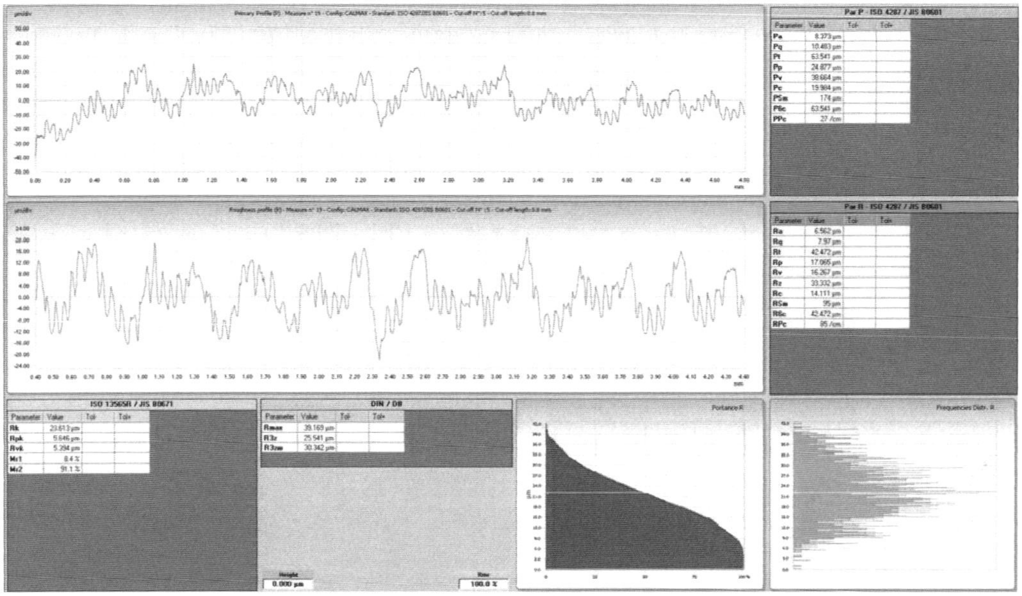

Figure 4. Indicative results from measurements taken on the 6th cutting zone (1st tool steel bar).

3. Experimental results and analysis

3.1. Results

The results obtained from the experiments are presented in Table 3. Analysis of variance (ANOVA) for the results was performed to evaluate the importance of machining parameters and estimate the experimental error. ANOVA for the results of main cutting force – *Fz* and arithmetic surface roughness average – *Ra* are given in table 4.

IOP Conf. Series: Materials Science and Engineering **968** (2020) 012012 doi:10.1088/1757-899X/968/1/012012

Table 3. Experimental results for main cutting force (*Fz*) and arithmetic surface roughness average (*Ra*).

No.	n (rpm)	s (mm/rev)	a (mm)	F_Z (N)	R_a (μm)
1	1500.00	0.050	0.5000	128.476	1.011
2	2000.00	0.050	0.5000	104.210	0.981
3	1500.00	0.200	0.5000	208.923	3.999
4	2000.00	0.200	0.5000	179.380	3.801
5	1500.00	0.050	1.5000	181.104	1.473
6	2000.00	0.050	1.5000	157.742	1.045
7	1500.00	0.200	1.5000	381.403	4.082
8	2000.00	0.200	1.5000	369.989	3.760
9	1750.00	0.125	1.0000	241.514	2.069
10	1750.00	0.125	1.0000	236.655	1.971
11	1750.00	0.125	1.0000	346.399	1.689
12	1750.00	0.125	1.0000	327.738	1.332
13	1342.00*	0.125	1.0000	330.974	1.566
14	2158.00*	0.125	1.0000	285.807	1.294
15	1750.00	0.025*	1.0000	83.848	0.980
16	1750.00	0.250*	1.0000	404.271	8.946
17	1750.00	0.125	0.1800*	22.128	1.717
18	1750.00	0.125	1.8200*	266.242	1.896
19	1750.00	0.125	1.0000	236.270	1.870
20	1750.00	0.125	1.0000	241.126	1.763

Table 4. ANOVA results for main cutting force (*Fz*) and mean arithmetic surface roughness (*Ra*).

ANOVA results for main cutting force, Fz regression model				
	Seq SS	Adj SS	Adj MS	% contribution
Model	194669	194669	19466.9	92.10
Linear	147807	147807	49269.0	69.93
Square	38449	38449	12816.3	18.19
Interactions	8302	8302	2767.5	3.93
Residual error	16706	16706	1856.2	7.90
Pure error	194669	194669	19466.9	2.10
Lack-of-fit (p value)			0.729	
ANOVA results for surface roughness, Ra regression model				
	Seq SS	Adj SS	Adj MS	% contribution
Model	61.7461	61.7461	6.8607	93.64
Linear	43.9142	43.9142	14.6381	66.60
Square	17.7683	17.7683	5.9228	26.95
Interactions	0.0637	0.0637	0.0212	0.10
Residual error	4.1930	4.1930	0.4193	6.36
Pure error	0.3375	0.3375	0.0675	0.51
Lack-of-fit (p-value)			0.770	

IOP Conf. Series: Materials Science and Engineering **968** (2020) 012012 doi:10.1088/1757-899X/968/1/012012

Analysis of the experimental data obtained from the central composite design was carried out on *MINITAB® R17* software using the full quadratic response surface model as given by equation 1.

$$y = \beta_0 + \sum_{i=1}^{k} \beta_i x_i + \sum_{i=1}^{k} \beta_{ii} x_i x_i + \sum_{i<j} \sum \beta_{ij} x_i x_j \quad (1)$$

Y is the response -main cutting force *Fz* (*Nt*) and surface roughness *Ra* (μm) - and x_i is the i_{th} parameter. Probability of *F*-value greater than calculated *F*-value due to noise is indicated by p-value. If *p*-value < 0.05, significance of corresponding term is established. For lack of fit, *p*-value > 0.05. An insignificant lack of fit is desirable because it indicates that any term excluded by the model is insignificant and that the developed model fits well. *Anderson–Darling* normality test is used to verify the suitability of the models corresponding to the main cutting force and surface roughness, for practical applications. If *p*-value for the *Anderson–Darling* test is lower than the chosen significance level, i.e. 0.05, it is concluded that the data do not follow a normal distribution. In this work, ANOVA indicates that both quadratic models are suitable for predicting the main cutting force *Fz* (*Nt*) and surface roughness *Ra* (μm) with high contributions, i.e. 92.10% and 93.64% and *p*-value for lack-of-fit far beyond 0.05 (*p*-value > 0.25) - see figure 5a and figure 5b.

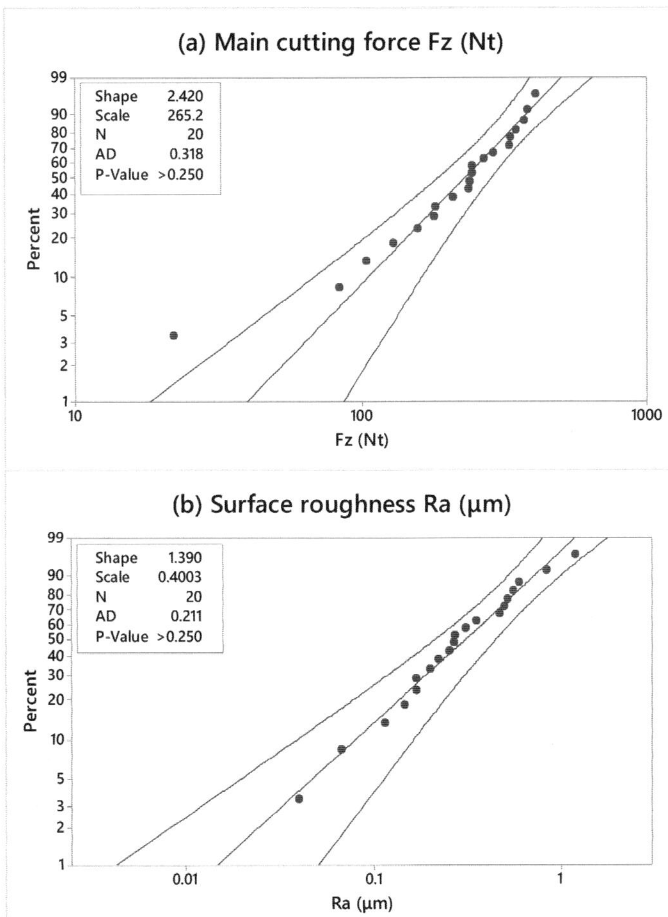

Figure 5. Probability plots for regression models: (a) main cutting force, *Fz*; (b) surface roughness, *Ra*.

3.2. Statistical analysis

Based on *p*-value, it has been concluded that the main cutting force *Fz* (*Nt*) and surface roughness *Ra* (*μm*) are mainly influenced by the linear terms and square terms followed by interaction terms. The individual significance of each term is calculated by t-test at 95% confidence level, thus; terms having *p*-value less than 0.05 are significant. The coefficient of determination (R^2) which indicates the percentage of total variation in the response explained by the terms in the model has been found equal to 92.10% and 93.64% for main cutting force *Fz* (*Nt*) and surface roughness *Ra* (*μm*) respectively. Specifically the linear terms in the case of main cutting force *Fz*, have a 69.93% contribution whereas square terms reach 18.19% as a contribution percentage. Interaction terms contribute to main cutting force as much as 3.93%. In the case of surface roughness *Ra*, linear terms have a 66.60% contribution whereas square terms reach 26.95% as a contribution percentage. Interaction terms contribute to surface roughness as much as 0.10%.

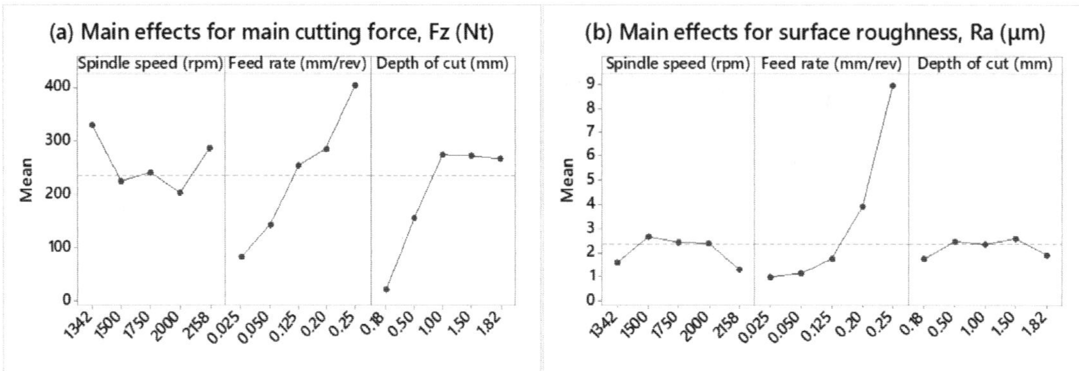

Figure 6. Plots for the main effects: (a) main cutting force, *Fz*; (b) surface roughness, *Ra*.

From figure 6a it is evident that feed rate *f* (*mm/rev*) and depth of cut *a* (*mm*) have the largest effect on the response of main cutting force *Fz* (*Nt*). Main cutting force gradually increases with the increse of feed rate. Main cuting force reaches its highest mean under the feed rate value equal to 0.25 *mm/rev*. A noticeable increase in main cutting force is observed for the three first levels of depth of cut whilst the two higher leves equal to 1.50 *mm* and 1.82 *mm* do not seem to further increase main cutting force. Spindle speed reduces the mean of main cutting force in general, however its highest level tends to increase it. From figure 6b it is evident that feed rate *f* (*mm/rev*) and spindle speed *n* (*rpm*) have the largest effect on the response of surface roughness *Ra* (*μm*). Surface roughness gradually increases with the increse of feed rate whilst it reaches its highest mean under the feed rate value equal to 0.25 *mm/rev*. Spindle speed comes as the second variable to affect surface roughness. Its lower level maintains low surface roughness, however surface roughness increases when moving towards the second level for spindle speed. The next spindle speed levels gradually reduce surface roughness where the lowest result is observed for spindle speed equal to 2158 *rpm*. Depth of cut is the third parameter to affect surface roughness and based to its rank, it can be advocated that it does not significantly affect surface roughness. Nevertheless, lowest and highest levels for depth of cut seem to be advantageous for surface roughness, whereas intermediate level deteriorate it. Contour plots allow to depict a three-dimensional surface on a two-dimensional plane. They graph two predictors-variables on *X* and *Y* axes whereas the response variable is appeared in *Z* axis as a contour. Corresponding contour plots for the responses of main cutting force and surface roughness were generated to show how their values change as functions of different pairs of independent variables. Figure 7 shows the resulting changes in main cutting force when altering spindle speed, feed rate and depth of cut, as a fucntion of two inputs, i.e. $Fz = f(n, f)$, $Fz=f(n, a)$, $Fz = f(f, a)$ whereas figure 8 shows the resulting changes in surface roughness when altering spindle speed, feed rate and depth of cut, as a function of two inputs, i.e. $Ra = f(n, f)$, $Ra=f(n, a)$, $Ra = f(f, a)$.

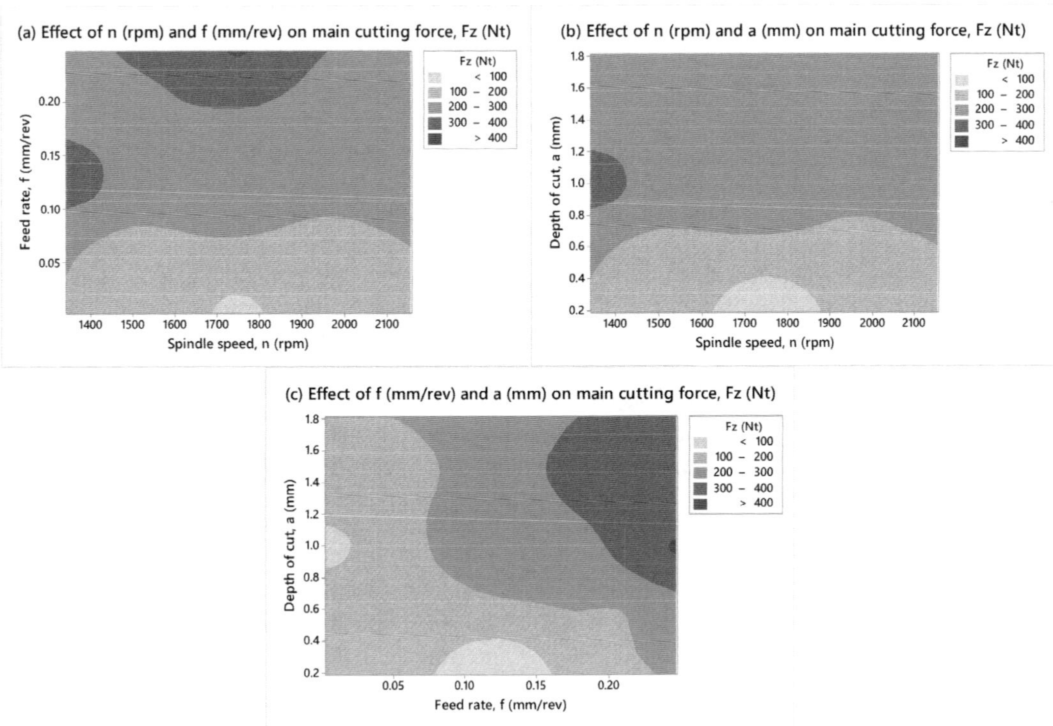

Figure 7. Contour plots for *Fz*: (a) $Fz = f(n, f)$, (b) $Fz = f(n, a)$, (c) $Fz = f(f, a)$.

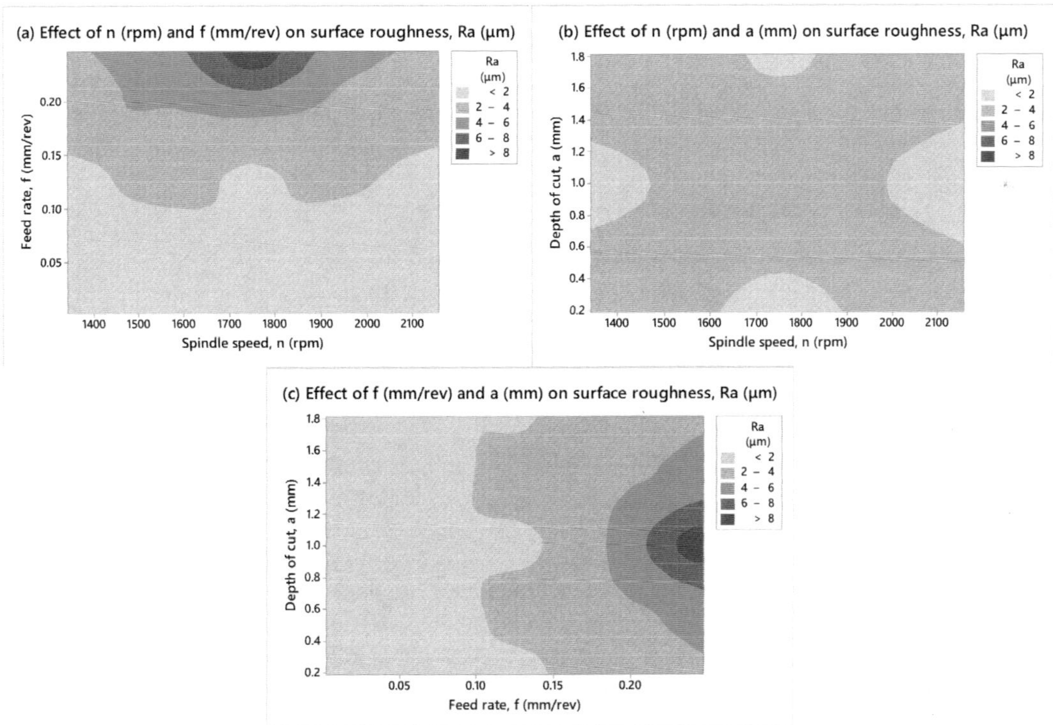

Figure 8. Contour plots for *Ra*: (a) $Fz = f(n, f)$, (b) $Fz = f(n, a)$, (c) $Fz = f(f, a)$.

4. Experimental results and analysis

To investigate whether the prediction for the responses can be improved an artificial neural network (ANN) with back propagation algorithm was adapted to model the CNC turning process. For this analysis, the three independent variables spindle speed *n*, feed rate *f* and depth of cut *a* were considered as the input parameters for the ANN. Each of these parameters has been characterized by one neuron, thus, the input layer in the ANN structure had three neurons. The database has been built considering the series of CNC turning experiments at the limit ranges of each independent variable. Experimental results for main cutting force and surface roughness were used to train the ANN to understand the input–output correlations. The database was then divided into two categories, namely the training category which has been exclusively used to adjust the network weights and the test category, which corresponds to the set that validates the results of the training protocol. Factorial runs were used for training, whilst the rest of experimental runs were used for testing. Experimental runs corresponding to center points have the same factor level values, hence mean of main cutting force and surface roughness were used as representative values related to these experiments. Neural network training involved updating the weights of the connections in such a manner that the error between the network's outputs and the actual output is minimized. To determine the number of neurons in the single hidden layer selected, different network structures with varying number of neurons in the hidden layer were tested. After training, the topology *3-10-2* was finally selected as the optimum. The activation level of neurons was determined by tan–sigmoid transfer function except for output layer neurons for which linear output transfer function was used so that output is not limited to small values. Figure 9 gives the schematic representation of ANN used in present work. The ANN was simulated in *MATLAB® 2014b*. To evaluate the competence of this trained network, the training data set was presented to the trained network.

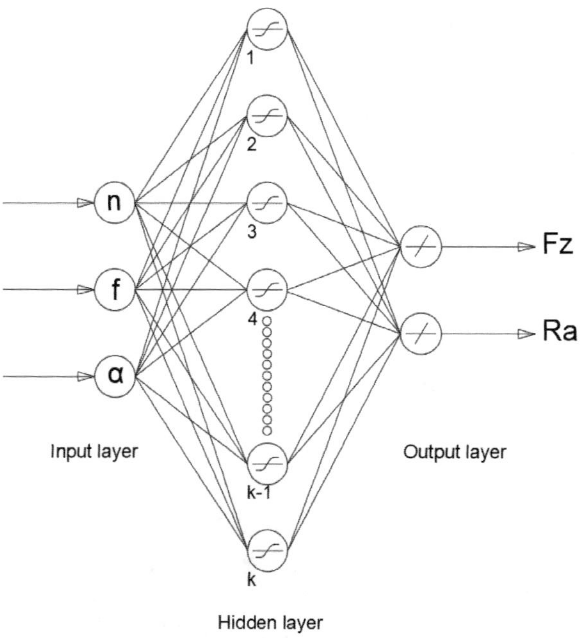

Figure 9. ANN architecture with three inputs, one hidden layer (10 neurons) and two outputs.

Mean squared error (MSE) in ANNs is the average squared difference between network output values and target values. For the network implemented in the study, best validation performance was equal to 185.4179 at epoch 11; see figure 10.

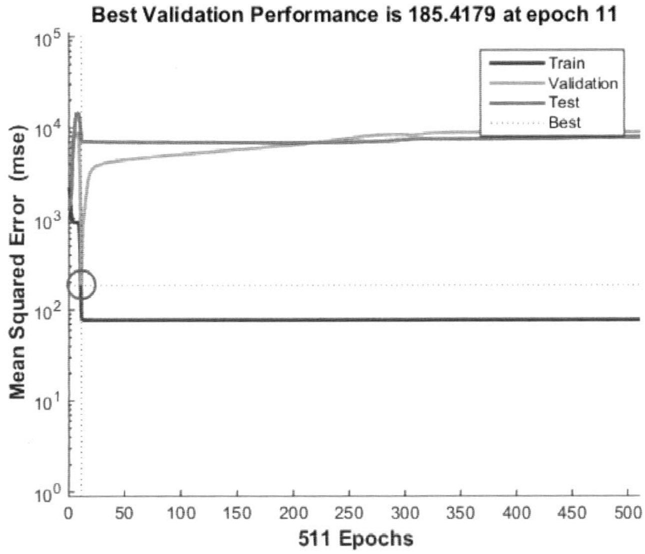

Figure 10. Best validation performance for the ANN implemented.

Figure 11 shows the regression analysis results between the network response and the corresponding targets. High correlation coefficient R^2 between predicted results (outputs) and targets validated the performance of the ANN implemented. Regression values measure the correlation between output values and targets. The results acquired for this study shown a good correlation among output values and targets during training (*R=0.981*), validation (*R=0.998*), and testing (*R=0.97956*).

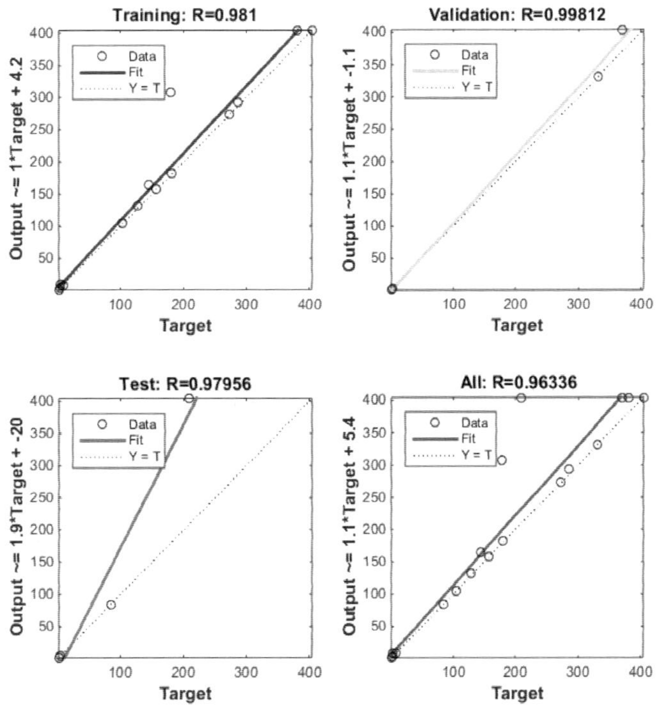

Figure 11. Regression plots for ANN performance verification.

The values predicted by ANN and regression models for main cutting force and surface roughness are depicted in figure 12. Figure 12a shows the comparison among experimental, predicted and ANN result whereas figure 12b shows the same comparison referring to the results obtained for surface roughness. In both figures, it can be observed that the trend corresponding to the results variation for the ANN matches more accurately the one corresponding to the experimental results. Based on these outputs it is proven that the ANN is better for modelling the non-linearity presented in the experimental domain.

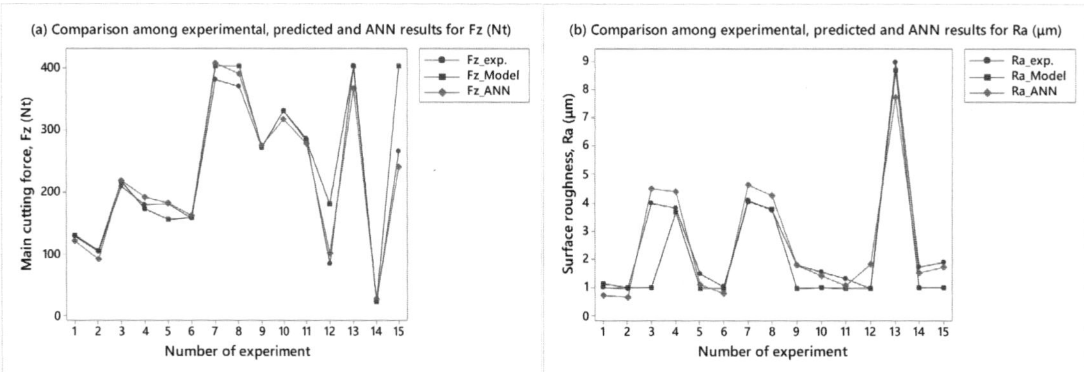

Figure 12. Comparisons among results obtained: (a) main cutting force *Fz (Nt)*, (b) surface roughness *Ra (μm)*.

5. Conclusions

In this work, the effect of cutting parameters (spindle speed-*n*, feed rate-*s* and depth of cut –a) on the main cutting force -*Fz,* and arithmetic surface roughness average -*Ra* were studied. Experimental results were obtained through the utilization of response surface methodology by establishing a central composite design. The analysis and interpretation of the respective results was carried out by performing analysis of variance using the full quadratic model and sequential SS for tests. Signal-to-noise and interaction plots were also exploited to study the effects of parameters on the responses. The experimental outputs were further used for the training, testing and validation of a feed forward back propagation artificial neural network model. It was shown that the regression models and the ANN can predict experimental results when CNC turning the commercially available CALMAX® tool steel, in terms of main cutting force and surface roughness with good correlation. To depict variations of quality objectives studied, contour plots were generated considering the different values (levels) of cutting parameters. As a future work authors plan to implement different evolutionary algorithms and optimize the CNC turning operation of CALMAX® tool steel by formulating a multi-objective optimization problem and having the regression models under the role of objective functions.

6. References

[1] Hong S Y, Markus I and Jeong W 2001 New cooling approach and tool life improvement in cryogenic machining of titanium alloy Ti-6Al-4V *International Journal of Machine Tools and Manufacture* **41** 2245–2260

[2] Prashanth P G, Venugopal K A and Paul S, 2003 Effect of cryogenic cooling with liquid nitrogen jets in turning titanium alloys *Recent Trends in Manufacturing ARTM*

[3] Venugopal K A, Tawade R, Prashanth P G, Paul S and Chattopadhyay A B 2003 Turning of titanium alloy with TiB2 – coated carbides under cryogenic cooling *Proceedings IMechE* B **217** 1697–707

[4] Ramana M V, Srinivasulu K, Krishna G and Mohana R 2011 Performance Evaluation and Selection of Optimal Parameters in Turning of Ti-6Al-4V Alloy Under Different Cooling Conditions *International Journal of Innovative Technology & Creative Engineering* **1**(5).
[5] Cotterell M and Byrne G 2008 Characterisation of chip formation during orthogonal cutting of titanium alloy Ti–6Al–4V *CIRP Journal of Manufacturing Science and Technology* **1** 81–85
[6] Ezugwu E O, Da Silva R B, Bonney J and Machado A R Evaluation of the performance of CBN tools when turning Ti–6Al–4V alloy with high pressure coolant supplies *International Journal of Machine Tools and Manufacture* **45** 1009–1014
[7] Armendia M, Garay A, Iriarte L M and Arrazola P J, Comparison of the machinabilities of Ti6Al4V and TIMETAL® 54M using uncoated WC–Co tools *Journal of Materials Processing Technology* **210** 197–203
[8] DeGarmo E P, Black J T and Kohser R A 2003 *Materials and Processes in Manufacturing, 9th Edition* (Wiley: Hoboken)
[9] Kalpakjian S, Schmid S R 2003 *Manufacturing Processes for Engineering Materials, 4th Edition* (New Jersey: Prentice Hall)
[10] Musfirah A H, Ghani J A and Che Haron C H 2017 Tool wear and surface integrity of inconel 718 in dry and cryogenic coolant at high cutting speed *Wear* 376-377 125-133
[11] Devillez A, Le Coz G, Dominiak S and Dudzinski D 2011 Dry machining of Inconel 718, workpiece surface integrity *Journal of Materials Processing Technology* **211** 1590-1598
[12] Xavior M A and Adithan M 2009 Determining the influence of cutting fluids on tool wear and surface roughness during turning of AISI 304 austenitic stainless steel *Journal of Materials Processing Technology* 2 **209** 900-909
[13] Lalwani D I, Mehta N K and Jain P K 2008 Experimental investigations of cutting parameters influence on cutting forces and surface roughness in finish hard turning of MDN250 steel *Journal of Materials Processing Technology* **206** 167-179
[14] Feng C X and Wang X F 2003 Surface roughness predictive modelling: neural networks versus regression *IIE Transactions* **35** 11–27
[15] Escamilla I, Torres L, Perez P and Zambrano P 2008 A Comparison between Back Propagation and the Maximum Sensibility Neural Network to Surface Roughness Prediction in Machining of Titanium (Ti 6Al 4V) Alloy *Advances in Artificial Intelligence* ed A Gelbukh and E F Morales (Berlin: Springer-Verlag)1009–1019
[16] Davim J P, Gaitonde V N and Karnik S R 2008 Investigations into the effect of cutting conditions on surface roughness in turning of free machining steel by ANN models *Journal of materials processing technology* **205** 16-23
[17] Myers R H and Montgomery D C 2002 Response Surface Methodology – Process and Product Optimization Using Designed Experiments 2nd Edition *Wiley Series in Probability and Statistics*

Acknowledgements
This work was funded by the "*Research & Management committee – E.L.K.E. ASPETE*" under the auspices of the Research program "*Parametric Analysis & Machining Parameters Optimization of Special Purposes Steels*" for the project "Research strength in ASPETE 2018-2020"-Ref.No.: 80156.

IOP Conf. Series: Materials Science and Engineering **968** (2020) 012013 doi:10.1088/1757-899X/968/1/012013

Study on the influence of technological parameters on the friction stir butt welding process of pure copper plates

M A Constantin[1], E L Nitu[1], D M Iordache[1] and C Badulescu[2]
[1] University of Pitesti, Faculty of Mechanics and Technology, Manufacturing and Industrial Management Department, Targul din Vale Street No.1, Pitesti, Romania
[2] ENSTA Bretagne, IRDL-UMR CNRS 6027, F-29200 Brest, France

E-mail: eduard.nitu@upit.ro

Abstract. Friction stir welding - FSW is a relatively new welding process, which is increasingly used in industry, due to the advantages which it has in relation to conventional processes (by melting and adhesions). The advantages of the FSW welding process (as a solid phase welding process) are all the more obvious in the case of joining copper and its alloys, because they have a high melting temperature and high thermal diffusivity. The influence of the technological parameters of the process, the tool rotational speed and the welding feed, on the temperature and the axial force, as well as on the quality of the joint surface is presented. The study shows that the stabilization of the process takes place after a certain time from the beginning of the advance stage and highlighted the major influence of the tool rotational speed on the process temperature, surface defects and the roughness of the joint surface.

1. Introduction

Copper and copper alloys offer unique combinations of electrical and thermal conductivity, mechanical strength, formability and corrosion resistance, nonmagnetic behaviour, which makes them used in a wide range of engineering applications.

In industry, the joining of copper and copper alloys is usually done by various classical processes of: fusion welding (arc shielding gas, acetylene flame, laser, etc.), adhesion joining (brazing, soldering) or solid-state welding (electrical resistance and pressure end-to-end, ultrasound, friction stir welding). Due to the problems generated by melt welding (cracks, pores, oxides) and the relatively low strengths obtained by adhesion joining, solid state welding processes are preferred in many situations.

Friction stir welding - FSW is a relatively new welding process [1], which is increasingly used to join copper structures and its alloys. The FSW process is able to perform thick welds, with mechanical properties superior to melt welding, reliably and reproducibly [2 - 5], which is why this welding process is in full growing.

Scientific research on the joining of copper and its alloys through the FSW process is mainly focused on the study of the influence of technological parameters and the shape of the tool (rotating active element) on the quality of the joint [6 - 11], the possibility of combining different structures and types of alloys, inclusive dissimilar [7, 12 - 15] or the possibility of using external heat sources - the application of hybrid processes [16 - 18]. For this purpose, different theoretical-experimental research techniques and/ or numerical modelling techniques and simulation were used.

The research on FSW joining of copper alloys has shown that due to the relatively high melting temperature and high thermal diffusivity of copper alloys, their FSW joining is limited to a narrow range of technological parameters. The temperature during the FSW process has a decisive role on the quality of the joint (appreciated by its microstructure and mechanical properties), and the axial pressure force influences the morphological aspect of the joint, but also the defects in the weld seam.

This paper presents the main results of the experimental research of the FSW joining process of a structure consisting of two similar pieces positioned butt-to-butt. The two welded parts are made of Cu-DHP, a material commonly used in the automotive industry, in the construction of radiators and air conditioning system elements. This study presents an analysis of the dependence of the process parameters, temperature and axial force, on the technological parameters of the process, the rotational speed of the rotating active element (tool) and welding feed. Also, the surface quality of the joints is evaluated by visual morphological analysis and through the roughness parameters.

2. Experimental procedure

2.1. Materials joined and experimentation

In the experiments, two plates of Cu-DHP (chemical composition: Cu 99.9% and P 0.015 - 0.04%), of dimensions 100x250x3 mm, having the mechanical characteristics from the table 1, were joined by FSW butt-welding process, figure 1.

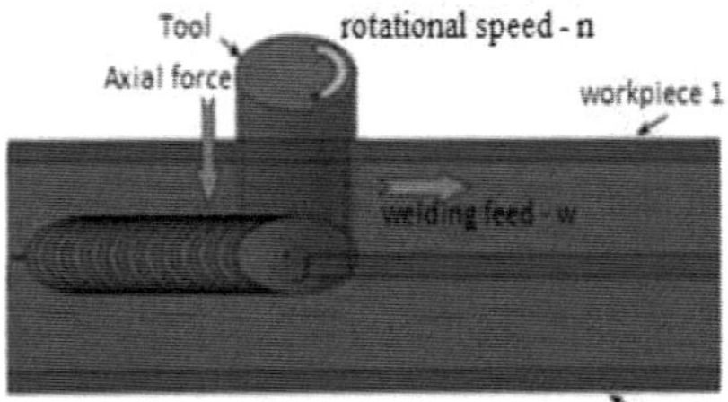

(a)

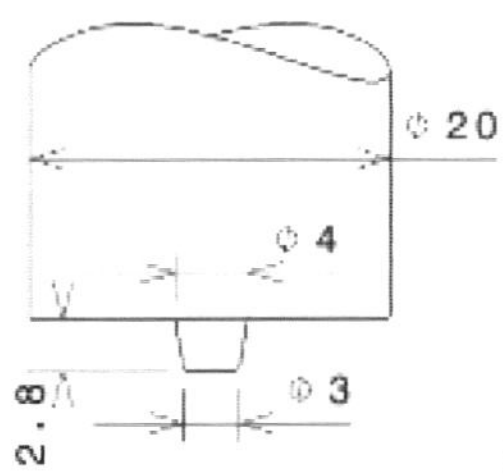

(b)

Figure 1. FSW butt-welding: (a) process scheme, (b) dimensions of the welding tool.

Table 1. Mechanical characteristics of Cu-DHP.

Mechanical characteristics	Tensile strength, Rm [MPa]	Yield strength, $Rp_{0.2}$ [MPa]	Vickers hardness, HV0.3 [N/mm^2]	Elongation, A5 [%]
Value	260 MPa	206 MPa	81	55

In order to analyse the influence of the technological parameters: rotational speed - n, and welding feed - w, on process characteristics, several experiments were performed, with different values of these technological parameters, table 2.

Table 2. Experimental plan of FSW of process.

Exp. no.	*1*	*2*	*3*	*4*	*5*	*6*	*7*
Rotational speed, n [rpm]	1200	1000	1200	1000	800	800	1000
Welding feed, w [mm/min]	90	90	150	150	90	150	120

2.2. Process parameters measurement and welds surfaces analysis

The experiments were performed on a specialized welding machine, FSW 4-10, figure 2. The welding process was monitored throughout its evolution, by recording the values of temperature and axial force.

The process temperature was measured using the infrared thermography method, using an infrared thermographic camera of the type FLIR A40M (temperature measurement range - 40^{0}C....+ 2000^{0}C, accuracy of data measurement ± 2°C, 76800 pixels, 60 fps). The thermographic data measured during the welding process were recorded and processed on a laptop, using the specialized software ThermaCAM^{TM} Researcher. The positioning and fixing of the thermographic camera were done with the help of a special device, which allowed its positioning behind the active element (tool), at a constant distance from it, monitoring the joint area in the immediate vicinity of the tool throughout the process evolution.

The axial process force was measured using a force sensor mounted on the main shaft of the welding machine head, a sensor that includes a WIKA compression transducer, model F1211 (force measurement range 0 - 20 kN and relative linearity error of ± 0.3% * Fnom). The acquired data were recorded and processed on a laptop, using the GRAPHTEC midi LOGGER GL240 system.

Figure 2. Experimental stand of Friction Stir Welding process.

The roughness of the FSW joints was determined on specimens taken from the end area of the joint, using a portable digital roughness-meter, type MAHR PS-10. The roughness measurement was performed in three areas of the specimen, in a direction parallel to the welding direction using a stylus method.

3. Results and discussions

The evolution of the temperature and the axial force during the FSW process were represented superposed on the same diagram, over the sampling scheme of the specimens from the welded structure. This allowed to highlight the moment when the process becomes stable: the temperature and the axial force remain approximately constant during the process. Equally, this helped to highlight local process conditions in which the joint was made, respectively, the conditions associated with the areas from which the different specimens used in the study were collected.

Figure 3 shows as an example such a diagram for experiment 6. It is found that the process becomes stable after a distance of about 70 mm from the beginning of the joint. On the whole of the experiments performed, the distances from which the FSW process becomes stable are between 60 and 90 mm.

The average values of temperature and axial force in the process stability area are presented in table 3 and are represented graphically, depending on the technological parameters n and w, in figure 4, respectively, figure 5.

Table 3. Average values of temperature and axial force in the process stability area.

Exp. no.	*1*	*2*	*3*	*4*	*5*	*6*	*7*
Technological parameters (n, w)	(1200,90)	(1000,90)	(1200,150)	(1000,150)	(800,90)	(800,150)	(1000,120)
Average temperature, [^{0}C]	693	466	580	550	476	500	596
Average axial force, [kN]	12.5	9.7	11.8	13.2	12.6	12.6	10.3

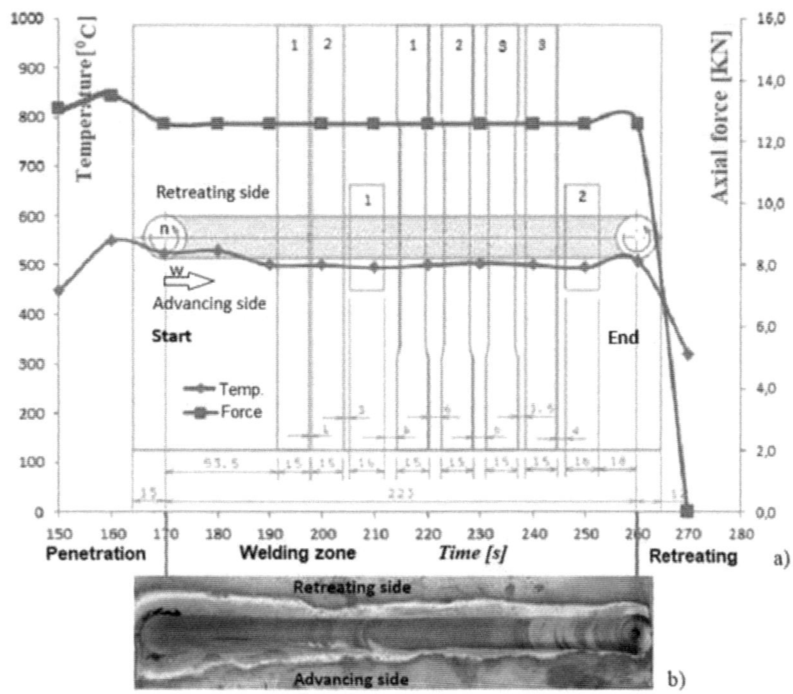

Figure 3. Experience no. 6 (n = 800 rpm, w = 150 mm/min): (a) evolution of temperature and axial force in the FSW process, (b) appearance of the welded joint.

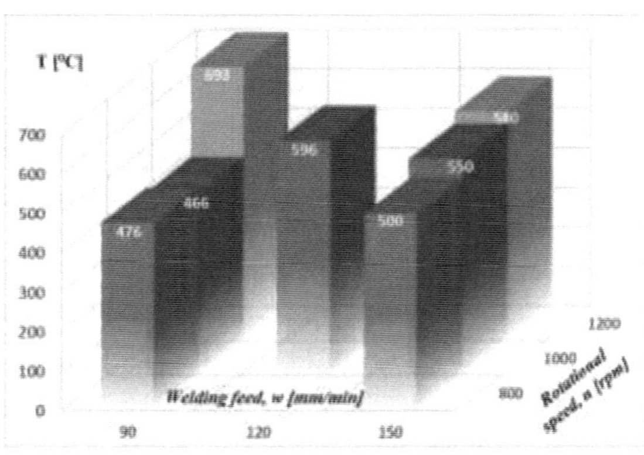

Figure 4. Dependence of average process temperatures on technological parameters.

From the analysis of the values in table 3 and of the graphs in figures 4 and 5, the following can be deduced:

- The average temperature values are between 466 and 693 ^{0}C and represent between 44% and 64% of the copper melting temperature (values similar to those found in other scientific studies);
- The process temperature is significantly influenced and in the same way by the tool rotational speed, while the welding feed has a smaller influence and depends on the value of the tool rotational speed. Thus, increasing the tool rotational speed produces a significant increase in the process temperature. Instead, increasing the welding feed causes a slight increase in the process temperature at low tool rotational speeds and a decrease in the process temperature at high tool rotational speeds. These influences of the technological parameters on the process temperature are due to the ratio in which are the velocities of the points on the periphery of the shoulder / tool pin produced by the two technological parameters: the velocities generated by the speed are between 50265 - 75400 mm/min at the shoulder periphery, and between 7539 - 11309 mm/min at the periphery of the pin, while the velocities generated by the welding feed are between 90 - 150 mm/min. Therefore, the effects produced by them (heat development due to friction at the shoulder-part interface and inside the joint by mixing of materials produced by pine) are different.
- In connection with the axial force in the process, it was not possible to establish its dependence on the two varied technological parameters in this study, n and w. This is caused by the complex mechanisms that influence the evolution of the axial force and which is done at the same time: on the one hand, the plasticization of the material (determined by the increase of the process temperature - which is influenced differently by the two technological parameters) leads to the decrease of the axial force, on the other hand, the increase of the friction between the tool and the welding material (also determined differently by the two technological parameters) leads to the increase of the axial process force.

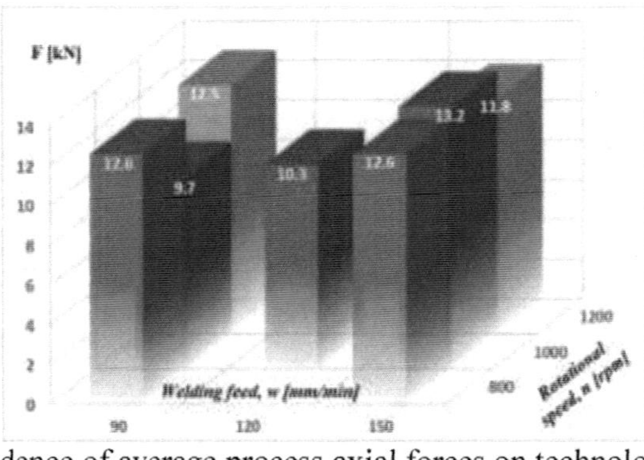

Figure 5. Dependence of average process axial forces on technological parameters.

The stabilization area of the welding process is also highlighted by the visual morphological analysis of the joint structure, figure 6. At the same time, this analysis provides important information on the production of surface defects of the joint, which are presented below:

- *Channel defect* (visible in all experiments): this occurs especially in the first part of the joint and is caused by insufficient reinforcement of the material on the advancing side. The appearance of this defect is due to insufficient plasticization of the material (low process temperature) and / or insufficient penetration of the tool in the joint structure (improper adjustment of the technological system). Once the FSW process stabilizes, the first cause is eliminated. The second cause can be highlighted on the joint structure by the existence of specific circular tracks, smaller than the diameter of the tool shoulder.
- *Excessive burr defect* (highlighted in exp. no. 1): this is caused by welding materials in too hot working conditions. When the process temperature is too high (above 650 ^{0}C), excessive

IOP Conf. Series: Materials Science and Engineering **968** (2020) 012013 doi:10.1088/1757-899X/968/1/012013

plasticization of the material and its mixing causes the appearance of these burrs, more prominent on the retreating side.

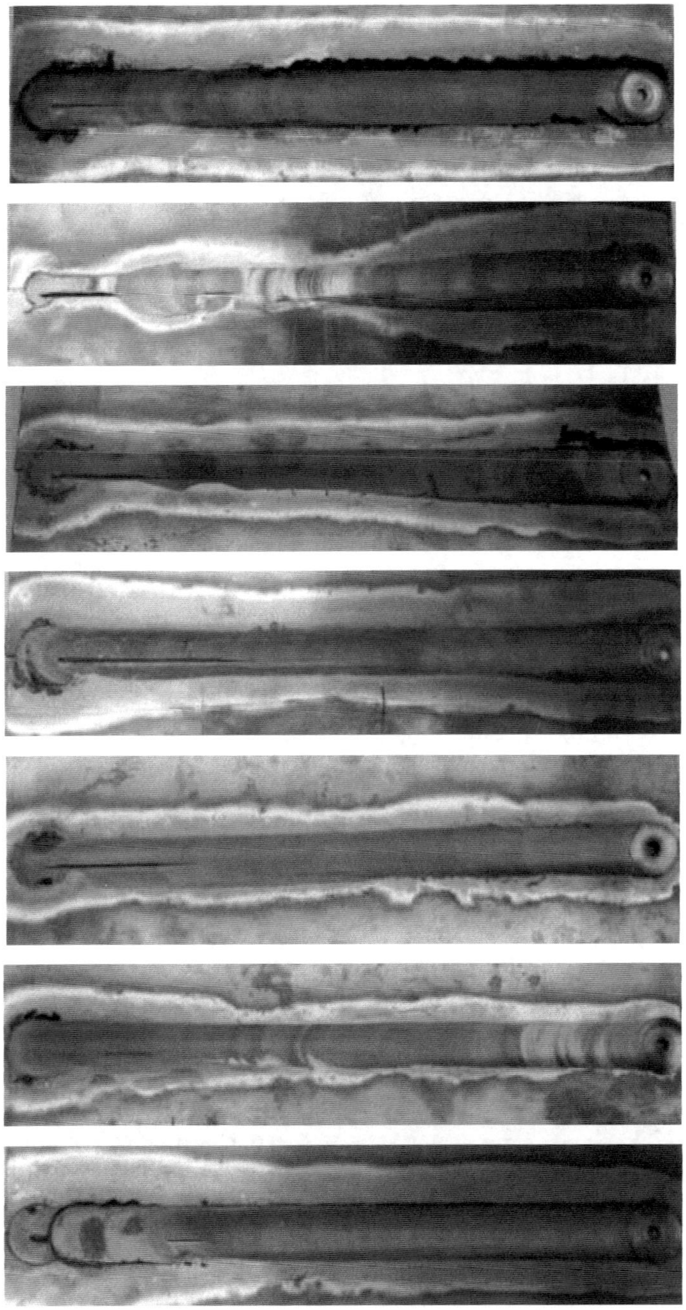

Exp. no. 1 (1200, 90)

Exp. no. 2 (1000, 90)

Exp. no. 3 (1200, 150)

Exp. no. 4 (1000, 150)

Exp. no. 5 (800, 90)

Exp. no. 6 (800, 150)

Exp. no. 7 (1000, 120)

Figure 6. Visual morphological aspect of the surface's joints.

The roughness of FSW joints is influenced by:

- the kinematics of the process that determines the appearance of "circular tracks": the spacing and height of the circular tracks specific to the FSW process depend on the tool rotational speed and the welding feed;

- the deformations caused by the friction of the tool shoulder with the upper surface of the parts and process temperature, which favours excessive plasticization of the material.

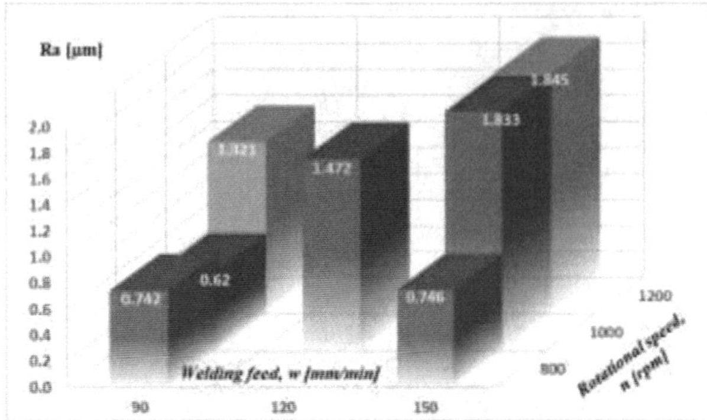

Figure 7. Dependence of average samples roughness on technological parameters.

The dependence of the average values of the roughness of the parts joined by the two technological parameters considered in the study, n and w, presented graphically in figure 7, highlights the fact that the roughness of the joint increases with increasing values of the two technological parameters:

- the increase of the feed rate, w, kinematically determines the increase of the surface roughness. This increase is more accentuated with the increase of the tool rotational speed, which causes the increase of the process temperature and, thus, the plasticization of the material;
- the increase of the tool rotational speed, n, leads to the increase of the process temperature and, thus, to the plasticization of the material, which also produces the increase of the surface roughness;
- the increase of the tool rotational speed influences the roughness of the joined surfaces to a greater extent than it influences the increase of the welding feed.

4. Conclusions

The main conclusions regarding the dependencies of the process parameters and of some characteristics of the welded joints by the technological parameters of the FSW process are summarized below.

The temperature in the FSW process depends on the two technological parameters, increasing with the increase of the tool rotational speed and with the reduction of the welding feed, the tool rotational speed having a major influence in this dependence. The axial force in the process does not show an explicit dependence on the two technological parameters, due to the two complex mechanisms that influence it and which is done at the same time: the material plasticization and friction between the tool and the material.

The FSW joints present two categories of surface defects: channels and excessive burrs. The channel type defect is determined by the insufficient penetration of the tool in the joining structure (improper regulation of the technological system) and by the low process temperature, while the excessive burr type defect is caused by too high process temperatures. Therefore, the temperature of the process is particularly important in relation to the occurrence of these defects:

- a high temperature ensures a better mixing of the material (creating the premises to avoid channel defects);
- increasing the temperature to values higher than 650 ^{0}C leads to excessive plasticization of the material and, thus, to the appearance of excessive burrs and to the increase of roughness.

FSW joints roughness depends on the two technological parameters, increasing as their values increase, the tool rotational speed having a greater influence in this dependence as well.

The study will continue through the structural and mechanical characterization of the joints and an analysis of the influence of defects on their mechanical behaviour.

5. References

[1] Thomas WM, Nicholas ED, Needham JC, Murch MG, Temple-Smith P and Dawes CJ 1991 Friction stir butt welding (International Patent Application No. PCT/GB92/02203 and GB Patent Application No. 9125978.8 and US Patent Application No. 5,460,317)

[2] Mishra RS and Mahoney MW 2007 *Friction Stir Welding and Processing* (Ohio: ASM International)

[3] Mitesh P, Utsav D and Mehul J 2015 Effects of FSW on Mechanical properties and Microstructure of copper at Weld Joint *Int. J. of Engineering Development and Research* **3**(4) 967-978

[4] Nandan R, DebRoy T and Bhadeshia H K D H 2008 Recent advances in friction-stir welding - process, weldment structure and properties *Progress in Materials Science* **53**(6) 980-1023

[5] Savolainen K 2012 *Friction Stir Welding of Copper and Microstructure and Properties of the Welds* (Doctoral Thesis, Aalto University)

[6] Mahgoub A, Bazoune A, Merah N, Al-Badour F and Shuaib A 2019 Effect of Welding Parameters on the Mechanical and Metallurgical Properties of Friction Stir Spot Welding of Copper Lap Joint *Arabian J. for Science and Engineering* **44** 1283-1292

[7] Mehta K P and Badheka V J 2017 Hybrid approaches of assisted heating and cooling for friction stir welding of copper to aluminum joints *J. of Materials Processing Technology* **239** 336-345

[8] Milicic M, Gladović P, Bojanić R, Savković T M and Stojic N 2016 Friction stir welding (FSW) process of copper alloys, *Metalurgija* **55**(1) 107-110

[9] Pashazadeh H, Teimournezhad J and Masoumi A 2017 Experimental investigation on material flow and mechanical properties in friction stir welding of copper sheets *Int J Adv Manuf Technol* **88** 1961–1970

[10] Sharma N, Siddiquee A N, Khan Z A and Mohammed M T 2018 Material stirring during FSW of Al–Cu: Effect of pin profile *Materials and Manufacturing Processes* **33**(7) 786-794

[11] Sun Y, He D, Xue F and Lai R 2018 Effect of Tool Rotational Speeds on the Microstructure and Mechanical Properties of a Dissimilar Friction-Stir-Welded CuCrZr/CuNiCrSi Butt Joint, *Metals* **8** 526

[12] Scutelnicu E, Birsan D and Cojocaru R 2011 Research on Friction Stir Welding and Tungsten Inert Gas assisted Friction Stir Welding of Copper *Recent Advances in Manufacturing Engineering* 97-102

[13] Sharma N, Khan Z A and Siddiquee A N 2017 Friction stir welding of aluminum to copper - An overview *Trans. Nonferrous Met. Soc. China* **27** 2113-2136

[14] Constantin M A, Boşneag A, Nițu E and Iordache M 2018 Comparative study on microhardness between friction stir welding and tungsten inert gas assisted friction stir welding of pure copper *MATEC Web of Conferences* **178** 178 03002

[15] Constantin MA, Boşneag A, Nițu E and Iordache M 2017 Experimental investigations of tungsten inert gas assisted friction stir welding of pure copper plates *IOP Conf. Ser.: Mater. Sci. Eng.* 252 012038

[16] Pashazadeh H, Masoumi A and Teimournezhad J 2013 Numerical modelling for the hardness evaluation of friction stir welded copper metals *Materials and Design* **49** 913-921

[17] Pashazadeh H, Teimournezhad J and Masoumi A 2014 Numerical investigation on the mechanical, thermal, metallurgical and material flow characteristics in friction stir welding of copper sheets with experimental verification *Materials and Design* **55** 619-632

[18] Sahlot P, Singh A K, Badheka V J and Arora A 2019 Friction Stir Welding of Copper: Numerical Modeling and Validation *Transactions of the Indian Institute of Metals* **72** 1339-1347

Acknowledgements
This work was supported by a grant of the Romanian Ministry of Research and Innovation, CCCDI-UEFISCDI, project number PN-III-P3-3.1-PM-RO-FR-2019-0048/01.07.2019

Methodology for improving production flows on an assembly line

E L Nitu, A C Gavriluta, N Belu and C A Gavriluta
University of Pitesti, Faculty of Mechanics and Technology, Manufacturing and Industrial Management Department, Targul din Vale Street No.1, Pitesti, Romania

E-mail: eduard.nitu@upit.ro

Abstract. Globalization of production and strong competition in the economic environment requires manufacturers from automotive industry to offer their customers a wide range of products, of the best quality and at lower prices. In order to achieve a competitive advantage, the automotive industry must to adapt their production systems to mass customisation, so that they can provide the variety demanded by the customers while limiting their costs and maintaining their profitability. In this context, research to increase the performance of assembly lines is increasingly numerous and use different techniques, as: layout design, mathematic modeling, dynamic simulation, Lean manufacturing etc. This paper presents a methodology for improving production flows on an assembly line, which was developed to provide the designer and manager of assembly lines in the automotive industry with a set of logically related steps and steps that would allow to achieve a high-performance assembly line. The stages and steps of this methodology consist in the use of methods from different fields, such as: layout design, modelling-simulation of production flows, labour study, lean manufacturing, and for their application different techniques and tools are used.

1. Introduction

Automotive industry faces many challenges in the era of globalization and therefore tries to continuously improve its production system in order to satisfy its clients demands with lower and lower costs.

The assembly is an important part of the production system, and therefore, to increase the performance of the assembly lines were used, along the years, different techniques, as: layout design, mathematic modeling, dynamic simulation, meta-heuristic approach, cost based approach, integrated approaches, Lean manufacturing etc. [1]. Some of the most relevant researches in the field are presented in the followings.

The first research direction concerning the organization of an assembly lines is the layout design because a good layout would contribute to the overall efficiency of operations. This layout must to facilitate the manufacturing process, and additional objectives include [2]: minimizing material handling, maintaining flexibility of arrangement and operation as needs change, promoting high turnover of work-in-process – keeping it moving, holding down investment in equipment, making economical use of floor space, remoting effective utilization of labor, providing for employees' safety, comfort and convenience.

Another important direction of research is to optimize the performance parameters of the assembly line. Hakami and al. [3] showed several mathematical models of different performance parameters of an assembly line. Chramcov and al. [4] proposed a mathematical model to reduce the assembly duration on an automated assembly line. Authors included, also, heuristics algorithms in their simulation model to determine the control of the assembly line. Lee and al. [5] studied the effect of monotony on the performance of operators to improve the productivity of the assembly line. Dao and al. [6] proposed a genetic algorithm to find an optimal solution for the performance of a line. Kia and al. [7] used genetic programing and discrete events simulation to study the influence of production programing rules on the performance of an assembly line.

In order to achieve a competitive advantage, the automotive industry must to adapt their production systems to mass customisation, so that they can provide the variety demanded by the customers while limiting their costs and maintaining their profitability [8]. In this context, research on the implementation of Lean manufacturing concepts in the organization of assembly lines is increasingly numerous. The core idea of Lean manufacturing is to work permanently on eliminating waste from the manufacturing process [9]. Lean facility layout means to arrange the physical equipment within a workshop to help the facility work in a productive way [10]. Using the same equipment and production line for the different models of a product, and standardized work are two of the Lean principles [9]. An empirical review of the Lean Manufacturing strategy is carried out by Chahal et al [11]. The result of the research done in this article show that Lean Manufacturing is the most profitable methodology of continuous improvement in the industrial area and in the research area.

Based on the literature overview and on discussions with experts in the field, this paper presents a methodology for improving production flows on an assembly line, which was developed to provide the designer and manager of assembly lines in the automotive industry with a set of logically related stages and steps that would allow to achieve a high-performance assembly line. The stages and steps of this methodology consist in the use of methods from different fields, such as: industrial systems design, modelling-simulation of production flows, labour study, lean manufacturing, and for their application different techniques and tools are used.

2. Methodology description

The methodology presented in this paper is part of a research project in which the authors are involved and is the result of several studies and research conducted in recent years [12-21].

The development of this methodology was based on a conceptual starting model [12], which was appreciated as having a level of technological training 2 - TRL2 [22]. This proposed model has 3 main interrelated investigations areas: layout design, modelling and simulation flows and Lean manufacturing, and includes both static activities and dynamic activities.

2.1. Methodology stages

The methodology was developed as an iterative process of design, implementation and exploatation, integrating the three main areas of the conceptual model of improving production flows. The application of the modern production requirements, expressed through the concept of Lean manufacturing, is found in the most activities within the methodology, especially in the design and exploatation of the assembly line. The stages and activities specific to the methodology for improving production flows on an assembly line are shown in figure 1 and there are briefly explained in the following subchapters.

The ***initial data*** are related to the product, the production process and the industrial system, as well as to the preset performance indicators.

The specific initial data of the *product* to be assembled are related to:

- the product to be assembled: shape, mass, overall dimensions, quality characteristics, specific requirements (brittleness, positioning during transport etc.), the product components and the joint mode of them to the product;

- the product components: shape, mass, overall dimensions, material and quality characteristics, specific requirements (brittleness, positioning during transport etc.), mode of realization (purchase from the supplier, execution in the company).

The specific initial data of the *production process* are related to:

- the general production flow, highlighting the position of the assembly process under it (connections with suppliers and customers) and takt time;

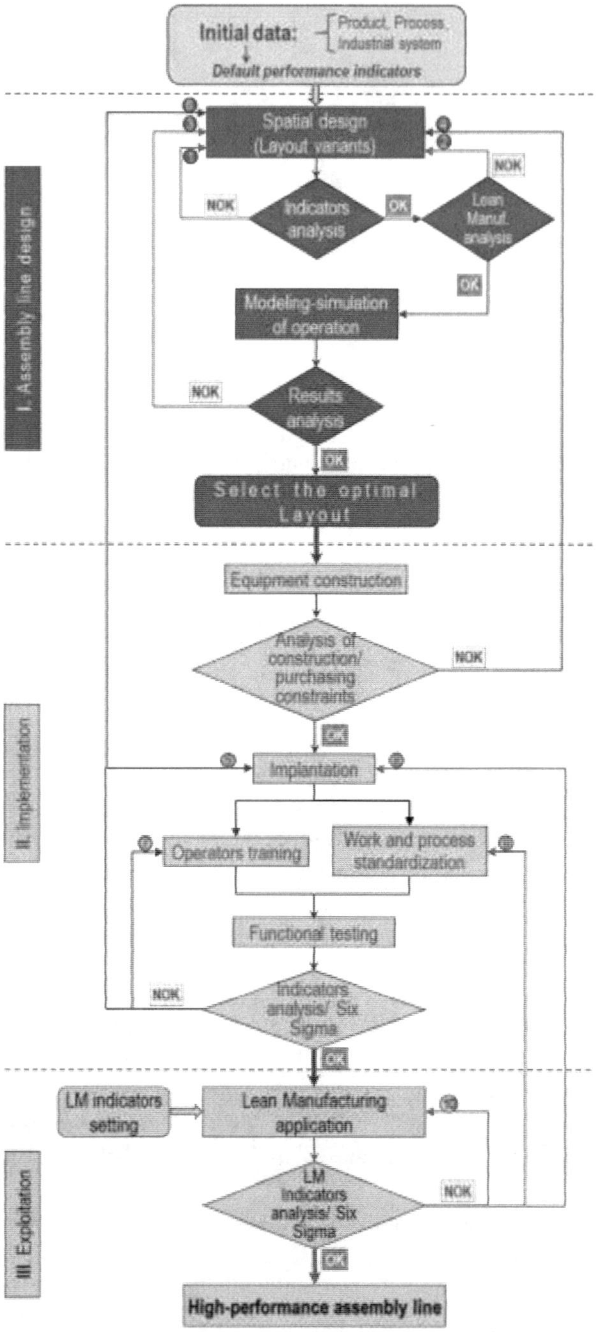

Figure 1. Specific stages and steps of the methodology for improving production flows on an assembly line.

- the assembly process, highlighting: the technological assembly process (operations and specific phases), the available means of production (assembly equipment and tools, control, marking, labeling, inter-operational transfer, supply of workstations etc.);

- the organizational elements of the process: production capacity and accepted limits for variation, possibilities for launching production.

The specific initial data of the *industrial system* are related to:

- the general plan of the industrial system, highlighting the positioning of the area intended for the assembly line under it and its characteristics (shape, dimensions, inputs-outputs etc.);
- the energy sources availabilities (compressed air, electricity etc.);
- the production environment characteristics (temperature, humidity, lighting etc.);
- the average dimensions of operators.

The performance indicators of the assembly line can relate to: the operation (cycle time, takt-line), the production achieved (production capacity, interoperation stocks size), the occupied areas (size, occupancy degree, production capacity/ surface), the NVA activities of the operators (displacements, handlings), the operation flexibility (possibility of variation to production capacity, possibility of attendance for several workstations by the operators), the cost (implantation, operation).

In order to ensure some preset objectives, the management of the company can impose for these indicators (or only for some of them) certain values, minimum or maximum, which the assembly line must to ensure in its operation, so that its performance is defined.

2.2. First stage: assembly line design

The objective of this stage is to develop the spatial organization plan of the production system associated with the assembly line (consisting of the assembly line and the related logistics system) and its operation mode, so that to ensure the achievement of preset performance indicators for the assembly line.

The first step of this stage consists in "***Design of the spatial organization of the assembly line***" [12, 13]. The purpose of that is to develop a several variants of spatial organization plans, which can realize, mainly, as a result of:

- the different design of the work space for the operators (area independent of the part supply method to the workstations);
- the use of different methods to supplying the workstations with parts;
- the use of different means to achieving the inter-operational transfer;
- the spatial organization of the logistics area;
- the spatial arrangement of the assembly line.

In this extensive step, on which can return for five times (through the 5 loops), it integrates specific Lean manufacturing concepts, such as:

- ensuring a high flexibility for workstations, so that a greater diversity of products of the same type can be realized;
- the realization of modular - unitary structures of the workstations, so as to decrease the costs of their realization and the operating times;
- the possibility of adapting the process to the client's request: the "one-piece flow" production in order to increase the reactivity to the client's request, which requires the achievement of "setup zero" times for the workstations;
- to minimise the activities without added value (especially displacements and handlings) so as to be meet the "NVA = 0" objective.

After developing of the variants for the spatial organization plans of the production system, in the second step of this stage - "***Indicators analysis***", the performance indicators for each variant are calculated and compared with the preset ones, being possible two situations:

- the values of some indicators are not appropriate: the design of the spatial organization (loop 1) is being re-examined, so that to ensure the obtaining of the desired values for the preset indicators, for at least one variant.
- the values of all indicators are appropriate for the analyzed organization variant: the following step may be taken.

The third step, "***Lean Manufacturing Analysis***", aims to analyze the application of Lean principles to the design of spatial organization variants with appropriate performance indicators, two situations being possible:

- the Lean principles were not applied for a variant of the analyzed spatial organization of the production system: the design of the spatial organization of that variant is being re-examined (loop 2).
- the Lean principles were applied for the analyzed spatial organization variant: the fours step may be taken.

"***Modeling - simulation of assembly line operation***" [14, 15] is the fourth step and consists in simulating the operation of the assembly line for each variant of spatial organization with appropriate performance indicators and meeting the requirements of Lean manufacturing. The operation of the assembly line is simulated for different production scenarios.

The results obtained by simulation for each scenario related to the performance indicators are analyzed in the fifth step, "***Results analysis***", two situations being possible:

- the results are not appropriate: the design of the spatial organization is being re-examined (loop 3).
- the results are appropriate: the following step may be taken (six).

The sixth step of the first stage, "***Selection of the optimal variant of spatial organization***", consists in selecting that variant of organization that leads to obtaining the "best" performance indicators of the assembly line. Depending on the tactical objectives of the management, one or more selection indicators can be chosen, in addition to those preset by the initial data. The results of this stage are represented by the location plan (implantation) of the assembly line, the operation mode of the assembly line and its set of performance indicators.

The location plan (implantation) of the assembly line is a complex document, which includes:

- the plan for the positioning of the production and logistical means,
- flow chart,
- the pedestrian traffic plan and the utilities connections plan etc.

The operating mode of the assembly line is a document that presents:

- the organizing of the logistic flow, respectively, supplying the line workstations with parts or other materials / equipment specific to the assembly process;
- the parts flow management, respectively: definition of the launch order, the launch point in execution and production declaration, presentation of the manufacturing change principle.
- the organizing of the production maintenance means;
- the mode to recover of the packaging and waste.

2.3. Second stage: implementation

The objective of this stage consists in the physical realization on the allocated location of the assembly line and the related logistics system also the production and work organization (work standardization and training of operators), so that the assembly line to ensure a stable process with the preset performance indicators.

In the first step of this stage, "***Equipment construction***", the components of the assembly line (workstations and logistic equipments) designed in "Spatial design" step are sent for execution. Thus, these equipments can be made within the company and/ or can be purchased from suppliers. The fabricants of these equipments analyze the possibilities of respecting the execution projects and, in case the execution requirements can be respected, they continue with the next step. If the execution requirements cannot be met, it returns to the step " Spatial design " in the first stage (loop 4), in order to redesign the equipment with problems and rebuild the layout.

The second step of this stage, "***Implementation***", consists in the physical location on the allocated site for all components of the assembly line (workstations and logistics area), according to the optimal spatial organization plan (established in stage I). It is recommended to carry out an installation project, in which the following main phases of the installation are highlighted: preparation (has the role to ensure the basic conditions for installing machinery and equipment on the allocated sites and refers to all participants from this process), moving (moving machinery, equipment and industrial furniture to the installation points), the actual installation (positioning and effective installation of machinery, equipment and industrial furniture on the allocated sites), start-testing (checking the correct operation of the equipment), general cleaning (general checks of the implantation and cleaning of the area).

The third step of the methodology includes two directions of action carried out in parallel, respectvely "***Work and process standardization***" and "***Operators training***". The work standardization aims to establish the working standards (working instructions) for each workstation of the line and the area of logistical preparation, in which are detalied: gestural rules for operations performing, self-maintenance activities, workplace health and safety rules, operation duration etc. The process standardization aims to establish the rules regarding to the development of all the processes necessary for the operation of the assembly line and their concatenation, respectively: the operations durations and supply processes with components (same as those of work standards), interoperation stock sizes, management of parts flow, maintenance of production means, recovery of packaging and wastes. To meet these standards, a series of techniques and tools of visual management are used, of different levels, such as: level 1 - communication of information to a large group of employees (production area information panels, announcements, process maps, skills / flexibility matrices etc.); level 2 - standards documentation (instructions for standard work on the workstation, checklist before work, overflow markings of containers etc.); level 3 - warning of the failures occurrence / abnormal conditions (panels and patterns with failures, cables and Andon plates, etc.); level 4 - preventing the failures occurrence - (motion sensors, Poka Yoke systems etc.).

The purpose of the operators training is to develop their skills, so that they reach the desired rate as quickly as possible. In this regard, specific techniques are used, such as: DOJO workshops, workstations rotation (versatility), communication skills training workshops (changing roles between the workshop leaders and operators), one off lesson etc.

In the fourth step, "***Functional testing***", the verification of the mode in which the assembly line works is carried out, through the implementation by the trained operators of the previously developed work and process standards.

The results obtained after testing the operation of the assembly line are analyzed in the fifth step, "***Indicators analysis/ Six Sigma***". The method used to analyze assembly line performance indicators is Six Sigma which uses quantitative data based on which statistical analyzes can be performed (Pareto, Analysis Hypothesis / Inference Tests, Correlation / Regression Analysis). These data are specific to the manufactured product (dimensions, deviations, tolerances, physical and / or mechanical properties, etc. obtained in different operations on the assembly line) and the assembly process (activities durations, stocks sizes, productivity, delivery times, etc.).

If the performance indicators are not appropriate, it is possible to intervene on:

- the implantation of the assembly line, through the appropriate modification of its spatial organization (loop 5). For example: the work standardization shows that the workstation from the assembly line is not well organized, in which case a new spatial organization of the workstation is proposed.
- the spatial design (loop 6). This situation can occur when there are large deviations of some indicators from those preset, which may have as a cause the non-compliance (during the implementation stage) of the spatial organization plan developed (for example, by using other machinery or equipment than those initially planned) or as a result of the constructive or technological modification of the product. This loop is one with a major impact on the whole process and longer duration.
- the training of operators and / or work and process standardization (loop 7). For example: it is found that there are major differences between the durations of identical operations performed by the same operator, in which case it is proposed to resuming the work standardization process.

If the values of the indicators resulting from this stage are in accordance with those preset through the initial data, it is considered that the assembly line is working apropriate and proceed to the exploitation stage of the assembly line.

IOP Conf. Series: Materials Science and Engineering **968** (2020) 012014 doi:10.1088/1757-899X/968/1/012014

2.4. Third stage: exploitation

The objective of this stage is to operate / use the assembly line and apply Lean Manufacturing techniques and tools, so as to ensure its operation at a high level of performance.

The implementation of the Lean Manufacturing concept at the company level, so as to ensure its successful and sustainable development, is significantly influenced by the good start of this process. In this regard, the company must ensure the realization of all the basic / fundamental elements of Lean Manufacturing, such as:

- the highest-level management demonstrates leadership and commitment to the LM project by developing a Lean policy to support the strategic direction of the organization and by setting Lean objectives. LM's objectives must be consistent with the company's overall policy.
- the company's management is responsible for setting up the "Lean team", which does not have to include many people, but must be a multifunctional team.
- the Lean team is trained. This training should be conducted by a consultant / expert Lean, providing a first impetus for the introduction of Lean culture within the team.
- the company's management must establish and maintain communication and feedback channels with the Lean team.

Within the proposed methodology, it is considered that all these basic elements have been achieved within the company. Thus, the first task of the Lean team is to delimit the perimeter of action, which can be done using the product / process matrix or a Pareto analysis. Proper selection of the initial scope of action is important for concentrating resources and maximizing gains. After selecting the initial perimeter, the team must develop a Lean implementation plan (master plan), which must include a timely planning of actions to be taken and a budget. Then, it can proceed to the first step in implementing the Lean Manufacturing methodology.

In the <u>first step</u> of this stage, "***Lean Manufacturing indicators setting***", the Lean performance indicators specific to the assembly line are defined, derived from the Lean objectives established at the company level. The initial values of these indicators are also determined to define the current state of assembly line performance. Lean performance indicators of the assembly line (whose value is to be achieved / increased by implementing the Lean Manufacturing methodology) can be related to:

- Productivity - the number of products assembled in an hour;
- Quality - the number of products without defects;
- Cost - unit cost with assembling the product on the assembly line;
- Delivery time - the period of time elapsed from the moment the customer requests the product and the moment when it is available for it.
- Safety and environment - possible risk situations regarding the working environment, together with the measures to prevent them, the actions of their correction and constant monitoring (in order to overcome the possible danger situations and to guarantee the conditions of safety).
- Morality - teamwork or optimal use of available human resources knowledge. The aim is to increase the skills of team members (to work in a team, lead a team, think clearly and logically, solve problems) and strengthen the self-confidence of team members.

"***Lean Manufacturing application***" is the <u>second step</u> and consists in analyzing and improving the performance of the assembly line by implementing methods specific to the concept of Lean Manufacturing. In this large-scale step, a set of specific methods and techniques are used, such as: Kaizen, 5S, Kanban/ Conwip, Jidoka/ Poka-Yoke, Hoshin etc. For their efficient use, the authors aimed to develop and integrate a specific methodology for their application.

The results obtained from the implementation of the Lean manufacturing concept are analyzed in <u>step three</u>, "***LM indicators analysis/ Six sigma***". As the name of this step suggests, the method used to analyze assembly line performance is SixSigma. If the results obtained are not adequate, it is possible to intervene on:

- implantation of the assembly line, by corresponding modification of its spatial organization (loop 8). For example: if the application of the Kaizen method shows that a workstation within the assembly line is not properly organized, a new spatial organization of the workstation is proposed.

- training of operators and/ or standardization of work and process (loop 9). For example: if a 5S audit reveals that labor standards are not being met, it is proposed to resume the operator training process.
- Lean Manufacturing continuous improvement process (loop 10): if the required LM indicators are not achieved, the specific Lean manufacturing methods will be applied again, so as to identify the solutions for achieving the desired indicators.

When the results obtained from the analysis of performance indicators are in line with the objectives set out in the first step of this stage, the assembly line is considered to operate at the desired level of performance.

3. Methodology application

The developed methodology was experimented and validated in the laboratory using an experimental demonstrator (TRL4 level) developed in a Lean Learning Factory laboratory, figure 2.

Figure 2. Overview of the Lean Learning Factory laboratory.

This laboratory [16, 17] includes integrated platform for research and learning, such as:
- for the *layout design direction*: modular and flexible equipments for configuring and supply workstations; equipments for interoperable transfer; digitized system for a workstation (PLC system, HMI, identification camera, scanner, pick to light, wireless communications); support elements for the analysis of the location of the means of production (support write board and video projector);
- for the *modeling - simulation direction of production flows*: guide for learning the use of modeling and simulation in the analysis of production systems;
- for the *work study direction*: platforms for learning and applying specific methods and techniques (DOJO, Standardization etc.);
- for the *Lean manufacturing direction*: platforms for learning and applying specific methods and techniques (VSM, Lean corner, 5S, Poka-Yoke, Kaizen, Visual management).

Along with the physical supports (research equipment and platforms), the laboratory also includes computer supports, being equipped with a server and computers interconnected in the network on which software specific to the research directions are installed: layout design (IMPACT, Tecnomatix - FactoryFlow), production planning (ABAS), modeling and simulation with discrete events of production flows (Rockwell Arena and Tecnomatix - PlantSimulation).

Thus, along with experimenting with the developed methodology, this laboratory also allowed research on layout design for an assembly line [12, 13], workspace and ergonomics analysis [18, 19], modeling and simulation of the production flow [14, 15], Six Sigma application [20] and a RFID based Kanban system implementation [21].

4. Conclusions

The methodology presented in this paper had as a starting point a level of technological training 2 - TRL2 with 3 main interrelated investigations areas: layout design, modelling and simulation flows and Lean manufacturing, and includes both static activities and dynamic activities. During the research project, this model was developed until TRL 4 and the developed methodology was experimented and validated in the laboratory, within an experimental demonstrator.

The methodology for improving production flows on an assembly line developed in this context consists of an iterative process in three stages (design, implementation and operation of the assembly line), with several steps each and return loops.

The data necessary for the application of this methodology are numerous and complex, coming from many fields of industrial activity: product design, technology design, industrial systems design, ergonomics, production management, and the variables involved in the process are diverse (products - through their diversity), process (through assembly technology), supply of components (methods of supply of workstations), organization of workstations (working methods used), location of workstations and means of transfer (various models and variants), production management and control (by the methods used).

The methods used in the activities are, in turn, very diverse: standards and norms specific to industrial fields (technical drawing, ergonomics, logistics), rules and principles specific to the design of assembly processes, spatial organization, ergonomics and work organization, and production management. The concepts of modern production, expressed through the concept of Lean manufacturing, are found in most of the activities within the methodology. The techniques and tools used are largely in the field of computer aided engineering (computer aided design, technological design, balancing production lines, optimization, modelling-simulation), but also in the field of flow analysis (flow charts) or comparative analysis (decision-making techniques, Six Sigma).

The results obtained by applying the methodology are quantified by the performance indicators of assembly line operation and specific Lean manufacturing and are influenced by the accuracy of input data, as well as the use of different methods, techniques and tools in the application of the methodology. Therefore, in order to facilitate the proper understanding and application of the methodology, the authors aim to develop a "Guide to using the methodology to improve production flows on an assembly line", which should be available to the main beneficiaries: students and companies.

Also, another direction of further research aims at integrating in this laboratory the specific methods and techniques of Industry 4.0 and the development of a Lean Learning Factory with Industry 4.0 at the University of Pitesti.

5. References

[1] Rane A B and Sunnapwar V K 2017 Assembly line performance and modeling *Journal of Industrial Engineering International* **13** 347–355

[2] Muther R and Hales L 2015 *Systematic Layout Planning - Fourth Edition* (Marietta, GA, USA: Management & Industrial Research Publications)

[3] Hakami N, Movahedi M M and Heidarinezhad M 2014 Comparative study of assembly line balancing methods in an automotive industry *Int J Sci Res* **05**(02) 1–21

[4] Chramcov B, Marecki F and Bucki R 2015 Heuristic control of the assembly line *Intelligent systems in cybernetics and automation theory (Advances in intelligent systems and computing)* **348**(2) pp 189–198

[5] Lee J, Kim J, Seo K, Roh S, Jung C, Lee H, Shin J, Choi G and Ryu H 2016 A case study in an

automotive assembly line: exploring the design framework for manufacturing gamification *Advances in ergonomics of manufacturing: managing the enterprise of the future* (New York: Springer International Publishing) pp 305–317

[6] Dao S D, Abhary K and Marian R 2017 Optimisation of assembly scheduling in VCIM systems using genetic algorithm *J Industrial Eng Int* **13** 275–296

[7] Kia H, Ghodsypour S H and Davoudpour H 2017 New scheduling rules for a dynamic flexible flow line problem with sequence-dependent setup times *J Industrial Eng Int* **13** 297–306

[8] Limere V 2012 Optimising part feeding in the automotive assembly industry: deciding between kitting and line stocking *Int J of Prod Res* **50**(15) 4046-4060

[9] Ohno T 1988 *Toyota Production System, Beyond Large-Scale Production* (Portland, Oregon, USA: Productivity Press)

[10] Wilson L 2015 *How to Implement Lean Manufacturing - Second Edition* (McGraw-Hill Education, USA)

[11] Chahala V and Narwalb M S 2017 An empirical review of lean manufacturing and their strategies, *Management Science Letters* **7** 321–336

[12] Gavriluta A, Nitu E, Gavriluta A, Belu N and Rizea A 2018 Methodology for designing the layout for an assembly line to the automotive industry using the Lean concept, *Proceedings of the 6th RMEE Management Conference*, ed. Abrudan I, Cluj-Napoca Romania, Todesco Publishing House pp 495-502

[13] Gavriluta A, Nitu E, Gavriluta A, Belu N and Rizea A 2018 Designing a layout for an assembly line used in the automotive industry, *Proceedings of the 6th RMEE Management Conference*, ed. Abrudan I, Cluj-Napoca Romania, Todesco Publishing House pp 646-654

[14] Gavriluta A, Gavriluta CA, Pascu I and Neacsu C 2019 The development of a methodology of learning to use simulation in the analysis of production system performances production system *IOP Conf. Ser.: Mater. Sci. Eng.* 564 012100

[15] Gavriluta A, Gavriluta C A, Neacsu C and Pascu I 2019 Experimentation of a methodology to use modelling and simulation in the analysis of performances of an automotive industry production system *IOP Conf. Ser.: Mater. Sci. Eng.* 564 012101

[16] Gavriluţă A, Niţu E L, Gavriluţă A, et al. 2018 The development of a laboratory system to experiment methods to improve the production flows, *Proceedings in Manufacturing Systems* 13 (3) pp 127-132

[17] Nitu E L and Gavriluta A C 2019 Lean Learning Factory at the University of Pitești *IOP Conf. Ser.: Mater. Sci. Eng.* 591 012095

[18] Anghel D C, Nitu E L, Rizea A D, Gavriluta A, Gavriluta A and Belu N 2019 Ergonomics study on an assembly line used in the automotive industry *MATEC Web of Conferences 290 12001*

[19] Belu N, Nitu E L, Gavriluta A C and Ionescu L M 2019 An approach with genetic algorithms to improve the workstation space planning *IOP Conf. Ser.: Mater. Sci. Eng.* 591 012002

[20] Belu N, Rizea A D, Niţu E L, Gavriluţă A C, Gavriluţă A C 2018 An application of Six Sigma to PPM reduction in the relationship with the external customer *IOP Conf. Ser.: Mater. Sci. Eng.* 400 062006

[21] Belu N, Niţu E L, Rizea A D, Gavriluţă A C, Ionescu L M, Gavriluţă A C 2018 Implementation of Single-Loop Kanban System Using Indoor Location Based on RFID, *Proceedings of the IVth International Congress of Automotive and Transport Engineering* ed. Filip N and Ghereş M (Cluj-Napoca: UT Press) pp 311-318

[22] DOE G 413.3-4 *Technology Readiness Assessment Guide 2009*, Department of Energy, U.S.A, available at: https://www.directives.doe.gov/directives-documents/400-series/0413.3-EGuide-04, accessed: 2018-08-01

Acknowledgements
This work was supported by a grant of the Romanian Ministry of Research and Innovation, CCCDI – UEFISCDI, project number PN-III-P1-1.2-PCCDI-2017-0446 / 82-PCCDI-2018, within PNCDI III.

IOP Conf. Series: Materials Science and Engineering **968** (2020) 012015 doi:10.1088/1757-899X/968/1/012015

Structural identification of the bearing manufacturing process – Case-study

C Afteni[1], G R Frumusanu[1], M Afteni[2] and V Paunoiu[1]
[1] "Dunarea de Jos" University, Manufacturing Engineering Department, Domneasca Street 111, 800201 – Galati, Romania
[2] Rulmenti S.A, Barlad, Republicii Street 320, 731108 – Barlad, Romania

E-mail: Cezarina.Afteni@ugal.ro

Abstract. In the manufacturing industry, globalization, shorter life cycles of products and changing customer needs, leads to high competitive pressure on companies. In addition to product quality and variety, flexibility, shorter processing times, and high-level compliance with delivery times have become essential factors for market success through efficient, efficacious and continually manufacturing processes optimization. To be successful in a highly competitive global production environment, a company must be able to deliver products that customers request at the requested time. In this paper is developed a novel method of structural identification of the bearing manufacturing process. This method allows the structuring of its activities, at all levels involved (order acceptance, production planning, product design, processes planning, and product processing), by elaborating the tree of the specific activities. The relations between the manufacturing process stages and related information circuit are revealed and the identification of the manufacturing process variants, at the level of each manufacturing activity is performed. Following the selection of the best alternatives from each level of the manufacturing activity, the optimal technological path is obtained for taking over an order for the bearing manufacture.

1. Introduction

In today's unpredictable production environment, companies are faced with a multitude of challenges to meet ever-increasing customer's desires and demands, to offer new and more customized products at the best price. Despite the benefits for consumers, global competition makes it difficult for manufacturers to select and configure their manufacturing processes.

In current manufacturing environment it is necessary to make diverse decisions, which mainly concern to all stages of the manufacturing activity: order acceptance, production planning, product design, processes planning, and product processing. Such a decision consists in selecting, at a given moment, the most suitable alternative among the potential ones. The selection can be made after a given criterion (e.g. cost, timespan, consumed energy etc.), or a combination of criteria [1]. The authors of this paper have already developed an original selection algorithm, based on comparative assessment [2]. Unlike the usual methods underpinning the comparative assessment, supposing the direct, separate evaluation of feature value for each alternative, before decision making, the developed algorithm of selection aims to establish a ranking of the potential alternatives set. The authors also suggest in [3], a new approach of optimizing the manufacturing process, based on both new concept (the holistic optimization), and method ("zoom & pick") for its implementation. According to the new concept, the optimization problem gets a new structure, which includes not only the optimal solution finding, but also the optimal formalization of the problem as well as the tooling for assessing the position of potential solutions relative to the optimal one. In paper [4], a methodology for preliminary process selection was created through its integration in a design process. Denno et. al presents in [5] a methodology, called production system identification, to produce a model of a manufacturing system from logs of the system's operation. Production system identification is similar to machine-learning methods of process mining in that they both use logs of operations. The paper [6] describes an object-oriented manufacturing process information model which comprises classes on the necessary manufacturing information, such as manufacturing activities, work piece, manufacturing equipment, estimated cost and time, and manufacturing process sequences. The model includes the representation of concurrent activities, alternative activities, and parallel activities. This model is capable of describing the hierarchical structure of the information representing manufacturing processes of a product. Gecevska et al describe in [7] a model for manufacturing activities that allows integration, having considers three dimensions of planning. The product development cycle may be seen as a set of answers to a series of actually simple questions: Why to produce? What to produce? How to produce? Where to produce? Who to produce? When to produce? The answers to these questions will identify what functions are necessary for the cycle from developing an idea to the realization of the final product. In the paper [8], experimental investigation, modeling and optimization of the drilling are performed using Taguchi Design of Experiments (DOE) and artificial neural networks (ANN) methods. In [9] is proposed a method of production scheduling oriented to energy consumption optimization for process industry, a production scheduling model is established whose goal is to obtain the minimum of total process energy consumption. The simulation results show that the production scheduling method orientated to energy consumption optimization is superior to the production scheduling method oriented to process time optimization, and it can realize the goal of reducing energy consumption. The paper [10] presents a manufacturing process management (MPM) solution that enables a real-time assessment of component manufacturability and a parallelization of product design and manufacturing processes. This solution provides an interactive and effective bridge between the Product Data Management systems that support design activities, and the Enterprise Resource Planning and Manufacturing Execution systems that support production planning and execution activities of complex products.

Within this paper, a method for structural identification of the manufacturing process is developed. It allows the structuring of the activities, at all levels involved (order acceptance, production planning, product design, processes planning, and product processing), by elaborating the tree of the specific activities (the relations between the manufacturing process stages and the related information circuit are revealed). The identification of the manufacturing process alternatives, at the level of each manufacturing activity is also performed.

The method aims to enable the selection of the best alternatives at each level of the manufacturing activity, according to different optimization criteria (such as, for example, cost, timespan, energy consumption, and other critical consumption or combination combinations thereof); the selection itself does not make the object of presented research.

Following the section of the best alternatives from each level of the manufacturing activity, the optimal technological path is obtained for the manufacturing of a considered product.

This method has been sampled by a case study in the case of bearing manufacturing. The aim was to draw the manufacturing graph, including the structuring of its activities, at all involved levels, according with the new proposed method. There were also identified the manufacturing process alternatives, at the level of each manufacturing activity stage.

The paper is organized as follows: the second section presents the manufacturing process levels. The next section describes the activities identification from each level of the manufacturing process. The fourth section shows the alternatives for activities accomplishment at a manufacturing process level. Section five deals with the application of this method through a case study on bearing manufacturing. The last section presents the paper's conclusion.

IOP Conf. Series: Materials Science and Engineering **968** (2020) 012015 doi:10.1088/1757-899X/968/1/012015

2. The manufacturing process levels

The manufacturing process will be further considered as composed of activities structured on five successive levels (order acceptance, production planning, product design, processes planning, and product processing, see table 1), [11].

Table 1. The manufacturing process levels [11].

Level	A	B	C	D	E
Manufacturing activity	Order acceptance	Production planning	Product design	Processes planning	Product processing
Object of the activity	Order	Product	Component	Operation	Phase

The tree of specific activities that reflect the relationships between the activities at different levels of the manufacturing process, which will be subjected to optimization, as well as the related information circuit are sampled in figure 1.

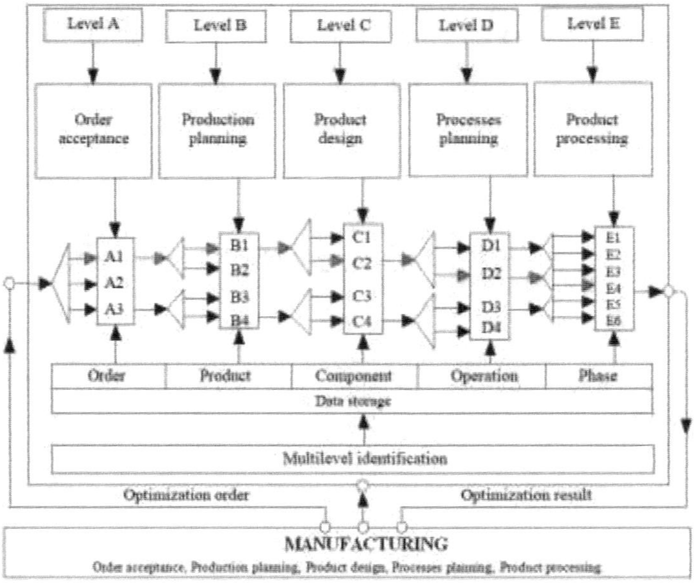

Figure 1. Levels for optimization of manufacturing process – block diagram – [11].

In figure 1 at level A there are three potential activities A1, A2 and A3, according to the method, are selected two activities which lead congruent to previous experience to the best results (lower cost, highest profit, etc.), namely A1 and A3. For each of the selected activities, at level B there are four potential activities B1, B2 for A1 and B3, B4 in case of activity A3, of which only two are selected, namely B1 and B4. Next, at level C, there are four potential activities: C1, C2 for B1, C3, C4 for B4. From these four activities, after a new evaluation, only two are selected, namely C2 and C4. Furthermore, for the selected activities, at level D there are four potential activities D1, D2 for C2 and D3, D4 for C4. From the four activities, three are selected which lead, according to previous experience, to the best results, namely D1, D2 and D3. Finally, for each of the selected activities, at level E, two activities E1, E2 for D1 and E3, E4 for D2, respectively E5, E6 for D3. From the six activities, it is selected, only one that leads, according to previous experience to the best results, namely E4.

Thus, following the successive evaluations and selections of the activities from the different levels of the manufacturing process, an optimal technological path is obtained for the considered manufacturing process, which includes activities A1, B1, C2, D2, and E4 (marked in red in figure 1).

3. Activities identification at each process level

The generic manufacturing process can be described by means of the activity objects (these can be, at the different levels of the process: order, product, component, operation, and phase).

The result of the manufacturing process identification consists in the proposed method case from the manufacturing process activity chain. Such a chain is exemplified in detailed representation in figure 2-a or synthetic in figure 2-b.

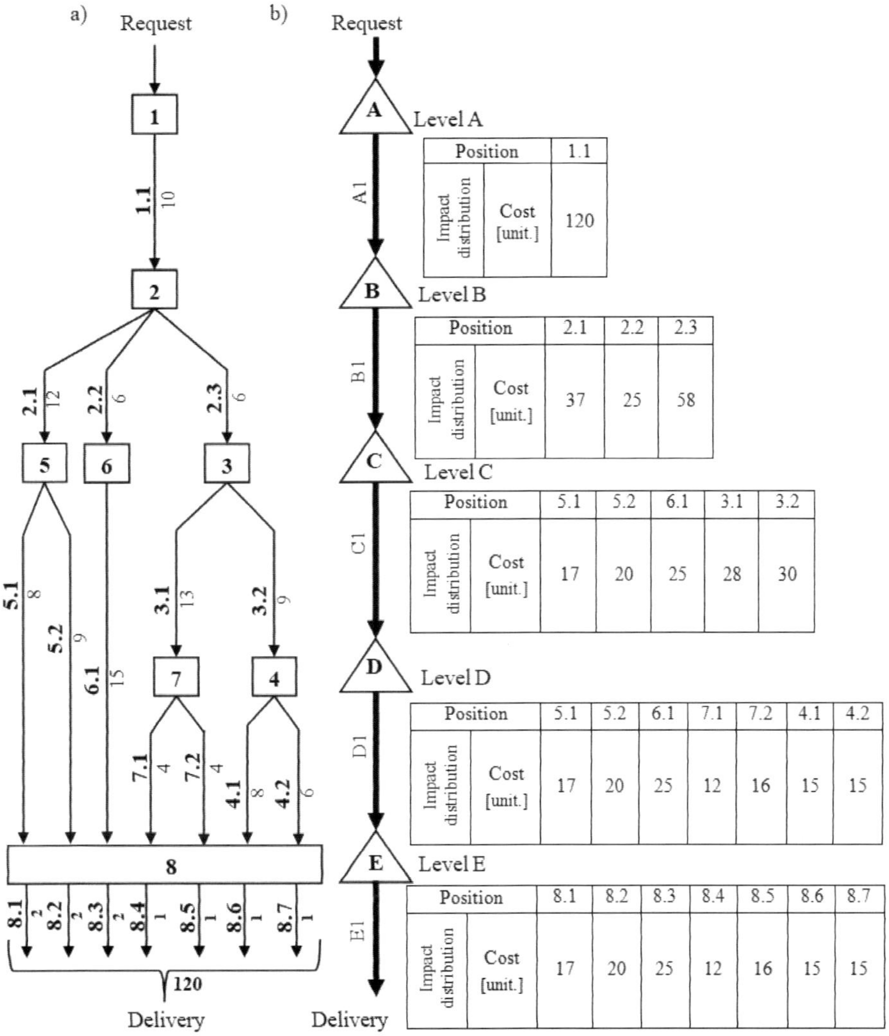

Figure 2. The activity chain for a manufacturing process variant [11].

In the example shown in figure 2 it was assumed that the manufacturing cost is 120 monetary units.

- At level A, activity 1 is carried out, having the cost of 10 monetary units, which will be divided into cost related to activities from level B.

- At level B, activity 2 is carried out, which is divided into three components 2.1, 2.2, and 2.3, having the following costs: 12 monetary units for 2.1, 6 monetary units for 2.2, 6 monetary units for 2.3. As mentioned above the cost of activity 1 is divided at this level so: 4 monetary units to 2.1, 2 monetary units to 2.2, 4 monetary units to 2.3. Both the cost of each activity and the cost assigned to itself are divided in turn to the next activities;

- At level C, there are three activities: 5, 6, and 3.

Activity 2 with component 2.1 is continued by activity 5, which is divided into two components, 5.1, having the cost of 8 monetary units, and 5.2 having the cost of 9 monetary units. The cost of activity 2.1, of 12 monetary units, as well as the cost assigned to itself of 4 monetary units is divided in two: 5 monetary units respectively 2 monetary units belong to 5.1, and 7 monetary units respectively 2 monetary units to 5.2. At the end of this level for each of the two components are added 2 monetary units. The cost of component 5.1 results from 17 monetary units, and that of component 5.2 from 20 monetary units.

Activity 2 with component 2.2 is continued by activity 6, which continues with component 6.1, having the cost of 15 monetary units, to which is added the cost of units from activity 2.2 and 2 monetary units from the end of the level. The cost of activity 6.1 results from 25 monetary units. Activities 5 and 6 ends at level E.

Activity 2 with component 2.3 is continued at level C with activity 3, which is divided into two components, 3.1, having the cost of 13 monetary units respectively 3.2, having the cost of 9 monetary units. Activity 3 ends at level D.

- At level D activity 3 is continued by two activities 7 and 4, which are each divided into two components, namely, 7.1, having the cost of 4 monetary units and 7.2, 4 monetary units, 4.1 – 8 monetary units and 4.2 – 6 monetary units. At the end of this level for each of the two activities is added 1 monetary unit.

- At the last level – level E activity 8 takes place which completes the above activities, it is divided into eight components namely 8.1 and 8.2 completes activity 5, 8.3 completes activity 6, 8.4 and 8.5 completes activity 7, 8.6 and 8.7 completes activity 4.

In the tables in figure 2 is given a synthetic presentation of the impact on the final cost of each direction of action within the graph.

The chain of activities identified to satisfy a given order showed in figure 2-a it includes both manufacturing activities applicable to the whole product (marked with numbers simple 1…8), as well as activities related to components of the product (whose symbols include the separator ".", for example, 2.1, 2.2, …, where the numbers before and after the dot refer to activities and to component, respectively). These correspond to the 5 levels marked A… E identified in the figure 2-b. An arrow symbolizes each activity, the notations under each arrow represent an attribute (for example, cost, timespan, etc.) of the activities (components of activities) performed for the transition from one activity level to the next. A determined set of activities (hence a succession of arrows), starting from process beginning point, corresponds to each component, no matter of component level. This set means the activities chain of the considered component.

The accomplishment of each activity is characterized by a resulting effect (be it cost, timespan and consumed energy etc.). The cumulated effect of all activities corresponding to the ways composing a path (when all the effects of the activities are of the same types) means the process effect corresponding to this path. More types of effects can be measured for the same path, by values of the corresponding effect-variables.

The problem addressed in this paper can be formulated now as the selection, from the manufacturing graph, of the optimal manufacturing path, according to a flexible criterion/set of criteria.

As it can be easily observed, the manufacturing graph may be very complex. Frequently, the structure of activities needing to be performed in order to cover the manufacturing path is also complicated. Moreover, the evaluation of path effect by decomposing the typical activities in basic nominal activities, finding the effect for each of them, and, finally, cumulating these effects is also a difficult task. For these reasons, the application of the combinatorial optimization method is not feasible for solving the problem, neither the application of other methods selecting the optimal path on the base of path effects direct evaluation.

4. Alternatives for activities accomplishment

At a manufacturing process level, in general, there may be several variants of the process. For example, suppose that at level A there are variants A1 and A2, this situation being symbolically presented in figure 3.

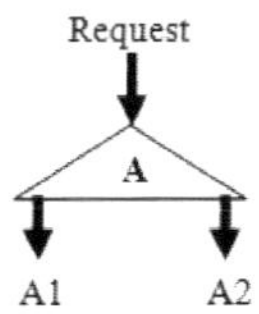

Figure 3. Variants of the manufacturing process at level A [11].

Figure 4 shows a manufacturing process for which these are several variants at each level. It should be noticed that for actually accomplishing the process, it is enough to follow a single path only. The ensembles of all potential paths that can be used for obtaining the product form the manufacturing process graph (figure 4).

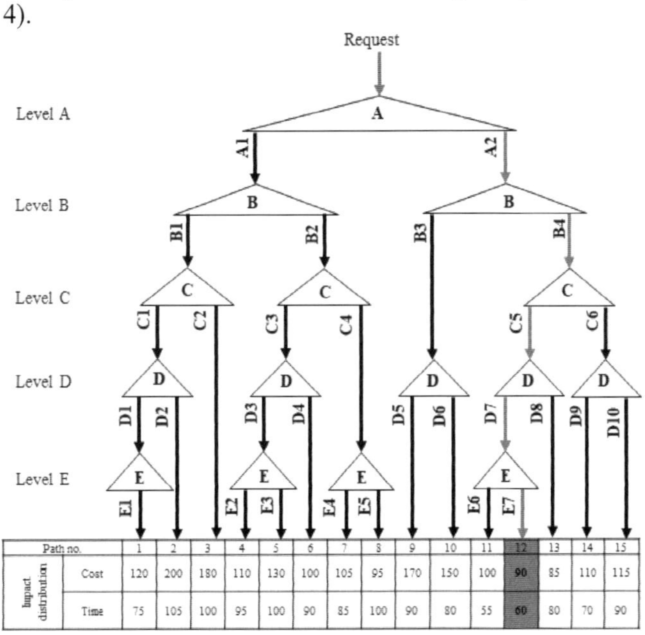

Impact distribution	Path no.	1	2	3	4	5	6	7	8	9	10	11	12	13	14	15
	Cost	120	200	180	110	130	100	105	95	170	150	100	90	85	110	115
	Time	75	105	100	95	100	90	85	100	90	80	55	60	80	70	90

Figure 4. Manufacturing process graph [11].

The activities required to be carried out between two consecutive identification levels will be called paths. The paths are represented by thick arrows and marked with letters according to their level starting point (for example, the paths from level B to level C are B1, B2, B3 and B4). Because the object manufactured can be found in different forms, which can be found in different ways, each of these forms must be represented separately. The paths that come from different forms of the object belonging to the same level are also different. Despite that in practice there can be several ways to bring the object from one level to another, for simplicity was considered a maximum of two paths to do this. You can highlight a path between the state of the object, from a certain level, and the endpoint of the process. The path to the realization of the object is the path between the beginning of the process and any among its endpoints. To define a route, a certain level must be selected a certain path, passing through a decision point. The optimal manufacturing path can be chosen according to the values obtained; thus, three possible variants are exemplified: *i)* variant 1: after minimum value for cost, *ii)* variant 2: after minimum value for time, and *iii)* variant 3: after a combination of the two criteria, the variant market with red in figure 4.

5. Case study

The case study was developed to exemplify the application of the structural identification method. To this end, the bearing manufacturing process was approached. The targeted optimization accuracy is set at 10%.

For the application of the structural identification method, the activities necessary for the manufacture of bearings and their components were identified, starting from the receipt of orders to the delivery of products to customers.

The bearing as a finished product consists of the following main elements: outer ring, inner ring, rolling elements (balls or rollers), and cage.

According to the proposed optimization method, the structural identification of the process of manufacturing was carried out along the optimal path. Thus identifying variants of manufacturing at each level was done only for the current point at the stage level (figure 5).

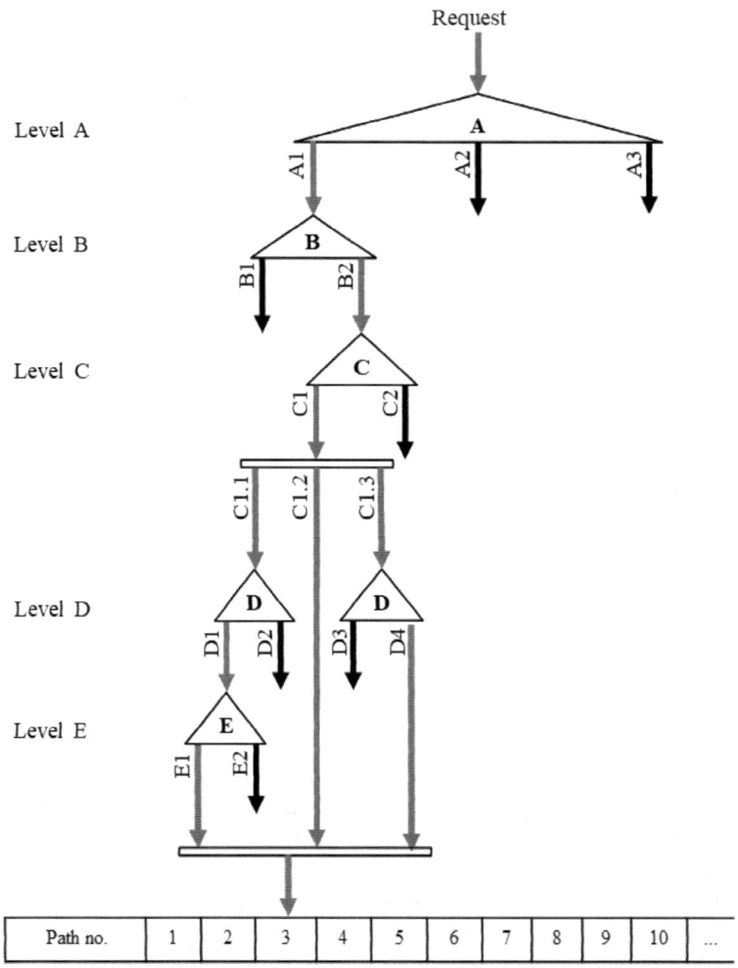

Figure 5. The optimal manufacturing path for the case study considered.

In table 2 are sampled the hypothetical results of the stage ″analysis of past activity″, while the ones of the stage ″assessment tooling″ in table 3. Table 4 summarized the results of the stage ″evaluation of the current processes″.

Table 2. Analysis of previous activity.

Identification level	Optimization object	Typical jobs
A	Order	A1 – Thrust bearing; A2 – Spherical bearing; A3 – Radial bearing
B	Assembly	B1 – Rolling element (ball); B2 – Rolling element (roller)
C	Subassembly	C1 – Thrust bearing with steel cage; C2 – Thrust bearing with brass cage
	Part	C1.1 – Outer ring; C1.2 – Inner ring; C1.3 – Roller
D	Plan	D1 – Plan I outer ring; D2 – Plan II inner ring; D3 – Plan I rolls; D4 – Plan II rolls
E	Operation	E1 – Turning on CNC lathes; E2 – Turning on classical lathes

Table 3. Assessment tooling.

Identification level	Optimization object	Typical features
P	Goal	P1- Efficiency; P2 - Productivity;
Q	Criterion	Q1 - Cost/product; Q2 - Cost/quantity; Q3 – Timespan; Q4 - Time/ quantity;
R	Function	R1 - Cost [Euro/dm^3]; R2 - Time [min/dm^3];
S	Arguments	(S1.1, S1.2, S1.3); (S2.1, S2.2); ... according to job causal model

Table 4. Evaluation of the current process.

Problem formalizing by			Assessment tooling by				Solution finding by		
			"identification"						
"identification"	*"selection"*		P	Q	R	S	*"identification & selection"*		
			"selection"						
Levels	Variables	Values	Goals	Criteria	Functions	Arguments	Results	Delta	Go to
A	x_A	A1, A3	P1	Q1	R1	S1.2	A1	35%	B-level
B	x_B	B1, B2	P2	Q3	R2	S2.1	B2	30%	C-level
C	x_C	C1, C2	P1	Q1	R1	S1.2	C1	25%	C-level
	x_{C1}	C1.1	P2	Q1	R1	S3.1	C1.1	17%	D-level
	x_{C2}	C1.2	P2	Q4	R2	S2.1	C1.2	8%	Stop
	x_{C3}	C1.3	P2	Q4	R2	S2.2	C1.3	15%	D-level
D	x_{D1}	D1, D2	P1	Q1	R1	S1.1	D1	11%	E-level
	x_{D4}	D3, D4	P1	Q1	R1	S1.2	D4	8%	Stop
E	x_{E1}	E1, E2	P1	Q2	R1	S1.2	E1	9%	Stop

Thus, at level A, is order acceptance, are three potential activities, namely: A1 – *Manufacture of thrust bearings*, A2 – *Manufacture of spherical bearings*, and A3 – *Manufacture of radial bearings*.

According to evaluation criterion Q1 – *cost*, activity A1 having the cost of 8 units/piece, activity A2 having the cost of 12 units/piece and activity A3 having the cost of 10 units/piece, two activities were analysed: A1 and A3. From these, the activity A1 was selected, which according to previous experience leads to best results (lowest cost), with the uncertainty of the evaluation of 35%. Consequently, the order for the *manufacture of thrust bearings* is accepted.

For this, at level B, corresponding to assembly design, these are two potential variants B1 – *Thrust ball bearing*, B2 – *Thrust roller bearing*. According to the evaluation criterion Q3 – *timespan*, in the case of thrust ball bearings, it is 10 units of time, while for thrust roller bearings it is 8 units of time. As a result, activity B2 is selected, the uncertainty of the evaluation being 30%, also higher that the accuracy level pursued.

Next, at level C, subassembly design, there are two potential activities C1 – *Manufacture of a steel cage* and C2 – *Manufacture of a brass cage*. According to criterion Q1, activity C1 (cost of 11 units/piece) and activity C2 (cost of 14 units/piece) were evaluated, being selected activity C1, with an evaluation uncertainty of 25%, also higher than the accuracy level pursued.

In terms of components at the same level, the following three have a significant impact on cost: C1.1 – *Outer ring*, C1.2 – *Inner ring*, C1.3 – *Rollers*. For these, the manufacturing cost is assessed C1.1 (cost of 2 units/piece) with an uncertainty of 17%, for component C1.2 (cost of 2 units/piece) with an uncertainty of 8% (in this case optimization stops) and for component C1.3 (cost of 3 units/piece) with an uncertainty of 12%. Furthermore, for the remaining components, at level D, process planning, there are four potential activities D1 – *Plan I – Outer ring* and D2 – *Plan II – Outer ring* for C1.1, respectively D3 – *Plan I – Rollers* and D4 – *Plan II – Rollers* for C1.3. Of the four activities, only two activities are selected, according to criterion Q1, for the manufacture of the bearing, namely D1 and D4, both activities having a cost of 4 units. In the case of activity D1 the evaluation uncertainty being 11%, the optimization is continued, while for the activity D4 evaluation uncertainty being 8%, the optimization stops.

Finally, for the selected operation plans, at level E, operation programming, two potential activities are identified E1 – *Turning on CNC lathes* and E2 – *Turning on conventional lathes*. These are evaluated according to the criterion Q2 – *cost/quantity*, the activity E1 having a cost of 4 units and the activity E2 a cost of 6 units. Activity E1 was selected as optimal. At this point, entire optimization process is stopped, because in all cases the resulting level of uncertainty drops below 10%.

Thus, following the successive evaluations and selections of activities at different levels of the manufacturing process, an optimal manufacturing path is obtained starting with an order for the bearings components manufacture, which includes activities A1, B2, C1 – C1.1, C1.2, C1.3, D1, D4 and E1 (marked in red in figure 5).

6. Conclusion

The method for structural identification of the manufacturing process allows the structuring of its activities, at all involved levels (order acceptance, production planning, product design, processes planning and product processing), by *i)* elaborating the tree of specific activities, the relations between the manufacturing process stages and the related informational circuit are highlighting, *ii)* identifying of the manufacturing process alternatives, at the level of each manufacturing activity, *iii)* selecting the best alternatives at each level of the manufacturing activity, according to different optimization criteria (such as, for example, cost, timespan, energy consumption, other critical consumption or combinations thereof). The exemplification regarding the application of the structural identification method in bearing manufacturing case, allowed the successive evaluation and selection of the optimal activities from different levels of the manufacturing process, thus obtaining the optimal manufacturing path starting with an order for the bearing manufacture.

7. References

[1] Frumusanu G, Afteni C and Paunoiu V 2020 Estimation of Roller Bearings Manufacturing Cost by Causal Identification and Comparative Assessment – Case Study Performed on Industrial Data *Int. J. Model. Optim.* **10**(4) pp 114–120

[2] Afteni C, Frumusanu G and Epureanu A 2018 Instance-based comparative assessment with application in manufacturing *IOP Conf. Ser. Mater. Sci. Eng. Pap.* **400**(042001) pp 0–8

[3] Afteni C, Frumusanu G and Epureanu A 2019 Method for Holistic Optimization of the Manufacturing Process *Int. J. Model. Optim.* **9**(5) pp 265–270

[4] Martínez-Rivero M D, Hernández-Castellano P, Marrero-Alemán M D and Suárez-García L 2019 Manufacturing process selection integrated in the design process: Test and results *Procedia Manuf.* **41** pp 827–834

[5] Denno P, Dickerson C and Harding J A 2018 Dynamic production system identification for smart manufacturing systems *J. Manuf. Syst.***48** pp 1–11

[6] Feng S C and Song E Y 2003 A manufacturing process information model for design and process planning integration *J. Manuf. Syst.* **22**(1) pp 1–15
[7] Gecevska V, Cus F, Dukovski V and Kuzinovski M 2006 Modelling of manufacturing activities by process planning knowledge representation *Int. J. Simul. Model.* **5**(2) pp 69–81
[8] Susac F and Stan F 2020 Experimental Investigation, Modeling and Optimization of Circularity, Cylindricity and Surface Roughness in Drilling of PMMA Using ANN and ANOVA *Mater. Plast.* **57**(1) pp 57–68
[9] Zhang L, Luo Y, Zhang Y and Song G 2015 Production scheduling oriented to energy consumption optimization for process industry based on self-adaptive DE algorithm *Int. J. Control Autom.* **8**(2) pp 31–42
[10] Huet G, Fortin C, Mcsorley G and Toche B 2009 Information Structures and Processes to Support Data Exchange between Product Development and Resource Planning Systems * *J. Oper. Logist.* **1**(514)
[11] Afteni C 2015 Holistic optimization of manufacturing process PhD Thesis *'Dunarea de Jos' University of Galati*, Series I 4: Industrial Engineering No. 70

Acknowledgement
This work was supported by the Romanian Ministry of Research and Innovation, CCCDI – UEFISCDI, project number PN-III-P1-1.2-PCCDI-2017-0446 / *Intelligent manufacturing technologies for advanced production of parts from automobiles and aeronautics industries* (TFI PMAIAA) - 82 PCCDI/2018, within PNCDI III.

Tool`s profiling for rotational volumetric deformation - analytical study

G A Costin, N Baroiu, V G Teodor, V Paunoiu and N Oancea
"Dunarea de Jos" University of Galati, Department of Manufacturing Engineering, Domneasca street, no. 111, Galati, Romania

E-mail: nicusor.baroiu@ugal.ro

Abstract. The paper develops in an analytical form an algorithm for the profiling of the active elements of the upper half-die forming which generates by means of plastic rotary volumetric deformation of the frontal teeth. The half die has the revolving axis inclined with respect to the axis of the deformed blank, which rotates around its own axis of symmetry. In the generation process, the generating half die also performs a movement of the blank. The constitutive surfaces of the tooth flanks to be generated are complex surfaces, formed of circular arcs with variable radius and disposed on a conical base surface, on the same axis with the half-die axis. Such teeth are used for front blades in the textile industry or drilling machines which performs, besides the rotational movement of the tool and an alternate rectilinear motion thereof, creating a percussion effect on the work piece. The surfaces of the upper half and the half die teeth may be regarded as reciprocal wrapping surfaces, so that the entire generation process is based on the analytical principles of surface winding. Starting from the constructive shape of the frontal tooth plate, in a classical scheme of the structure of the oscillating motion mechanism of the upper half matrix, the shape of the deformation teeth is determined based on a specific algorithm, based on the general theory of surface winding, and dedicated software products for it.

1. Introduction

The production of cold pressed steel pieces is widely used, even since the last decades of the twentieth century [2]. There are restrictions imposed by the qualities of the deformed material [7], the shape of the piece subjected to deformation processes, related to the type of deformation machine used [4], as well as the level of knowledge and analysis of metal deformation [3], [5], [6], [8], [9], which restricted the applicability domain of specific processes [1].

The types of specialized equipments, on which industrial applications can be developed, as well as the specific constructions of the tools used, constitutes continuous concerns, in the profile industries, in all the developed economies [1].

In the paper, an analysis is made based on which a dedicated algorithm is generated for profiling the shape of the generating tool when processing by volumetric plastic deformation, for a wheel with front teeth, bordered by conical surfaces. The constitutive surfaces of the flanks of the teeth to be generated were geometrically modeled and, based on the generating kinematics, the surfaces families generated in the processing process were determined, whose envelope is the composed surface of the flanks of the tool teeth.

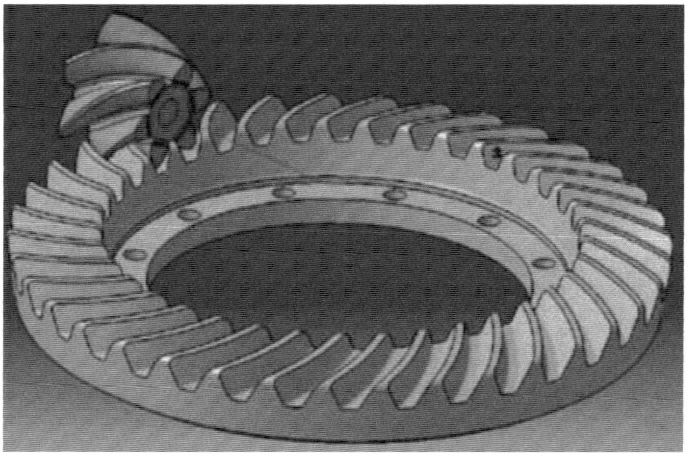

Figure 1. Pinion with conical flanks - constructive model [10].

2. Cam geometry with front conical flanks

Figure 2 shows the geometry of the teeth with a composed profile consisting of circle arcs and conical flanks.

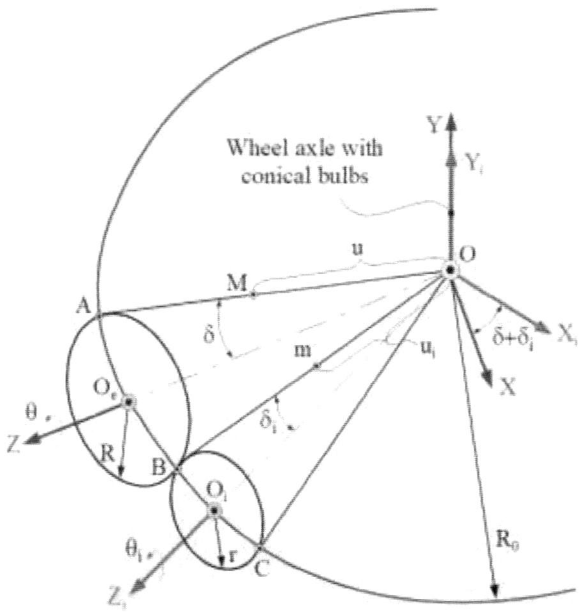

Figure 2. The flanks of the conical bulbs. Reference systems.

It is considered that the flanks of the conical bulbs have as directors, circle arcs of radii R, respectively r, with centers in O_e and O_i. The generators of the conical flanks are rectilinear, forming conical surfaces with the peak at the point O, a point on the symmetry axis of the wheel with conical bulbs. The radius on which the centers of the directing circles of the conical bulbs are located is defined as R_o.

Reference systems are defined, figure 2:

- XYZ – system associated with the front cam axis with the Z axis, given by the OO_e direction, right reference system, associated with the AB profile of the front bulb;

- $X_iY_iZ_i$ – system associated with the bottom, the BC arc of center O_i; the Z_i axis is defined by the OO_i direction, right system.

It is noted with δ, respectively δ_i the angles formed by the generators of the conical surfaces OM, respectively Om with the directions OO_e, OO_i. If it is noted with θ, respectively θ_i the angles that define on the circles of centers O_e, O_i, the current point on the circle AB, respectively BC, then, the conical flanks equations of the front teeth bulbs can be written in the forms:

AB arc of O_e center, see figure 2

$$AB\left| \begin{aligned} &X = -u \cdot sin\,\delta \cdot sin\,\theta; \\ &Y = u \cdot sin\,\delta \cdot cos\,\theta; \\ &Z = u \cdot cos\,\delta, \end{aligned} \right. \tag{1}$$

with θ and u independent variables ($tg\delta = R / R_0$).

Also, the conical surface that constitutes the bottom of the front tooth, the BC arc, the circle of radius r, with the center in O_i, is:

$$BC\left| \begin{aligned} &X_i = -u_i \cdot sin\,\delta_i \cdot sin\,\theta_I; \\ &Y_i = u_i \cdot sin\,\delta_i \cdot cos\,\theta_I; \\ &Z_i = u_i \cdot cos\,\delta_i, \end{aligned} \right. \tag{2}$$

with θ_i and u_i independent variables ($tg\delta_i = r / R_0$).

Note: The centers O_e and O_i are defined as points belonging to the circle of radius R_o with the center in O; the plane determined by O, O_e, O_i is the reference plane of the cam with front bulbs.

The coordinates transformation $X_iY_iZ_i \rightarrow XYZ$, see also figure 2, is given by the form:

$$\begin{pmatrix} X \\ Y \\ Z \end{pmatrix} = \begin{pmatrix} cos\,\alpha & 0 & sin\,\alpha \\ 0 & 1 & 0 \\ -sin\,\alpha & 0 & cos\,\alpha \end{pmatrix} \cdot \begin{pmatrix} X_i \\ Y_i \\ Z_i \end{pmatrix}, \tag{3}$$

where the X_i matrix has the meaning given by (2) and

$$\alpha = \delta + \delta_i. \tag{4}$$

Thus, the two conical surfaces can be defined in the same coordinate system, XYZ:

$$BC\ \begin{pmatrix} X \\ Y \\ Z \end{pmatrix} = \begin{pmatrix} cos\,\alpha & 0 & sin\,\alpha \\ 0 & 1 & 0 \\ -sin\,\alpha & 0 & cos\,\alpha \end{pmatrix} \cdot \begin{pmatrix} -u_i \cdot sin\,\delta_i \cdot sin\,\theta_I \\ u_i \cdot sin\,\delta_i \cdot cos\,\theta_I \\ u_i \cdot cos\,\delta_i \end{pmatrix} \tag{5}$$

or, obviously, in developed form:

$$BC\left| \begin{aligned} &X = \left[-u_i \cdot sin\,\delta_i \cdot sin\,\theta_I\right] \cdot cos\,\alpha + u_i \cdot cos\,\delta_i \cdot sin\,\alpha; \\ &Y = u_i \cdot sin\,\delta_i \cdot cos\,\theta_I; \\ &Z = -\left[-u_i \cdot sin\,\delta_i \cdot sin\,\theta_I\right] \cdot sin\,\alpha + u_i \cdot cos\,\delta_i \cdot cos\,\alpha. \end{aligned} \right. \tag{6}$$

The circular pitch of the teeth on the circle of radius R_o is defined as follows:

$$P_C = 2 \cdot R_o \cdot (\delta + \delta_i). \tag{7}$$

The maximum values of the linear parameters, u and u_i, measured along the generators of the conical surfaces are:

$$u_{max} = \sqrt{R_o^2 + R^2}, \text{ for the } AB \text{ lobe;} \tag{8}$$

$$u_{i_{max}} = \sqrt{R_o^2 + r^2}, \text{ for the conical surface at the bottom of the teeth, } BC. \tag{9}$$

The sizes R, r are constructive sizes.

The minimum value of the parameters u, u_i is defined for constructive considerations.

The variation limits of the parameters θ, respectively θ_i are:

$$\begin{aligned} \theta_{min} &= -\frac{\pi}{2};\ \theta_{max} = 0; \\ \theta_{i_{min}} &= \frac{\pi}{2};\ \theta_{i_{max}} = \pi. \end{aligned} \tag{10}$$

Correlations are also defined, see figure 2:

$$R = R_o \cdot sin\,\delta;\quad r = R_o \cdot sin\,\delta_i. \tag{11}$$

If it defined with Z_{teeth}, the number of teeth (Z_{teeth} - integer), then there is the relation:

$$2\pi \cdot R_o = Z_{teeth} \cdot P_C \text{ sau } P_C = \frac{2\pi \cdot R_o}{Z_{teeth}}. \tag{12}$$

3. The generating tool by volumetric deformation

Figure 3 shows the relative position of the reference systems associated with the cam with front conical flanks, XYZ, see also figure 1 and the reference system of the volumetric deformation tool, $X_2Y_2Z_2$ (Z_2 - rotation axis).

The process kinematics is also presented:

- I – rotational movement of the blank around its own axis (Y), of angle ϕ_1;
- II – rotational movement of the volumetric deformation tool around the Z_2 axis, of angle ϕ_2;
- III – radial feed movement.

The sizes are constructively defined:

- the rolling radius of the volumetric deformation tool, R_S, in correlation with the circular pitch of the teeth, see (12) and the number of tool teeth, Z_S (integer, constructive value);
- the inclination angle of the rotation axis of the volumetric deformation tool in relation with the reference frontal plane of the generated wheel (the frontal plane of the circle of radius R_o, the angle τ, see figure 3).

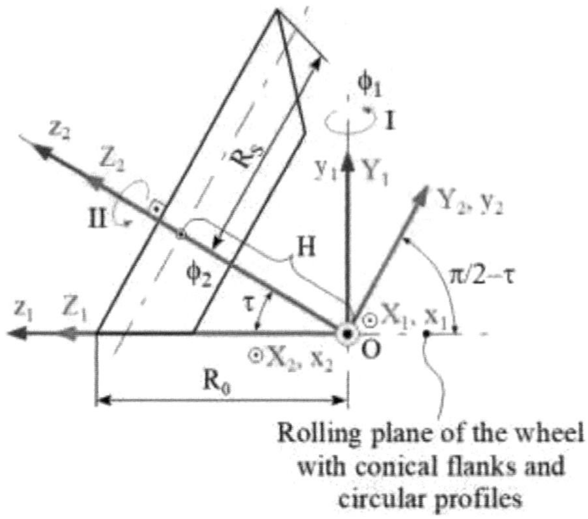

Figure 3. The kinematics of the generation process.

The reference systems are defined:
- $x_1y_1z_1$ – fixed reference system, y_1 axis superimposed on the rotation axis of the blank;
- $x_2y_2z_2$ – fixed system, z_2 axis superimposed on the rotation axis of the volumetric deformation tool;
- $X_1Y_1Z_1$ – mobile system associated with the blank;
- $X_2Y_2Z_2$ – mobile system associated with the deforming tool; the Z_2 axis is the rotation axis of the tool.

The process kinematics includes:
- the rotation of the $X_1Y_1Z_1$ system and, joined with it, of the blank:

$$x_1 = \omega_2^T(\phi_1)\cdot X_1; \tag{13}$$

- the rotation of the $X_2Y_2Z_2$ system joined with the volumetric deformation tool:

$$x_2 = \omega_3^T(\phi_2)\cdot X_2; \tag{14}$$

- the relative position of the fixed systems:

$$x_2 = \tau \cdot x_1. \tag{15}$$

where τ has the form:

$$\delta = \begin{pmatrix} 1 & 0 & 0 \\ 0 & \cos\tau & -\sin\tau \\ 0 & \sin\tau & \cos\tau \end{pmatrix}. \tag{16}$$

Thus, the relative movement of the space $X_1Y_1Z_1$ towards to $X_2Y_2Z_2$, meaning the space movement of the blank to be generated, in relation to the reference system of the volumetric deformation tool, $X_2Y_2Z_2$, is:

$$X_2 = \omega_3(\phi_2)\cdot\delta\cdot\omega_2(\phi_1)\cdot X_1 \tag{17}$$

or, developed:

$$\begin{pmatrix} X_2 \\ Y_2 \\ Z_2 \end{pmatrix} = \begin{pmatrix} cos\,\phi_2 & -sin\,\phi_2 & 0 \\ sin\,\phi_2 & cos\,\phi_2 & 0 \\ 0 & 0 & 1 \end{pmatrix} \cdot \begin{pmatrix} 1 & 0 & 0 \\ 0 & cos\,\tau & -sin\,\tau \\ 0 & sin\,\tau & cos\,\tau \end{pmatrix} \cdot \begin{pmatrix} cos\,\phi_1 & 0 & sin\,\phi_1 \\ 0 & 1 & 0 \\ -sin\,\phi_1 & 0 & cos\,\phi_1 \end{pmatrix} \cdot \begin{pmatrix} X \\ Y \\ Z \end{pmatrix}. \tag{18}$$

In the transformation (18), the (X) matrix has successively the meanings given by (1) and (6), for the two conical surfaces defined by the directors of the surfaces AB, respectively BC and with the common peak of the conical slopes, point O, see figure 1.

Following the development of the matrix product (18), it reached at the forms:

$$\Sigma_{AB,BC}\ {}_{\phi_1} \quad \begin{aligned} X_2 &= \left[cos\,\phi_1 \cdot cos\,\phi_2 - sin\,\phi_2 \cdot sin\,\tau \cdot sin\,\phi_1\right] \cdot X_1 - sin\,\phi_2 \cdot cos\,\tau \cdot Y_1 + \\ &+ \left[sin\,\phi_1 \cdot cos\,\phi_2 + sin\,\phi_2 \cdot sin\,\tau \cdot cos\,\phi_1\right] \cdot Z_1; \\ Y_2 &= \left[cos\,\phi_1 \cdot sin\,\phi_2 + cos\,\phi_2 \cdot sin\,\tau \cdot sin\,\phi_1\right] \cdot X_1 + cos\,\phi_2 \cdot cos\,\tau \cdot Y_1 + \\ &+ \left[sin\,\phi_1 \cdot sin\,\phi_2 - cos\,\phi_2 \cdot sin\,\tau \cdot cos\,\phi_1\right] \cdot Z_1; \\ Z_2 &= -cos\,\tau \cdot sin\,\phi_1 \cdot X_1 + sin\,\tau \cdot Y_1 + cos\,\tau \cdot cos\,\phi_1 \cdot Z_1. \end{aligned} \tag{19}$$

In further developments, it is accepted that the matrices formed with the coordinates of the two conical surfaces defined by (1) respectively (6), satisfy the condition:

$$\begin{Vmatrix} X \\ Y \\ Z \end{Vmatrix}_{AB,BC} = \begin{Vmatrix} X_1 \\ Y_1 \\ Z_1 \end{Vmatrix}_{AB,BC} \tag{20}$$

Equations (19), with (1), respectively (2), represent the family of surfaces generated by the blank flanks in relation to the reference system of the volumetric deformation tool.

The envelope of the composed surface family $\Sigma_{AB,BC}\ {}_{\phi_1}$ (19), in which, successively, Σ_{AB} depends on the independent variable parameters θ and u and also Σ_{BC} depends on the parameters θ_1 and u_1, is obtained by associating the surfaces family (19), (1), for the arc AB and (19), (6) for the arc BC, see figure 2, the enwrapping conditions:

- for AB:

$$\begin{vmatrix} \dot{X}_{2\phi_1} & \dot{Y}_{2\phi_1} & \dot{Z}_{2\phi_1} \\ \dot{X}_{2u} & \dot{Y}_{2u} & \dot{Z}_{2u} \\ \dot{X}_{2\theta} & \dot{Y}_{2\theta} & \dot{Z}_{2\theta} \end{vmatrix} = 0; \tag{21}$$

- for BC:

$$\begin{vmatrix} \dot{X}_{2\phi_1} & \dot{Y}_{2\phi_1} & \dot{Z}_{2\phi_1} \\ \dot{X}_{2u_1} & \dot{Y}_{2u_1} & \dot{Z}_{2u_1} \\ \dot{X}_{2\theta_1} & \dot{Y}_{2\theta_1} & \dot{Z}_{2\theta_1} \end{vmatrix} = 0. \tag{22}$$

The partial derivatives from (1) are:

$$\begin{aligned}&\dot{X}_{2u} = -\sin\delta\cdot\sin\theta;\\&\dot{Y}_{2u} = \sin\delta\cdot\cos\theta;\\&\dot{Z}_{2u} = \cos\delta;\end{aligned} \tag{23}$$

and also,

$$\begin{aligned}&\dot{X}_{2\theta} = \dot{X}_{1\theta} = -u\cdot\sin\delta\cdot\cos\theta;\\&\dot{Y}_{2\theta} = \dot{Y}_{1\theta} = -u\cdot\sin\delta\cdot\sin\theta;\\&\dot{Z}_{2\theta} = \dot{Z}_{2\theta} = 0.\end{aligned} \tag{24}$$

Also, for the conical surface that has as directories the BC arc see also (6), it results:

$$\begin{aligned}&\dot{X}_{2\theta_i} = -u_i\cdot\sin\delta_i\cdot\cos\theta_i\cdot\cos\alpha; && \dot{X}_{2u_i} = -\sin\delta_i\cdot\sin\theta_1\cdot\cos\alpha + \cos\delta_i\cdot\sin\alpha;\\&\dot{Y}_{2\theta_i} = -u_i\cdot\sin\delta_i\cdot\sin\theta_i; && \text{and } \dot{Y}_{2u_i} = \sin\delta_i\cdot\cos\theta_1;\\&\dot{Z}_{2\theta_i} = u_i\cdot\sin\delta_i\cdot\cos\theta_i\cdot\sin\alpha && \dot{Z}_{2u_i} = \sin\delta_i\cdot\sin\theta_1\cdot\sin\alpha + \cos\delta_i\cdot\cos\alpha.\end{aligned} \tag{25}$$

The partial derivatives of the surface family (19) are also defined in relation with the parameter ϕ_1:

$$\begin{aligned}\dot{X}_{2\phi_1} = &\left[-\sin\phi_1\cdot\cos\phi_2 - \cos\phi_1\cdot\sin\phi_2\cdot\frac{d\phi_2}{d\phi_1} - \frac{d\phi_2}{d\phi_1}\cdot\cos\phi_2\cdot\sin\tau\cdot\sin\phi_1 - \sin\phi_2\cdot\sin\tau\cdot\cos\phi_1\right]\cdot X_1 -\\&-\cos\phi_2\cdot\frac{d\phi_2}{d\phi_1}\cdot\cos\tau\cdot Y_1 +\\&+\left[\cos\phi_1\cdot\cos\phi_2 - \frac{d\phi_2}{d\phi_1}\cdot\sin\phi_1\cdot\sin\phi_2 + \frac{d\phi_2}{d\phi_1}\cdot\cos\phi_2\cdot\sin\tau\cdot\cos\phi_1 - \sin\phi_2\cdot\sin\tau\cdot\sin\phi_1\right]\cdot Z_1;\\\dot{Y}_{2\phi_1} = &\left[-\sin\phi_1\cdot\sin\phi_2 + \frac{d\phi_2}{d\phi_1}\cdot\cos\phi_1\cdot\cos\phi_2 - \frac{d\phi_2}{d\phi_1}\cdot\sin\phi_2\cdot\sin\tau\cdot\sin\phi_1 + \cos\phi_2\cdot\sin\tau\cdot\cos\phi_1\right]\cdot X_1 -\\&-\frac{d\phi_2}{d\phi_1}\cdot\sin\phi_2\cdot\cos\tau\cdot Y_1 +\\&+\left[\cos\phi_1\cdot\sin\phi_2 + \frac{d\phi_2}{d\phi_1}\cdot\sin\phi_1\cdot\cos\phi_2 + \frac{d\phi_2}{d\phi_1}\cdot\sin\phi_2\cdot\sin\tau\cdot\cos\phi_1 + \cos\phi_2\cdot\sin\tau\cdot\sin\phi_1\right]\cdot Z_1;\\\dot{Z}_{2\phi_1} = &-\cos\delta\cdot\cos\phi_1\cdot X_1 - \cos\tau\cdot\sin\phi_1\cdot Z_1.\end{aligned} \tag{26}$$

Note: In the expressions of the partial derivatives (26), $X_1Y_1Z_1$ have the meanings given, successively, by (1), respectively (6), see also the specification (20).

The size of the radius R_s of the volumetric deformation tool is defined:

$$2\overline{u}\cdot R_s = Z_s\cdot P_C, \tag{27}$$

where:
- R_s – the rolling radius of the deformation tool, see figure 3;
- P_C – the circular pitch of the teeth on the radius circle R_o, see (7);
- Z_s – the number of teeth of the deforming tool (constructive).

The inclination angle of the deforming tool axis, δ, is defined:

$$sin\,\delta = \frac{R_s}{R_o}. \tag{28}$$

(R_o – see figure 2; R_o – constructive, as the size of wheel with front teeth)

For the numerical representation of the surfaces of the deforming tool lobes, the condition $Z_S = H$ is associated, H - variable between the limits:

$$\begin{cases} H_{\max} = R_o \cdot \cos\delta; \\ H_{\min} = u_{\min} \cdot \cos\delta, \end{cases} \tag{29}$$

depending on the searched surface element: AB or BC.

4. Conclusions

The analytical study was developed for a conical tooth with a circular profile, based on the general analytical theorems (Gohman).

The algorithm allowed the determination of a specific enwrapping condition based on three independent variable parameters.

Obviously, a numerical solution to such a problem involves imagining a specific software product.

5. References

[1] Ayano M and Kumakura I 1984 The development of a rotary forging machine and is practical applications *Proceeding of the 3-rd International Conference Rotary Metal Working Processes* (Kyoto: IFS Publications) pp 71-80

[2] Bârzescu A 1987 Considerații tehnice privind aplicarea procedeului de deformare a metalelor prin forjare orbital (Technical considerations regarding the application of the metal deformation process by orbital forging) *Construcția de mașini* **39**(2-3) pp 126-130

[3] Grzeskowak J and Piszewski M 1987 Orbital forging on cocking die press in forging manufacture *ICFG, Meeting* (Warszawa, Poznan) pp 331-320

[4] Grozav S, Achimaș G and Chis I 1993 Contribuții privind cinematica preselor pentru deformare prin forjare orbital (Contributions on press kinematics for deformation by orbital forging) *Construcția de mașini* **45**(10-11) pp 21-24

[5] Stoian C and Stoian E V 1997 Contribuții privind simularea fizică a proceselor de deformare plastic (Contributions on the physical simulation of plastic deformation processes) *The Annals of "Dunărea de Jos" Univeristy of Galați*

[6] Teodorescu M, Stoian C, Oprescu G, Levcovici S and Nicoară D 1990, Cercetări privind procesul de deformare volumică orbitală la rece (Research on the process of cold orbital volumetric deformation) *Metalurgia*

[7] Wei J and all 1984 New development of rotary forging in China *Procedings of the 3-rd Conference on Rotary Metal Working Processes* (Kyoto: IFS Publications) pp 61-69

[8] Radzevich S P 2007 On the Accuracy of Precision Involute Hobs: An Analytical Approach *Journal of Manufacturing Processes* **9**(2) pp 121-136

[9] Radzevich S P 2018 Theory of Gearing: Kinematics, Geometry, and Synthesis (Boca Raton: CRC Press)

[10] Wang J, Kong L, Liu B, Hu X, Yu X and Kong W 2014 The Mathematical Model of Spiral Bevel Gears - A Review *Journal of Mechanical Engineering* **60**(2) pp 93-105

Acknowledgments

This work was supported by a grant of the Romanian National Authority for Scientific Research and Innovation, CNCS-UEFISCDI, project number PN-III-P1-1.2-PCCDI-2017-0446 - TFI PMAIAA, within PNCDI III.

New developed software for processing and analysing of the cutting forces records verified when machining by modern circle-segment end mills

A Polzer, T Vancura and T Trcka
Institute of Manufacturing Technology, Faculty of Mechanical Engineering, Brno University of Technology. Technicka 2896, 616 69 Brno, Czech Republic

E-mail: polzer@fme.vutbr.cz

Abstract. At present, 5-axis machining is often realized with ball end mills that do not achieve the advantages of the modern circle-segment end mills. In recent studies, these tools have been compared concerning the quality of the machined surfaces, machining times and cutting forces. An essential part of the research was an analysis of the force loading of the used tools. To get accurate data about force acting during the machining process, it was necessary to undergo experimental measuring using a cutting force dynamometer. For a force records analysis, the new software was developed based on the MATLAB platform. Evaluated results obtained by the new software are presented in the paper and compared with recent studies.

1. Introduction

Demands for increasing productivity in 5-axis machining of complex components lead to extensive research in the way of cutting tools design. The novel developed circle-segment tools are attracting considerable interest due to their advantages compared to conventionally used ball-end cutters. The circle-segment tools have a significantly increased radius of the cutting edge, which allows having larger path distance accompanied by the same or even better surface quality compared to ball-end cutters (see figure 1) [1].

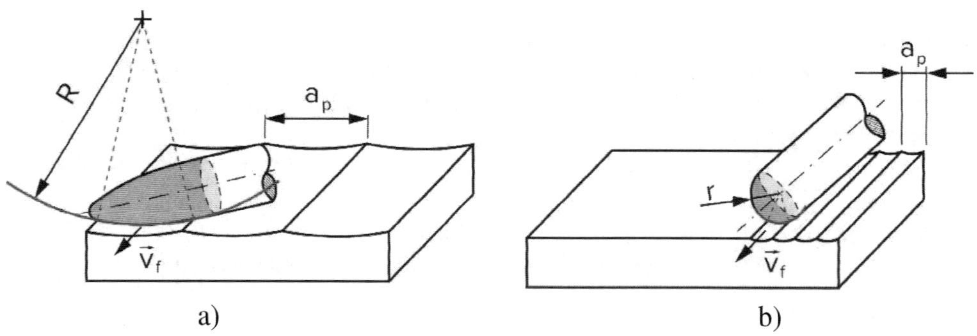

Figure 1. a) A circle-segment end mill, b) a ball end mill.

In the new researches carried out by Trcka [2] and Sebesta [3, 4] these tools are compared concerning the quality of the machined surfaces, machining times and cutting forces. An essential part of the research was also an experimental cutting force measuring of the tools. The uses and benefits of the experimental measurement are well known [5]. Force measurement can be performed by indirect or direct method, which is more accurate. The most used types of dynamometers for measuring cutting forces during machining are based on the piezoelectric effect. Rotary dynamometers (e.g. type 9170A by company Kistler) have an advantage in direct measurement of force components $\mathbf{F_c}$ (cutting force) and $\mathbf{F_{cN}}$ (normal cutting force) over stationary dynamometer. This is due to the rotation of the dynamometer coordinate system together with the tool. This characteristic becomes more effective in continuous 5-axis CNC milling [5].

The approach used for the processing of measured data is not well established and differs depending on the experience of the author. Filtering and evaluation are often performed in the Microsoft Excel. To establish a less time-consuming approach to cutting force signal processing and analysing, the software presented in this paper has been developed. For verification of the functionality, the results obtained by the new software are compared with the published results of experiments.

2. Development of the data processing system

Developing the software for automatisation of processing and analysing of forces records requires an understanding of the whole process of experimental forces measuring. A brief overview of the process is described in figure 2.

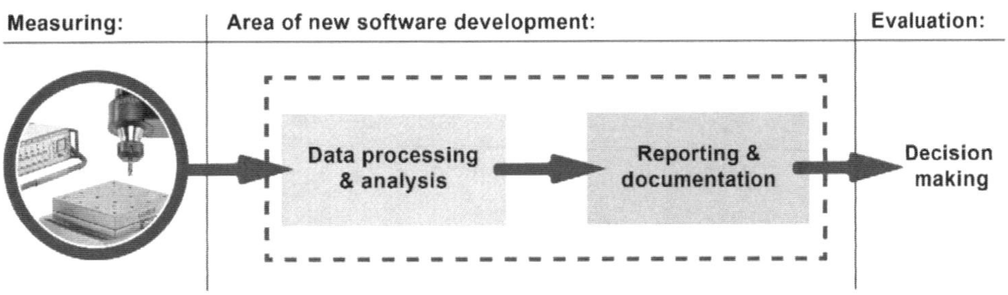

Figure 2. Structure of the experimental force measurement process.

Input data for the processing are cutting forces records saved in the file with the *txt* suffix. As an example of these data it is possible to use data from the program DYNOWARE by the company KISTLER. However, it is also possible to process data from other sources with the following features:

- The text file has an identical format with standard *ASCII*
- The separator of decimal space is a dot.
- The order of measured data columns must be observed: time-$\mathbf{F_x}$-$\mathbf{F_y}$-$\mathbf{F_z}$-$\mathbf{M_z}$ when the first four columns must be filled. Moment $\mathbf{M_z}$ is optional.
- The length of each measured force record must be identical.

At the same time, the following mathematical relationship for total force **F** and measured forces $\mathbf{F_{xm}}$, $\mathbf{F_{ym}}$, $\mathbf{F_{zm}}$ (forces in the coordinate system of the dynamometer), must be validated:

$$F = \sqrt{(F_{xM})^2 + \left(F_{yM}\right)^2 + (F_{zM})^2} \quad (1)$$

For the creation of the software an interactive developing environment was used within the program MATLAB App Designer. This Matlab toolbox allows creating a Graphic User Interface (GUI) with use of the UIControl objects which are set in the integrated editor into the form of default app.

Structure of the new software consists of four modules and GUI (see figure 3). Together with the default MATLAB toolbox, a toolbox for the signal processing (Signal Processing Toolbox), statistic evaluation of data (Statistics and Machine Learning Toolbox) and wavelet transformation (Wavelet Toolbox) were also used.

Data import module: this module is responsible for the loading of the measured force's records in the text format. Furthermore, it verifies completeness of input data and in the case of incompleteness, asks the user via a dialogue window for the data re-entering.

Data processing module: this module processes data before the analyses. Processing methods include, for example, numeric filtering of a cutting forces record, drift compensation or automatic detection of a cutting process signal area.

GUI: Graphic User Interface connects all modules and allows operating of the program both via the MATLAB interface and the stand-alone application with the support of the Runtime toolbox. Another task is to handle the input/output data, enabling their quick edits and displays the results of the analysis.

Data analysis module: this module allows analysing the input data. The module consists of three parts: statistical analysis, frequency analysis and time-frequency analysis of the measured data.

Data export module: this module enables the export of the analysed data. Graphs can be exported into the formats *JPG, GIF* or *EPS* and if it is needed, statistical data can be exported to the format *XLS.*

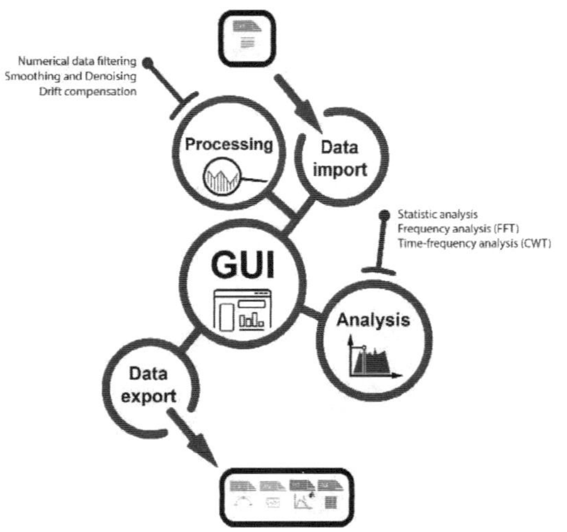

Figure 3. Concept of the software.

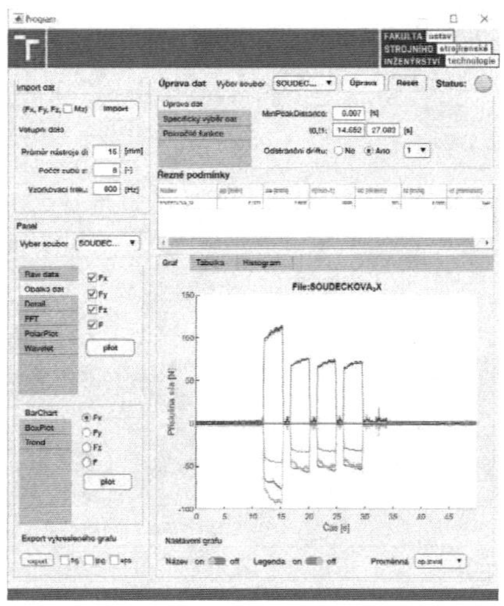

Figure 4. Graphical user interface of the program.

3. Cutting force data processing and analysing

This part aims to describe how the program works. After the successful uploading the files and entering the necessary cutting conditions, the software automatically processes the input data. During this phase, the signal is numerically filtered by the function *findpeaks* [6] to find the local maximums of the force records. The next step is the automatic detection of the signal area related to the machining process. For this task, the program uses an algorithm based on two functions: *findchangepts*, which detects change points in the signal and *midcross,* which detects the time instants where each transition of the signal crosses the 50 % reference level [6]. Once the signal is processed, the user has an option to edit the selected area of a cutting process signal or choose multiple intermittent segments of the signal. The next possibility of processing is signal smoothing and denoising. In this case, the user can use the Savitzky-Golay Filter [6] or the Wavelet denoising method [7]. The program also has a function for automatic drift compensation. After the processing, the user has several options on how to analyse the data:

- statistical analysis in the time domain,
- frequency analysis,
- time-frequency analysis.

3.1. Statistical analysis

The statistical analysis is an essential method of a cutting forces study. The main goal of the statistical analysis is to display the statistical characteristics of the analysed process. To get these characteristics, the program uses function *datastats* [8]. Statistical parameters which might be calculated are number of analysed values, minimum and maximum value, mean value, median, range of variation and standard deviation. For graphic interpretation of the statistical data, the user has at his disposal three types of graphs: a bar chart, boxplot and histogram. In the section "Normality test", normality of analysed data through the Anderson-Darling test can also be investigated [8]. In the case of the normally distributed data, the user can use the Grubb´s test for filtration of outliers.

3.2. Analysis in the frequency domain

To transfer the signal from a time domain to a frequency domain, the program uses the algorithm of Fast Fourier Transformation (FFT). The FFT output is the graph determining the power of each frequency in the force record. Based on analyses of the frequency spectrum of the signal, it is possible to recognise the dominant frequency of a force record or filter unwanted frequencies. In the case of machining, the dominant frequencies are usually related to the tooth passing frequency or the spindle speed frequency.

3.3. Analysis in the time-frequency domain

A disadvantage of the signal transforming to the frequency domain by Fourier transformation is the loss of time information. Wavelet transform (WT) decomposes a time signal into the time-frequency domain. In the time-frequency domain, it is possible to monitor the progress of cutting tool wear or analyse the cause of the vibration. Wavelet transformation can also be used for signal denoising. For the purpose, the program uses the Matlab function *wdenoise* [7].

4. Verification of functionality of the software

For verification of the software, the experiments with the circle-segment cutters were evaluated. The results obtained by the software have been compared with published results of these experiments. The presented experiments were performed at the Institute of Manufacturing Technology of the Faculty of Mechanical Engineering BUT. The measuring instruments used in the both cases consist of the table piezoelectric dynamometer KISTLER 9257B, multichannel charge amplifier KISTLER 5070A11000 and DYNOWARE software.

4.1. Trcka´s experiment

a) Design of the experiment

The first experiment carried out by Trcka [2] deals with the milling of a workpiece made by the additive manufacturing (AM) technology SLM (selective laser melting) from the powder material Ti-6Al-4V. For the experiment, solid carbide circle-segment end mill (EMUGE-FRANKEN 3540L.12250A) was used. The primary interest was an investigation of the cutting speed influence on the cutting forces and the roughness of the machined surface. During the experiment, two measurements were made at cutting speeds of 65 and 80 m.min^{-1}. The published results (total forces indexed as $\mathbf{F}_{pr}$) for the cutting speeds are stated in table 1.

Table 1. Published results.

	($\bar{F}_{pr}$) [N]	s(F_{pr}) [N]
v_c 65 m.min^{-1}	173.2276	1.371993
v_c 80 m.min^{-1}	146.9638	1.54645

b) Data processing

In figure 5 all input data managed by the program can be seen. Both the cutting force records were automatically processed by numerical filtration and searched by an algorithm to find the relevant cutting process area for statistical analysis. The results of the data processing can be seen in figure 6.

c) Results

The mean value of the total cutting force compared to the published results differs in the case of the first measurement by 0.12 % (0.21 N) and the second case by 0.39 % (0.57 N). The choice of different criteria of numerical filtration of the data resulted in different values of the standard deviations. The results of the total cutting force are graphically illustrated via a bar chart and boxplots (see figure 7). In figure 8 and figure 9, frequency and time-frequency representation of the signal can be seen.

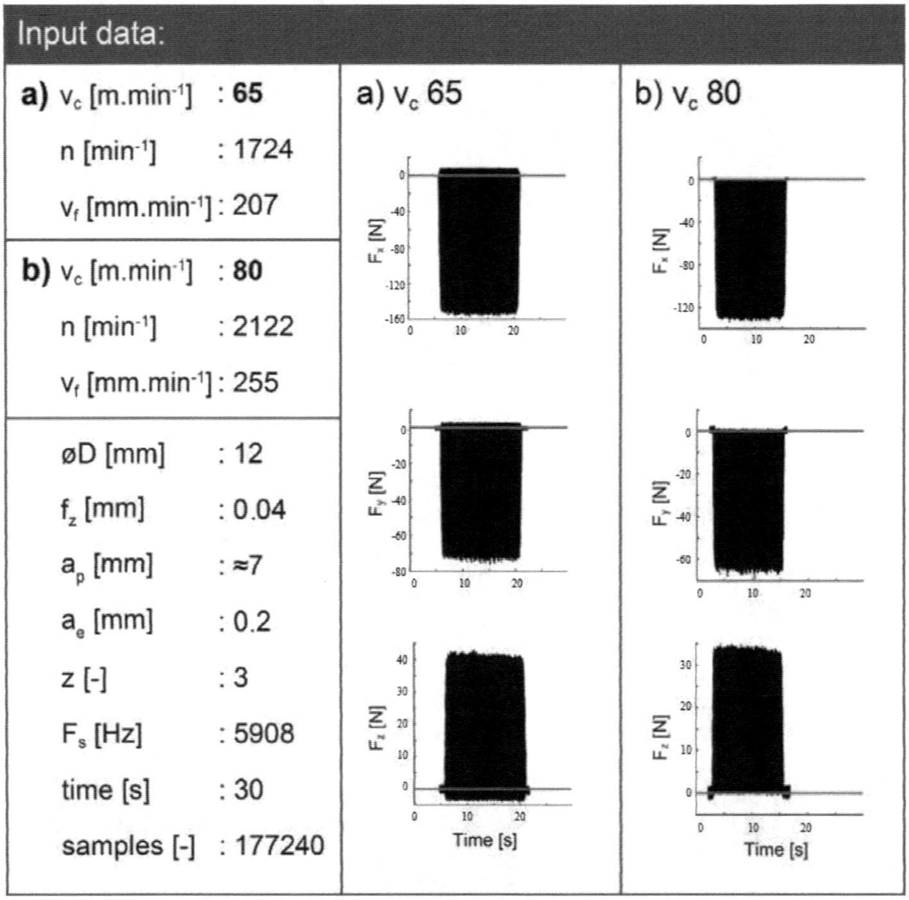

Figure 5. Program inputs.

IOP Conf. Series: Materials Science and Engineering **968** (2020) 012017 doi:10.1088/1757-899X/968/1/012017

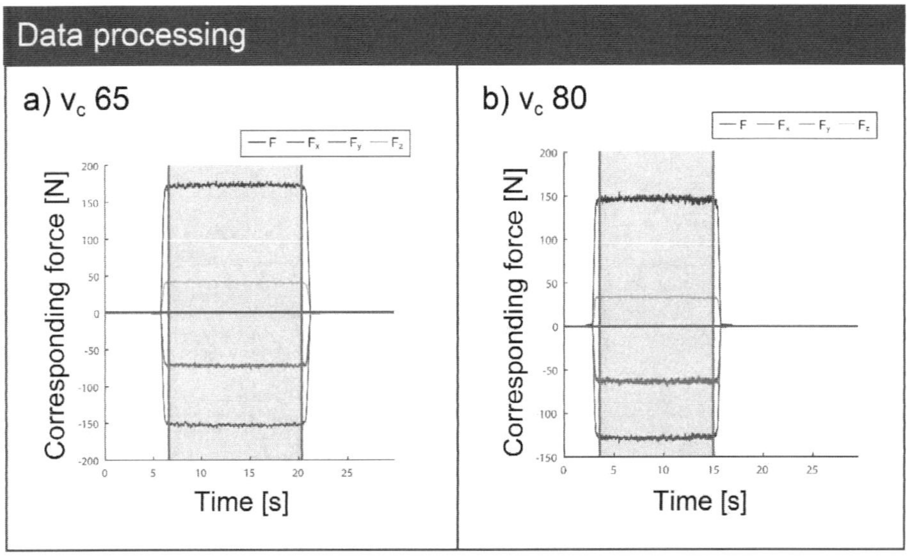

Figure 6. Data processing.

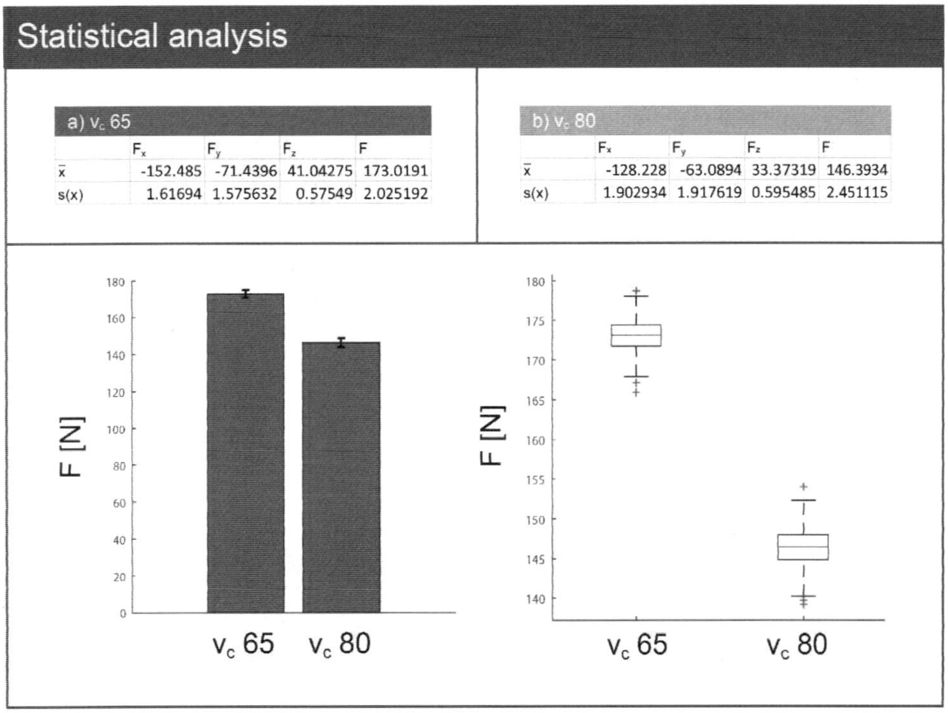

a) v_c 65

	F_x	F_y	F_z	F
$\bar{x}$	-152.485	-71.4396	41.04275	173.0191
s(x)	1.61694	1.575632	0.57549	2.025192

b) v_c 80

	F_x	F_y	F_z	F
$\bar{x}$	-128.228	-63.0894	33.37319	146.3934
s(x)	1.902934	1.917619	0.595485	2.451115

Figure 7. Statistical analysis.

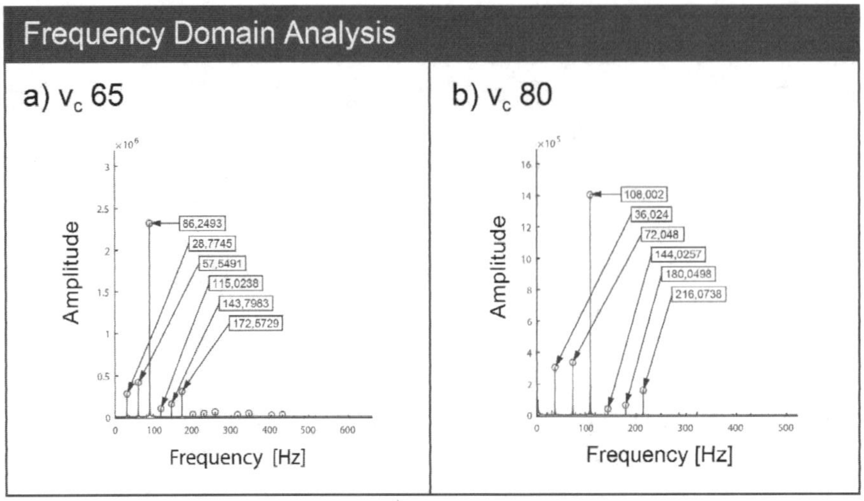

Figure 8. Frequency domain analysis: FFT.

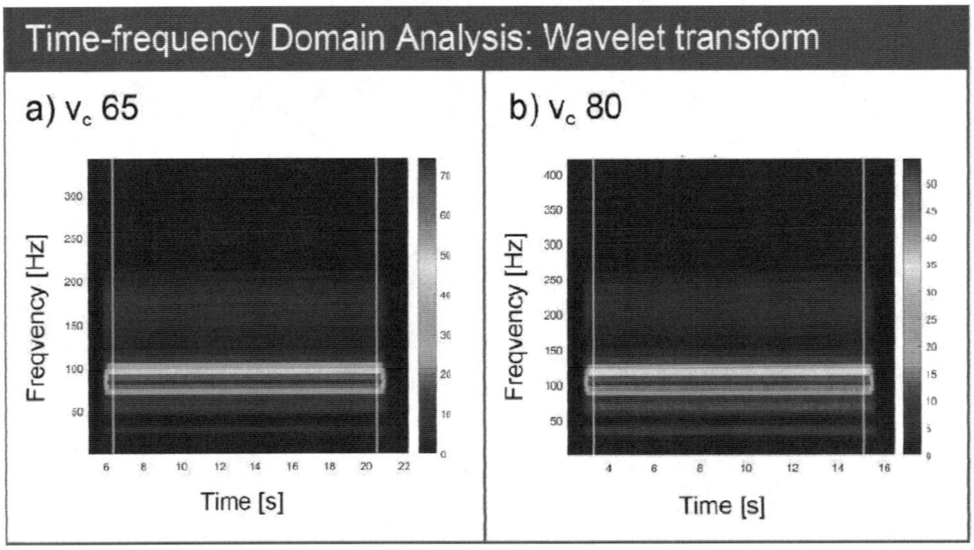

Figure 9. Time frequency domain analysis (CWT).

4.2. Sebesta's experiment

a) Design of the experiment

The second experiment carried out by Sebesta [3, 4] deals with the 3+2 and 5-axis milling of the bottom of the pocket. The workpiece material was aluminium alloy EN AW 7075-T6. The main goal was to investigate the benefits of the circle-segment end mill (EMUGE-FRANKEN 3540L.1615AA) in comparison with the ball end mill (EMUGE-FRANKEN 2550A.008). In this paper, only 3+2-axis milling with an inclination angle of the circle-segment end mill 20° takes apart. In ŠEBESTA´s research, there is no published statistical data, so only the data processing results can be compared.

b) Data processing

Two main issues of the measured signal can be seen in figure 10 a). The first issue is a signal drift and the second is a signal noise level. In figure 10 b), results of the automatic algorithm for the drift compensation can be seen. In the first step, the straight line between the starting and ending area where no cutting process occurs is computed. In the second step, the line is subtracted from the measured signal. This process repeated for all measured channels. In the last step, relevant areas for statistical analysis were chosen (see figure 10 c).

c) Results

The statistical analysis has been performed for two cases: a) a signal with noise and b) a denoised signal. The wavelet denoising was performed by using the db2 wavelets, with five-level of decomposition. The results for the both cases are presented in figure 11. As can be noted from the tables, the mean values of the signal with noise and the signal after wavelet denoising are different about 2 N.

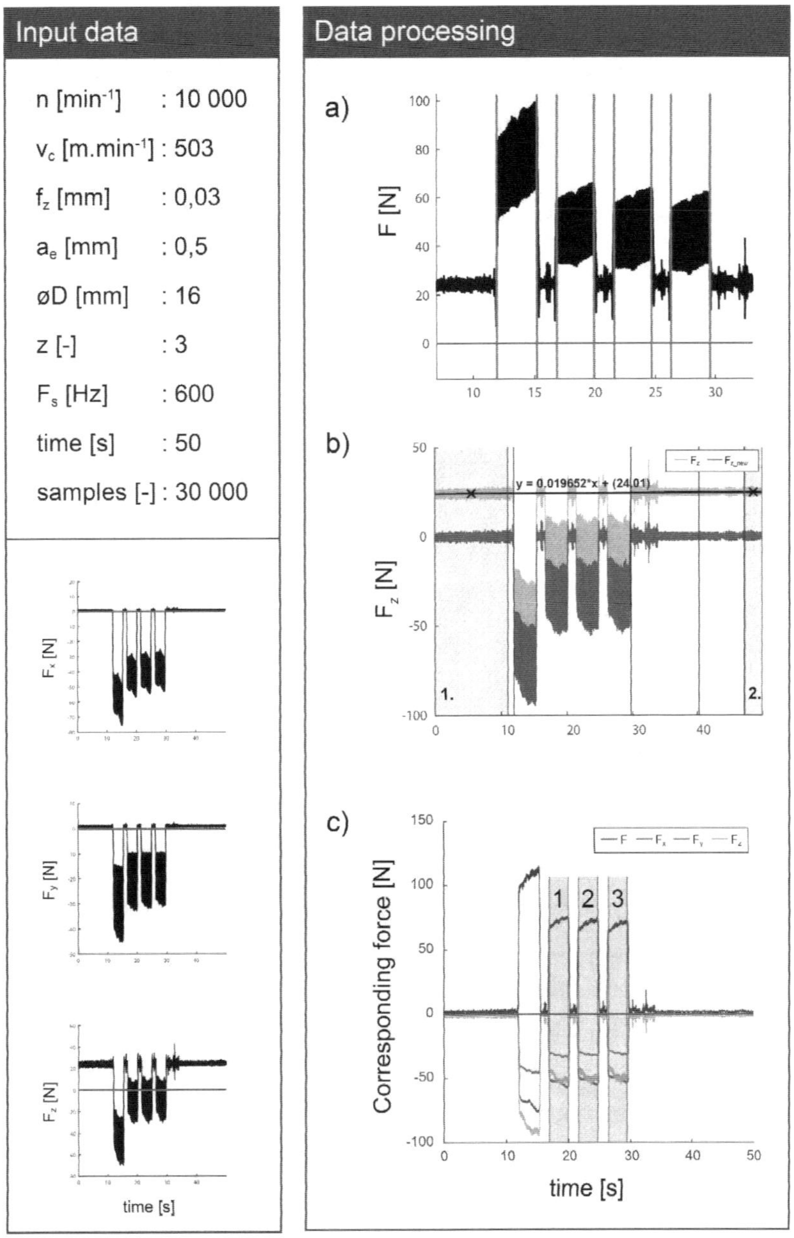

Figure 10. Program inputs & data processing.

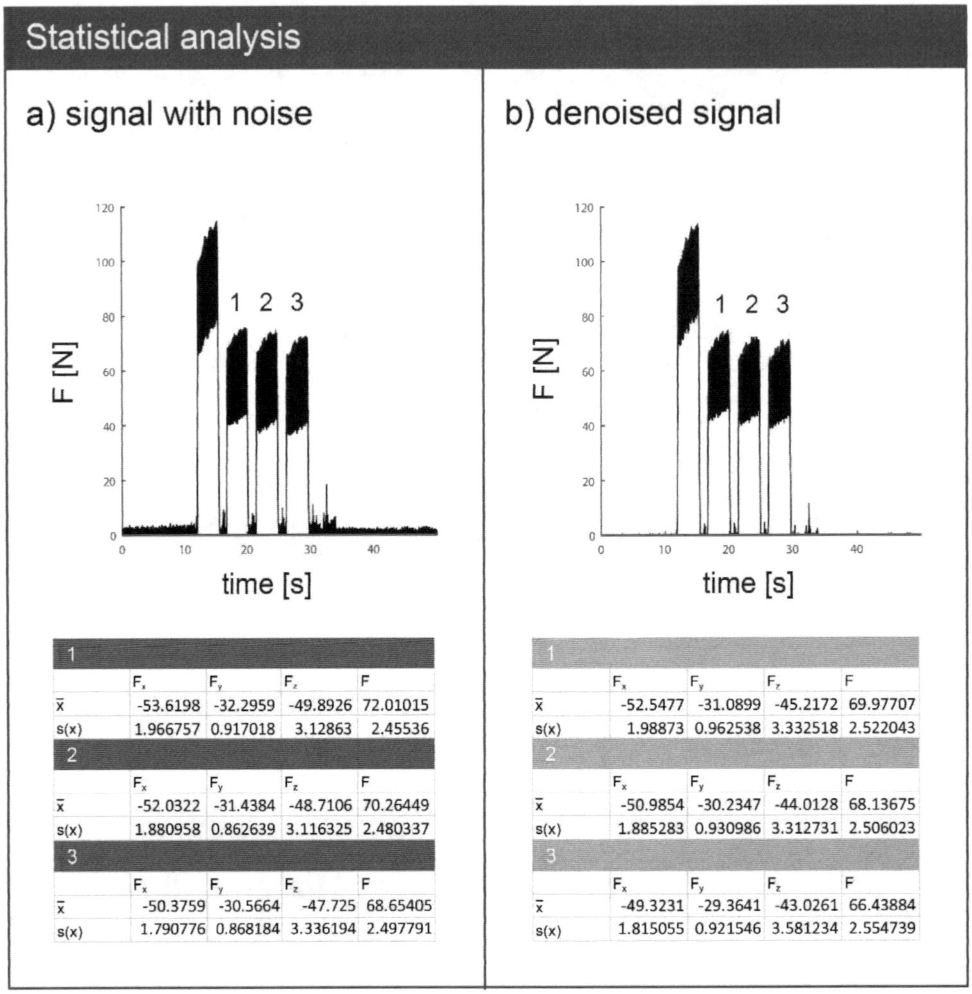

Figure 11. Statistical analysis.

5. Conclusion

The software for processing and analyzing of the cutting force´s records has been developed and verified with evaluating the already published experiments. The obtained results are demonstrating software functionality with the following conclusions:

(1) Results obtained by the new software differ only slightly with the published results of the experiments.

(2) The software is suitable for evaluating more cutting force records simultaneously with the possibility of comparing results with each other.

(3) In contrast with the use of conventional tools, the new software enables automatic data processing which significantly shortens the time that is needed to evaluate the experiment.

(4) All required tools for data processing and analysis are integrated into one software, which makes it easier for data handling.

IOP Conf. Series: Materials Science and Engineering **968** (2020) 012017 doi:10.1088/1757-899X/968/1/012017

6. References

[1] EMUGE-FRANKEN 2020 *Circle Segment End Mills* https://www.emuge.com/products/end-mills/circle-segment-cutters

[2] Trcka T, Polzer A and Sedlak J 2020 Analysing the performance of circle segment end mill with PCD inserts with laser-machined integral chip breaker when dry milling of additive manufactured Ti-6Al-4V titanium alloy *MM Science Journal (article in publishing process)*

[3] Sebesta V 2019 Application CAD/CAM software hypermill for machining *Brno University of Technology* (Brno) Diploma Thesis

[4] Polzer A, Sebesta V a Trcka T 2020 5-axis dry milling of bottom of pocket by ball and circle-segment end mills *MM Science Journal* **1** 3740-44

[5] Kistler C 2019 *Cutting force measurements in research and development: High-precision measurement systems for metal-cutting production.* Kistler Group. Winterthur https://www.kistler.com/files/download/960-002e.pdf?callee=frontend

[6] MATHWORKS 2020 *Statistics and Machine Learning Toolbox: User's Guide (R2020a)* https://www.mathworks.com/help/pdf_doc/stats/stats.pdf

[7] MATHWORKS 2020 *Signal Processing Toolbox: User's Guide (R2020a)* https://www.mathworks.com/help/pdf_doc/signal/signal.pdf

[8] Cabral G F 2016 Modelling and simulation of tool engagement and prediction of process forces in milling *RWTH Aachen University* ISBN 978-3-86359-427-5

Acknowledgments

This article was supported by the FME BUT Brno, Czech Republic as a specific research FSI-S-19-6014 called “Research of perspective manufacturing technologies”.

IOP Conf. Series: Materials Science and Engineering **968** (2020) 012018 doi:10.1088/1757-899X/968/1/012018

Influence of aluminium alloys relative positioning on dissimilar friction stir lap welds properties

S Sabari[1], I Galvão[1,2], C Leitão[1] and D M Rodrigues[3]
[1] CEMMPRE, Department of Mechanical Engineering, University of Coimbra, Portugal
[2] ISEL, Department of Mechanical Engineering, Polytechnic Institute of Lisbon, Portugal
[3] ISISE, Department of Mechanical Engineering, University of Coimbra, Portugal

E-mail: sree.sabari@dem.uc.pt

Abstract. Very thin sheets of heat treatable (AA6082-T6) and non-heat treatable (AA5754-H22) aluminium alloys were used to produce dissimilar joints by friction stir lap welding. Joint strength and mechanical heterogeneity were assessed by performing lap tensile-shear and microhardness testing, respectively. Optical microscopy was used to analyse the weld morphology. Welding machine outputs, such as torque and axial force, were analysed in order to capture differences in the alloys strength during dissimilar welding. From the process outputs analysis, it was concluded that higher Z-Forces were necessary while positioning the AA5754 alloy at the top of lap configuration, indicating that this alloy offered higher resistance to the tool stirring than the AA6082 alloy. These results were associated with the flow softening of the AA6082 alloy during welding, which was explained using finite element analysis to illustrate the welding thermal cycles and the precipitation behaviour of this alloy. The lap welds defect formation was also related to the base materials plastic properties at high temperatures, which was inferred from process outputs data analysis. Based on the results it was shown that the lap welds defects might be supressed by a proper positioning the dissimilar base materials in the lap joint.

1. Introduction

Friction stir welding (FSW) is a prominent technique that is already used for the successful joining of similar aluminium materials. However, compared to similar welding, dissimilar FSW is still a challenge since the interaction of base materials with different properties is required [1]. Simar et al., 2010, for example, welded AA2017 to AA6005A and found a similar kind of precipitate-related softening during welding, since both alloys are heat treatable [2]. However, when welding heat treatable and non-heat treatable aluminium alloys, the exposure of the dissimilar materials to high temperatures may leads to different microstructure-related softening responses [3]. It is widely known that recovery is the primary softening mechanism in non-heat treatable aluminium alloys, whereas precipitate dissolution or coarsening are the primary softening mechanism in heat treatable alloys [4, 5]. Considering the differences in the softening mechanisms between the heat treatable and non-heat treatable alloys, differences in flow behaviour during welding may be also expected.

The dissimilar flow behaviour of the base materials may have important influence on weld morphology, especially for lap joint configuration, where the upward material flow results in hook and cold lap defects formation [6]. Cederqvist & Reynolds, 2001 reported that the hook defect reduces the effective joining thickness [7] and Chen, 2012 referred that increments in hook height reduce the joint strength [8]. Many researchers, including Cao & Jahazi, 2011 [9] and Yazdanian et al., 2012 [10], adapted the process parameters and tool profiles for regulating the material flow in order to suppress the hook defect. But, from the studies made by Leitão, 2013 [11] and Galvão et al., 2013 [12] it was inferred that the inherent base material plastic properties are the key factor deciding the material flowability. Considering this aspect, Costa et al., 2015 explored the idea of changing the position of the aluminium alloys in the lap configuration to control the upward material flow and thereby to control the hook formation [13].

However, although the influence of the base materials positioning on weld properties was already addressed in a large number of works on dissimilar butt joining, the analysis of base materials positioning on lap welds quality, especially for the joining of very thin sheets, is scarce. Moreover, the relationship between the morphological and the mechanical properties of lap welds and the softening mechanisms of the base materials is not defined yet. This way, the aim of the present work was to analyse the influence of the relative positioning of the base materials on the properties of very thin AA6082/AA5754 friction stir lap welds. The softening mechanisms and their influence on the material flow were especially addressed. A coupled experimental and numerical approach was followed in this study.

2. Experimental procedure

One mm-thick sheets of the heat treatable AA6082-T6 and the non-heat treatable AA5754-H22 aluminium alloys were used as base materials to produce dissimilar friction stir lap welds. The alternative positioning of the base materials in the lap joint was tested, i.e. while some welds were produced with the AA6082 alloy positioned in the top of the joint, other welds were produced with the reverse positioning of the materials. In the text, the dissimilar welds made by placing the AA5754 alloy in the top of the joint will be labelled as D56, whereas the dissimilar welds made with the reverse base materials positioning will be designated as D65. In order to better understand the role of the dissimilar base materials flow on welds properties, three tools with different pin profiles and dimensions were used. Details on the used pin profiles, which were labelled according to their design (cylindrical - CL; conical - CN) and diameter, are displayed in Table 1. Following Costa et al., 2018, the tool rotational and traverse speeds were set equal to 600 rpm and 300 mm/min, respectively, the tool tilt angle was 2° and the tool plunge depth was 1.2 mm (positioning control). During the welding trials, the torque and force evolution was recorded [14].

Table 1. Geometries and dimensions of the pin profiles.

Tool designation	Cylindrical-CL6	Conical-CN6	Conical-CN8
Dimensions [mm]	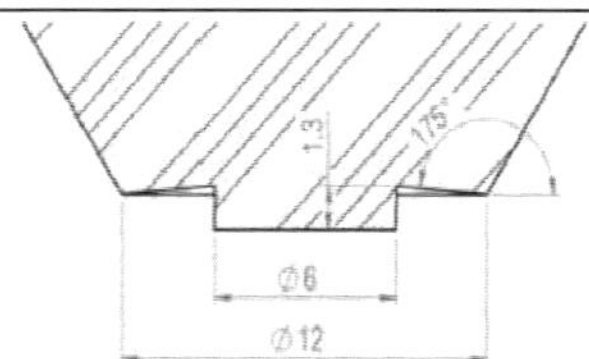	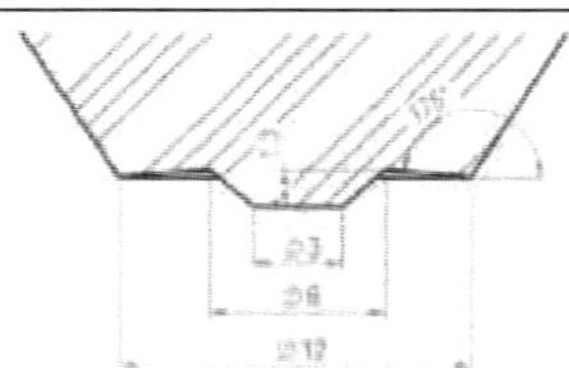	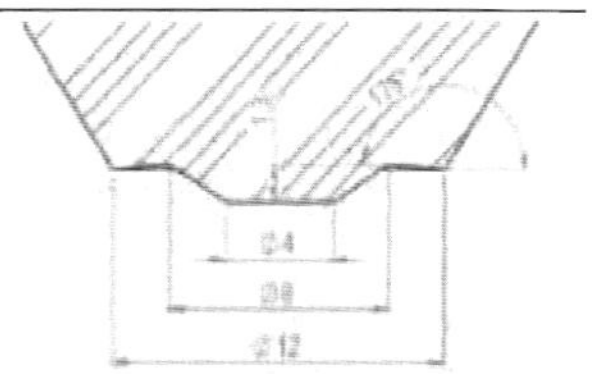

The cross-section of the welds was analysed in order to characterise the weld morphology. Keller's etchant was used to differentiate the flow of the different base materials. The mechanical properties of the dissimilar welds were evaluated by performing lap tensile-shear tests and microhardness measurements. The tensile-shear specimens were collected transversely to the welding direction. Advancing (AS) and retreating (RS) side loading schemes, defined based on Costa et al., 2018, were tested to assess the lap weld strength asymmetry [14].

3. Numerical simulation details

A thermal analysis of the lap welding process was done using finite element (FE) numerical simulation in order to correlate the temperature distribution in the similar joints with the precipitation behaviour of the AA6082 alloy. A simple FE FSW modelling approach was handled herewith by considering the AA6082 alloy as the only base material since the softening at the heat treatable alloy was the main phenomena of interest. FE modelling was done according to the procedures reported by Dialami et al., 2017 [15] and Andrade et al., 2020 [16]. The AA6082 alloy precipitation kinetics under the thermal conditions developed during welding was found using Thermocalc metallurgical software and correlated with the FE results.

4. Results and discussion

4.1. Mechanical properties

In order to understand the weld performance, tensile-shear testing was conducted by loading AS and RS specimens. The peak load values of the welds and the base materials are shown in Figure 1. From the figure it can be observed that all the welds were in under match when compared to both base materials. However, the D65 welds displayed higher load values than the D56 welds, which is especially noticed for the AS-loaded specimens. It can also be observed that the asymmetry in strength between the AS and the RS-loaded specimens is more evident for the D56 welds than for D65 welds. The RS-tested D56 specimens presented higher load values than the AS-tested specimens for all the used tools. On the other hand, the D65 weld produced with the CN8 tool performed well in both AS and RS loading conditions.

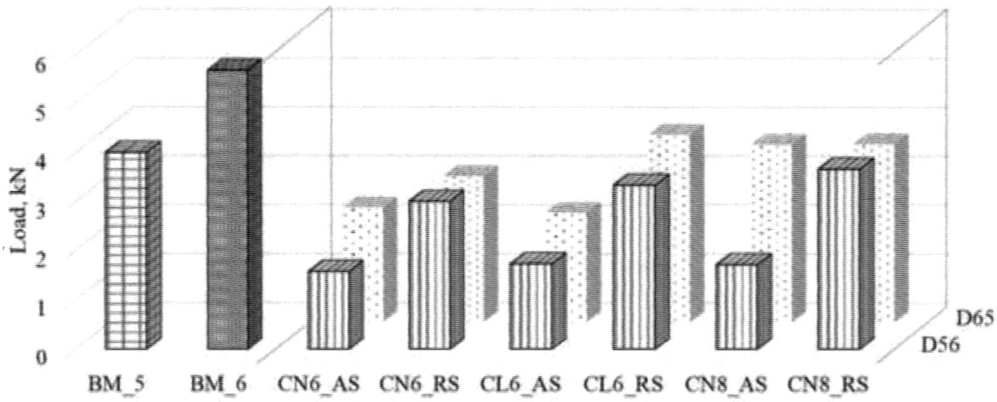

Figure 1. Tensile-shear properties of the welds and the base materials.

4.2. Force and torque analysis

Figure 2 displays the average values of Torque (Figure 2a) and Z-Force (Figure 2b) registered during welding. These values were computed considering only the steady-state stage of the process, i.e. discarding the initial and final instabilities. From Figure 2a, it can be observed that no important differences exist in the Torque values registered for the welds produced with different tools. As the part of the tool that drags most material during welding is the shoulder, which was similar for all the tools, it is easy to understand the non-variation in Torque among the different weld series.

Figure 2b shows that the Z-Force values varied according to both the tool and the base material positioning. Regarding the influence of the tool, lower Z-Force values were registered for the welds produced with the CN8 tool. This observation is agreeing with the previous study reported by Costa et al., 2015, where, axial force increases linearly with increasing shoulder plunge depth of different pin profiles [13]. So, with reference to that, it can say that the lower Z-Force of CN8 tool is attributed to the lower shoulder plunge depth. In which concerns to the positioning of the base materials, it can be observed that the D65 welds required lower Z-Force values than the D56 welds. Arora et al., 2011 related the evolution of the Z-Force with the material properties and reported the higher flow strength of the material as the reason for recording higher Z-Force values [17]. According to Leitão et al., 2012, the 5XXX aluminium alloys present higher flow strength at high temperature and strain rates than the 6XXX alloys [11]. Since the AA6082 is the most dragged material in the D65 welds, the lower Z-Force values agree-well with the mechanical behaviour of this alloy.

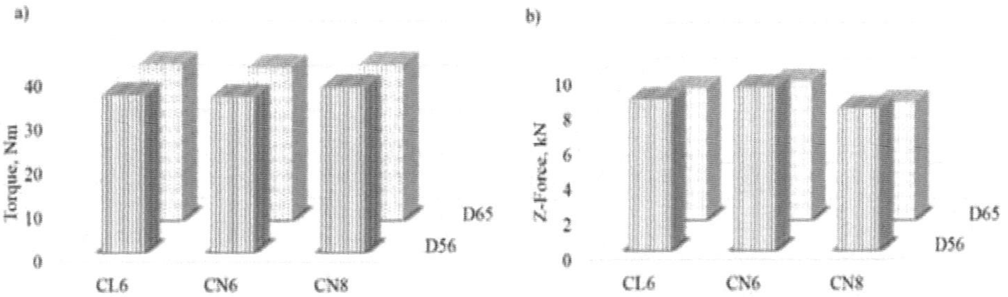

Figure 2. FSW machine process outputs: a) Torque; b) Z-Force.

4.3. Morphological characterisation

It is important to understand the influence of the flow strength and the positioning of the alloys on the material flow during welding. Figure 3 illustrates macrographs of the transverse cross-sections of the D65 (Figure 3a) and D56 (Figure 3b) welds produced with the CN8 tool. From Figure 3a, it can be observed that the cross-section of the D65 welds is free of defects. In turn, Figure 3b shows that hook formation occurred in the AS of the D56 welds. This agrees-well with the very low strength presented by these joints during AS loading, and therefore, with the strong asymmetry in strength between the AS and the RS. On the other hand, the smaller asymmetry in strength observed for the D65 welds is explained by the non-formation of defects.

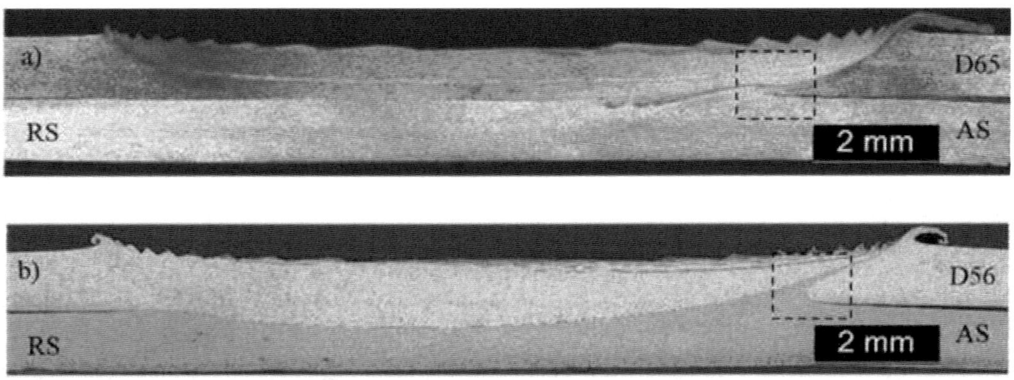

Figure 3. Macrostructure of the weld cross-sections: a) D65-CN8 weld; b) D56-CN8 weld.

Liu et al., 2016 explained that hook forms due to the upward material flow occurring at the AS [18]. Therefore, the lower hook height of the D65 welds clearly indicates that the placement of AA5754 plate at the bottom of the joint hinders the upward material flow. On the other hand, when the AA6082 alloy, which presents lower flow strength, is located at the bottom, the upward material flow is intense and hook formation is facilitated. The orientation of the hook along the thickness direction reduces the effective thickness of the top plate, and consequently, the load bearing area of the D56 AS-tested specimens.

4.4. Influence of base materials

In order to understand the differences in material flow between both dissimilar welds, the base materials softening mechanisms during welding should be understood. Figure 4a shows the hardness distribution along the longitudinal section of the top plate of the CN8 dissimilar welds. The central region of the graph, in which no microhardness values are displayed, corresponds to the pin zone, whereas the left and the right regions correspond to the trailing and the leading sides of the tool, respectively. From the figure, it can be observed that the microhardness values of the AA6082 alloy are under matched relative to the base material at both the trailing and the leading (pre-heat zone) sides of the tool. On the other hand, the microhardness of the AA5754 alloy is similar to that of the base material, or even higher, in some localised zones. These results agree-well with the previously-reported softening tendency of the AA6082 alloy with the temperature. Unlike the AA5754 alloy, the AA6082 alloy gets well softened before entering the processing zone, which may be one of the reasons for its higher flowability. So, placing the softer AA6082 at the bottom of the joint promotes the upward material flow, leading to a high hook height (Figure 3b), and consequently, affecting the strength of the welds (Figure 1).

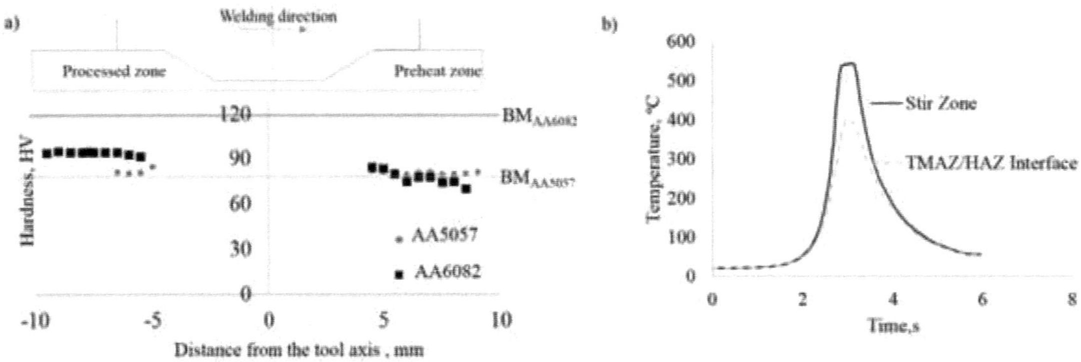

Figure 4. Thermomechanical characterisation: a) Microhardness profile along the welding direction; b) Thermal cycles obtained by FE simulation.

Table 2. Simulated dissolution temperatures of the precipitates.

Precipitate	GP Zone	β"	β'	β
Dissolution Temperature, °C	133 °C	325 °C	468 °C	530 °C

As illustrated in Figure 4b, where are shown the thermal cycles obtained by FE simulation, the peak temperatures in the stir zone and TMAZ/HAZ interface were of 553 °C and 400 °C, respectively. To get more insight about the softening mechanism of the AA6082 alloy, the possible precipitate evolutions and their respective dissolution temperatures were simulated and are displayed in Table 2. By correlating the thermal cycles and the precipitate dissolution temperatures, it is understood that the processed zone undergoes β dissolution whereas the temperature attained in the preheat zone or HAZ/TMAZ interface is not sufficient for β' dissolution and thereby it is assumed that β' coarsening may be occurred and yet it undergone β" dissolution. The precipitate dissolution/coarsening is exactly agreeing with the lower hardness regions of AA6082 alloy.

5. Conclusions

A coupled experimental and numerical research was developed to analyse the influence of the relative positioning of the base materials on the properties of very thin AA6082/AA5754 friction stir lap welds. From this research, the following conclusions can be drawn:

- The heat treatable AA6082 aluminium alloy exhibits lower flow strength and higher dynamic softening effect than the non-heat treatable AA5754 aluminium alloy during welding.
- The softening of the AA6082 aluminium alloy promotes upward material flow when this alloy is located at the bottom of the joint, which results in hook defect formation, and consequently, in the deterioration of the tensile-shear properties of the welds.
- Welds with a good mechanical behaviour can be achieved by positioning the AA5754 aluminium as the bottom plate in the joint. The higher flow strength and lower softening of the AA5754 alloy supress hook formation in these welds.

6. References

[1] Leitao C, Leal R M, Rodrigues D M, Loureiro A and Vilaça P 2009 *Mater. Des.* **30** 101–8
[2] Simar A, Jonckheere C, Deplus K, Pardoen T and De Meester B 2010 *Sci. Technol. Weld. Join.* **15** 254–9
[3] Jamshidi Aval H, Serajzadeh S and Kokabi A H 2011 *Mater. Sci. Eng. A* **528** 8071–83
[4] Wu L and Ferguson W G 2009 *Mater. Sci. Forum* **618–619** 203–6
[5] Summers P T, Chen Y, Rippe C M, Allen B, Mouritz A P, Case S W and Lattimer B Y 2015 *Fire Sci. Rev.* **4** 3
[6] Costa M I, Leitão C and Rodrigues D M 2017 *J. Mater. Process. Technol.* **250** 289–96
[7] Cederqvist L and Reynolds A P 2001 *Weld. J. (Miami, Fla)* **80** 281–287
[8] Z.W. Chen S Y 2012 *J. Achiev. Mater. Manuf. Eng.* **55** 629–37
[9] Cao X and Jahazi M 2011 *Mater. Des.* **32** 1–11
[10] Yazdanian S, Chen Z W and Littlefair G 2012 *J. Mater. Sci.* **47** 1251–61
[11] Leitão; C 2013 *Influence of Base Material Plastic Properties and Process Parameters on Friction Stir Weldability* (University of Coimbra, Portugal)
[12] Galvão I, Verdera D, Gesto D, Loureiro A and Rodrigues D M 2013 *J. Mater. Process. Technol.* **213** 1920–8
[13] Costa M I, Verdera D, Leitão C and Rodrigues D M 2015 *Mater. Des.* **87** 721–31
[14] Costa M I, Leitao C and Rodrigues D M 2018 *Soldag. Inspeção* **23** 32–42
[15] Dialami N, Cervera M, Chiumenti M and Agelet de Saracibar C 2017 *Int. J. Mech. Sci.* **122** 215–27
[16] Andrade D G, Leitão C, Dialami N, Chiumenti M and Rodrigues D M 2020 *Int. J. Mech. Sci.* **182** 105725
[17] Arora A, De A and DebRoy T 2011 *Scr. Mater.* **64** 9–12
[18] Liu H, Hu Y, Peng Y, Dou C and Wang Z 2016 *J. Mater. Process. Technol.* **238** 244–54

Acknowledgements

This research is sponsored by FEDER funds through Portugal 2020 (PT2020), by the Competitiveness and Internationalization Operational Program (COMPETE 2020) and national funds through the Portuguese Foundation for Science and Technology, under the projects: UID/EMS/00285/2020, POCI-01-0145-FEDER-00763 and Friction 4.0(POCI-01-0145-FEDER-032089). All supports are gratefully acknowledged.

Regression functions for determining the expression of surface Roughness at plane grinding of ceramic

H Bulea
Transilvania University of Brasov, B-dul Eroilor nr.29, 500036 Brasov, Romania

E-mail: bulea@unitbv.ro

Abstract. The aim of the work is to determine, through mathematical modelling using regression functions, the dependency expression between the surface roughness Ra at grinding with diamond wheels of aluminum oxide ceramic (Al2O3) for plain grinding with longitudinal advance, and the values of the grinding processing.

1. Mathematical model

Mathematical modelling of the data gathered through e experiments is used to determine the expression of roughness Ra=f(Sl,t,Vd), at plain grinding with longitudinal advance where: Sl (m/min) -longitudinal advance; t (mm) - depth of cut; Vd (m/s) - peripheral speed of diamond wheel, tested on the parameters of grinding, as function of multiple linear regression.

For determination of roughness expression at plain grinding, tested on the parameters of grinding, as a power function of the form like:

$$Y = e^{c} \cdot Sl^{x} \cdot t^{y} \cdot Vd^{z} \quad (1)$$

where: c an constant, x, y, z the exponents of the variables, the expression (1) is logarithmic, obtaining:

$$\ln Y = \ln\left(e^{c} \cdot X1^{x} \cdot X2^{y} \cdot X3^{z}\right) \quad (2)$$

which leads to obtaining the relationship:

$$\ln Y = c \cdot \ln e + x \cdot \ln X1 + y \cdot \ln X2 + z \cdot \ln X3 \quad (3)$$

which by identification with the general expression of multiple linear regression [1]:

$$Y = b0 + b1 \cdot X1 + b2 \cdot X2 + b3 \cdot X3 \quad (4)$$

leads to the relation:

$$c \cdot \ln(e) = bo \text{ which give } c = b0,\ x = b1,\ y = b2,\ z = b3 \quad (5)$$

where: b0, b1, b2, b3, Xl, X2, X3 [1]have the meanings given previously, and the function can be written as:

$$Y = e^{b0} \cdot X1^{b1} \cdot X2^{b2} \cdot X3^{b3} \quad (6)$$

IOP Conf. Series: Materials Science and Engineering **968** (2020) 012019 doi:10.1088/1757-899X/968/1/012019

2. Experimental components

The material used is a ceramic material with a high content of aluminum oxide, whose properties are [2]:Al_2O_3:99.5(%),density: 3450 (kg/m3),porosity: 1 (%), breaking resistance: 250 (MPa),compression strength: 1180 (MPa),flexion resistance 175 (MPa),reziliens: 2.9 (N/mm^2),hardness: 2000(Vikers), modulus of elasticity: 210 (MPa), thermal expansion module:8.3..-12.00 (10-6K^{-1}).

Processing was carried out on a grinding machine RPO-200, and measurements was performed on a roughness score FORM-Talysurf-120, and the diamond disk was 1-A1-150-10-4-D125-R-C75. Given the possibility of kinematic machine tool, the parameters of the system have changed in next order: [3] processing depth t (mm), longitudinal advance Sl (m/min), the longitudinal peripheral speed of the disc Vd (m/s), the values are[2]:

- depth of cut t: 0.02(mm); 0.03(mm); 0,04(mm); 0.05(mm); 0.06(mm); 0.07(mm)
- longitudinal advance Sl:1(m/min); 2(m/min); 2,5(m/min); 3(m/min)
- peripheral speed of diamond wheel Vd: 15(m/s); 23(m/s); 28(m/s); 31(m/s).

Table 1. Roughness R_a

t (mm)	S_l (m/min)	R_a (µm)			
		V_d=15 (m/s)	V_d=23 (m/s)	V_d=28 (m/s)	V_d=31 (m/s)
0.02	1	0.8431	0.7485	0.6524	0.6475
0.03	1	0.8564	0.7623	0.6595	0.6491
0.04	1	0.8827	0.7761	0.6687	0.6511
0.05	1	0.8912	0.7922	0.6732	0.6572
0.06	1	0.8953	0.7981	0.6996	0.6832
0.07	1	0.9058	0.8024	0.7036	0.6934
0.02	2	0.8614	0.7628	0.6727	0.6398
0.03	2	0.8725	0.7745	0.6812	0.6413
0.04	2	0.8963	0.7896	0.6983	0.6487
0.05	2	0.8975	0.7951	0.7075	0.6528
0.06	2	0.9012	0.8076	0.7181	0.6669
0.07	2	0.9427	0.8154	0.7234	0.7104
0.02	2.5	0.8763	0.7717	0.6793	0.6654
0.03	2.5	0.8824	0.7897	0.7021	0.6797
0.04	2.5	0.8979	0.7984	0.7153	0.6821
0.05	2.5	0.9233	0.8057	0.7334	0.7169
0.06	2.5	0.9357	0.8238	0.7645	0.7483
0.07	2.5	0.9421	0.8345	0.7892	0.7651
0.02	3	0.8862	0.7851	0.6859	0.6712
0.03	3	0.8998	0.7978	0.6908	0.6748
0.04	3	0.9371	0.8026	0.6993	0.6875
0.05	3	0.9547	0.8264	0.7367	0.7247
0.06	3	0.9772	0.8415	0.7782	0.7323
0.07	3	0.9964	0.8597	0.7956	0.7767

The parameter of surface roughness Ra was measured in three different areas of the workpiece surface processed, and the values taken into account is the average of the three measured values.

With each longitudinal advances Sl, and depth of machining t, have grinding to the four peripheral speed Vd of diamond disc, every time the measured parameter Ra under the conditions present in the previous chapter, the results are presented in Table 1 (roughness values representing the average of the measurements at the three points of measurement) [3].

IOP Conf. Series: Materials Science and Engineering **968** (2020) 012019 doi:10.1088/1757-899X/968/1/012019

3. Determination the Expression of Roughness at Plain Grinding Regression

To determine the expression of roughness Ra μm depending on the parameters of the cutting regime as a linear function of multiple regression as:

$$Ra=e^{c}\cdot Sl^{x}\cdot t^{y}\cdot Vd^{z} \tag{7}$$

where: x,y,z, Sl,t, Vd have the meaning mentioned, it solves the normal equations , are constituting the matrix B[4] of variables at the following form:

the expression (7) is logarithmed, obtaining:

$$\ln(Ra) = \ln(e^{c}\cdot Sl^{x}\cdot t^{y}\cdot Vd^{z}) \tag{8}$$

which leads to the relation

$$\ln(Ra) = c\cdot\ln e + x\cdot\ln(sl) + y\cdot\ln t + z\cdot\ln(Vd) \tag{9}$$

which by identification with the general expression of the multiple linear regression (4) leads to the relation:

$$c\cdot\ln(e)=b0 \text{ which gives } c=b0,\ x=b1,\ y=b2,\ z=b3 \tag{10}$$

where: b0, b1, b2, b3, Xl, X2, X3[4]: have the meanings given previously, and the function can be written as:the matrix of the variables being of the form[4]:

$$B = \begin{vmatrix} \ln(Sl11) & \ln(t11) & \ln(Vd11) & \ln(Ra11) \\ \ln(Sl21) & \ln(t21) & \ln(Vd21) & \ln(Ra21) \\ \dots & \dots & \dots & \dots \\ \ln(Sln1) & t\ln(n1) & \ln(Vdn1) & \ln(Ran1) \end{vmatrix} \tag{11}$$

With next notation:
X1:=B<0> - column with the values of logarithm of longitudinal advance Sl in the Table 1
X2:=B<1>- column with the values of logarithm of depth of cut t in the Table 1
X3:=B<2>- column with the values of logarithm of peripheral speed of wheel Vd in the Table 1
Y:=B<3>- column with the values of logarithm of roughness Ra in the Table 1

Determination of $Ra = e^{c}\cdot Sl^{x}\cdot t^{y}\cdot Vd^{z}$ have been made for next peripheral speed of the diamond disc:Vd = 15 (m/s),Vd = 23 (m/s),Vd = 28 (m/s),Vd = 31 (m/s) and are presented in the following subsections.

The term roughness $Ra = e^{c}\cdot Sl^{x}\cdot t^{y}\cdot Vd^{z}$ to the plane of the correction Vd = 15 (m/s).

For the determination of multiple linear regression coefficients, with the data are made matrix of the logarithm of variables and by solving the normal equation [1] were obtain the following values for multiple linear regression coefficients [5]:

$$\begin{bmatrix} b_0 \\ b_1 \\ b_2 \\ b_3 \end{bmatrix} = \begin{bmatrix} 1.943 \\ 0.054 \\ 0.070 \\ -0.680 \end{bmatrix}$$

and the regression expression would have the form:

$$Ra = e^{1.943}\cdot Sl^{0.054}\cdot t^{0.07}\cdot Vd^{z-0.68}\ (\mu m). \tag{12}$$

The term roughness $Ra = e^{c}\cdot Sl^{x}\cdot t^{y}\cdot Vd^{z}$ to the plane of the correction Vd=28 (m/s).

For the determination of multiple linear regression coefficients, with the data from Table 3 are made matrix of the logarithm of variables and by solving the normal equation [1] were obtain the following values for multiple linear regression coefficients[5]:

$$\begin{bmatrix} b_0 \\ b_1 \\ b_2 \\ b_3 \end{bmatrix} = \begin{bmatrix} 0.014 \\ 0.040 \\ 0.061 \\ -0.019 \end{bmatrix}$$

and the regression expression would have the form:

$$Ra = e^{0.014} \cdot Sl^{0.04} \cdot t^{0.061} \cdot Vd^{z-0.19}(\mu m). \tag{13}$$

The term roughness $Ra = e^c \cdot Sl^x \cdot t^y \cdot Vd^z$ to the plane of the correction Vd=28 (m/s).

For the determination of multiple linear regression coefficients, with the data from Table 3 are made matrix of the logarithm of variables and by solving the normal equation [1] were obtain the following values for multiple linear regression coefficients[5]:

$$\begin{bmatrix} b_0 \\ b_1 \\ b_2 \\ b_3 \end{bmatrix} = \begin{bmatrix} 0{,}152 \\ 0.074 \\ 0.091 \\ -0.730 \end{bmatrix}$$

and the regression expression would have the form:

$$Ra = e^{0.152} \cdot Sl^{0.074} \cdot t^{0.091} \cdot Vd^{z-0.73} (\mu m). \tag{14}$$

The term roughness $Ra = e^c \cdot Sl^x \cdot t^y \cdot Vd^z$ to the plane of the correction Vd = 31 (m/s).

For the determination of multiple linear regression coefficients, with the data from Table 1 are made matrix of the logarithm of variables and by solving the normal equation [1] were obtain the following values for multiple linear regression coefficients[5]:

$$\begin{bmatrix} b_0 \\ b_1 \\ b_2 \\ b_3 \end{bmatrix} = \begin{bmatrix} 2.600 \\ 0.064 \\ 0.109 \\ -0.768 \end{bmatrix}$$

and the regression expression would have the form:

$$Ra = e^{2.6} \cdot Sl^{0.064} \cdot t^{0.019} \cdot Vd^{z-0.768} (\mu m). \tag{15}$$

4. Evaluation of the roughness

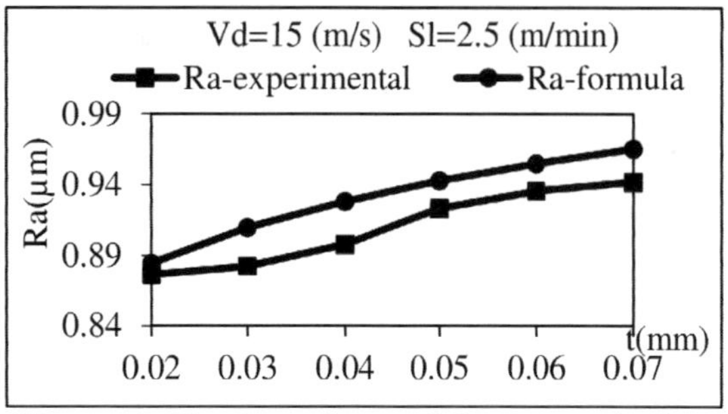

Figure 1. The dependence of the Ra at grinding with Vd =15 (m/s), and Sl =2.5 (m/min)

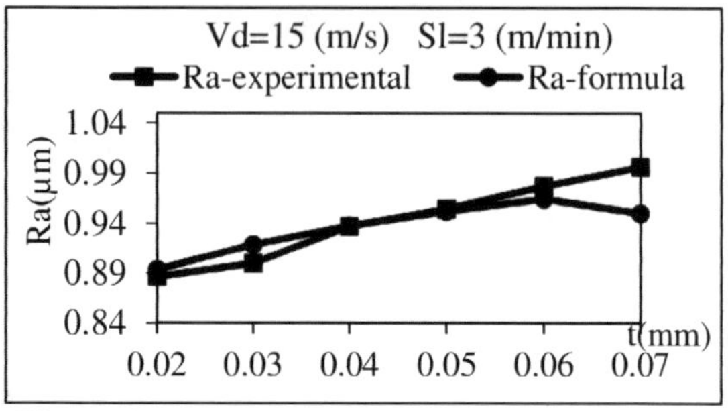

Figure 2. The dependence of the Ra at grinding with Vd =15 (m/s), and Sl =3 (m/min)

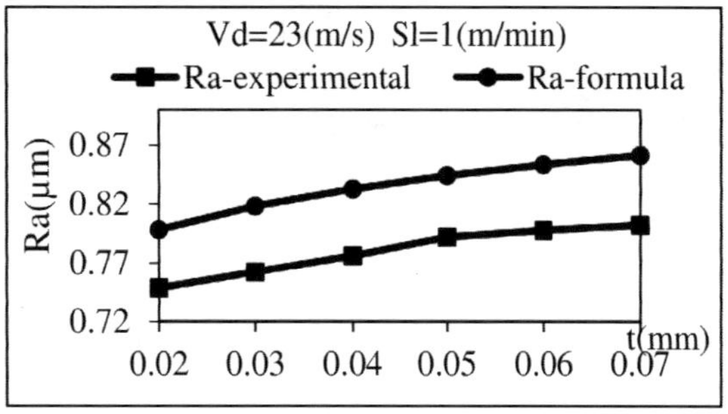

Figure 3. The dependence of the Ra at grinding with Vd =23 (m/s), and Sl =1 (m/min)

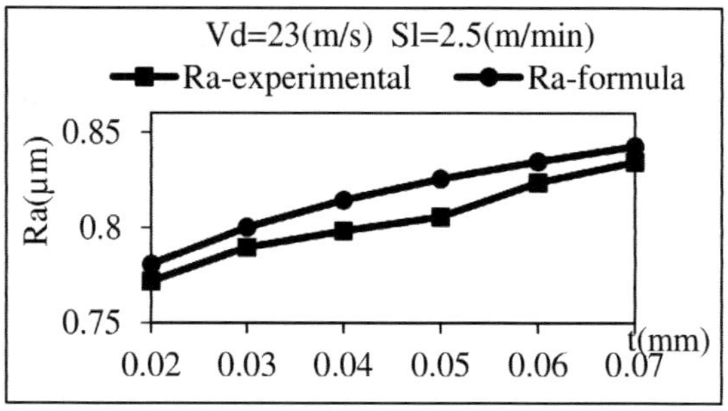

Figure 4. The dependence of the Ra at grinding with Vd =23 (m/s), and Sl =2.5 (m/min)

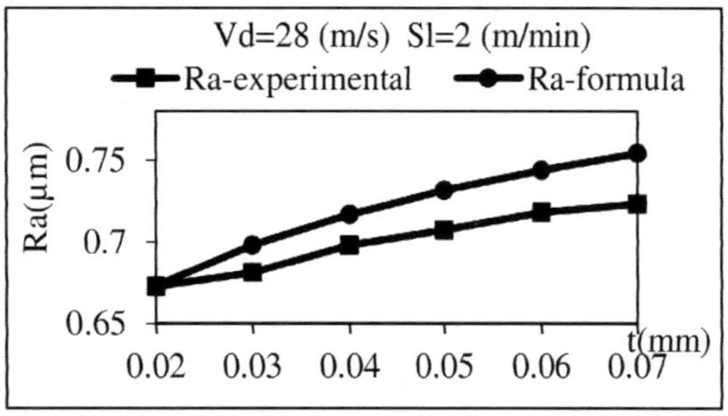

Figure 5. The dependence of the Ra at grinding with Vd =28 (m/s), and Sl =2(m/min)

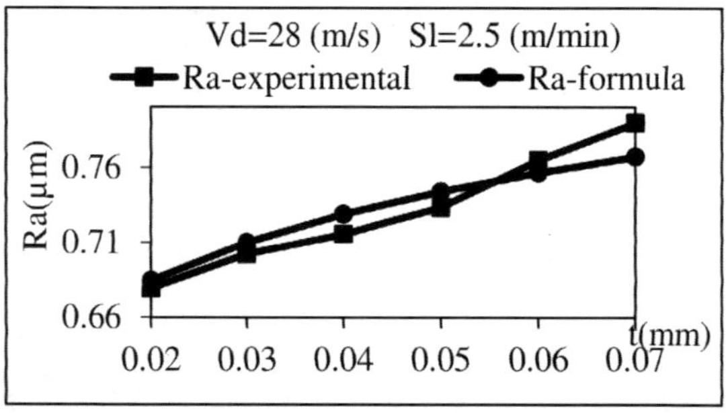

Figure 6. The dependence of the Ra at grinding with Vd =28 (m/s), and Sl =2.5(m/min)

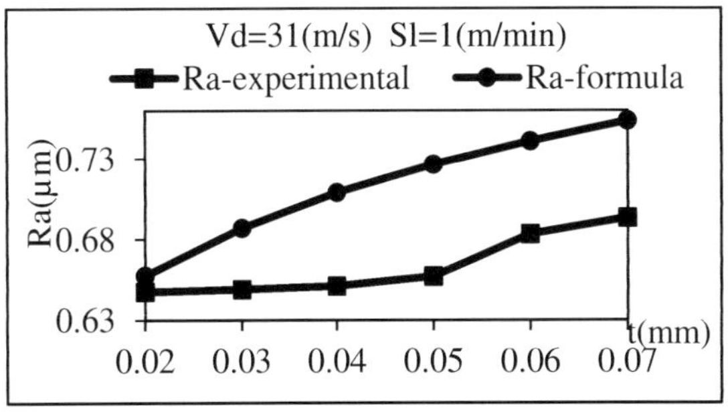

Figure 7. The dependence of the Ra at grinding with Vd =31 (m/s), and Sl =1(m/min)

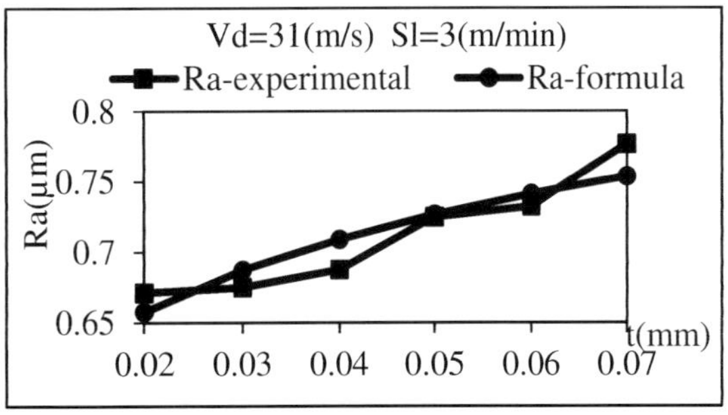

Figure 8. The dependence of the Ra at grinding with Vd =31 (m/s), and Sl =3(m/min)

5. Conclusions

The analysis of the diagrams presented in Figures 1 … 8 can lead to the following conclusions:

1. The variation curves of the roughness Ra analytically deducted, approximate better the experimental results for the peripheral speed of the diamond wheel of Vd=28(m/s) and Vd =15 (m/s) than for the speeds of Vd=31(m/s) Vd =23 (m/s);
2. The variation curves of the roughness Ra analytically deducted, approximate better the experimental results for the longitudinal advance Sl =2 (m/min) and Sl =2.5 (m/min);
3. The variation curves of the roughness Ra analytically deducted, approximate better the experimental results for the depth of cut t = 0.02 mm;
4. The variation of the roughness for the analytical curves is between 0.87-0.99 µm for the peripheral speed of the diamond wheel of Vd = 15 (m/s) is between 0.8-0.88 µm for the peripheral speed of the diamond wheel of Vd = 23 (m/s), between 0.65-0.78 µm for the peripheral speed of the diamond wheel of Vd = 28 (m/s) and between 0.66-0.77 µm for the peripheral speed of the diamond wheel of Vd = 31 (m/s);
5. The variation of the roughness of the analytical roughness decreasing when the peripheral speed of the diamond wheel increase from Vd = 15 (m/s) to Vd = 31 (m/s);
6. The variation of the analytical roughness increases steadily with increasing of depth of cut of processing;
7. The variation of the analytical roughness increases steadily with increasing of longitudinal advance Sl of processing;
8. It is recommended to be used the following expressions to determine the roughness at the plain grinding of the aluminum oxide:

$Ra = e^{1.943} \cdot Sl^{0.054} \cdot t^{0.07} \cdot Vd^{z-0.68}$ (µm) to the plane of the correction Vd = 15 (m/s);
$Ra = e^{0.014} \cdot Sl^{0.04} \cdot t^{0.061} \cdot Vd^{z-0.19}$ (µm) to the plane of the correction Vd = 23 (m/s);
$Ra = e^{0.152} \cdot Sl^{0.074} \cdot t^{0.091} \cdot Vd^{z-0.73}$ (µm) to the plane of the correction Vd = 28 (m/s);
$Ra = e^{2.6} \cdot Sl^{0.064} \cdot t^{0.019} \cdot Vd^{z-0.768}$ (µm) to the plane of the correction Vd = 31 (m/s).

6. References

[1] Montgomery D C 2012 Introduction to linear regression analysis pp 13-20.
[2] Bulea H 2019 Mathematical Model Using Regression Functions for Determining the Expression of Surface Roughness at Plain Grinding with Diamond Wheels of Aluminum Oxide Ceramic Al_2O_3 *Jurnal BDI Recent* pp 64-69
[3] Bulea H 1999 Optimization Technology for processing of minerals and ceramic parts materials *Transilvania University of Brasov* (Braşov) p144-150
[4] Constantinescu L 1980 Processing of experimental data with numerical computersnother reference (București: Ed Tehnică) pp 121-123
[5] *** 1998 Mathcad 7 Professional *Chapter 1*

Influence of vertical step on forces and dimensional accuracy of SPIF parts – a numerical investigation

M O Popp, G P Rusu, V Oleksik and C Biris
Lucian Blaga University of Sibiu, Department of Industrial Machinery and Equipment, Emil Cioran Street No. 4, Romania

E-mail: gabriela.rusu@ulbsibiu.ro

Abstract. Single point incremental forming (SPIF) is a new flexible sheet metal forming process characterized by low costs and the possibility to produce prototype parts without the need for a specific die. This is one of the reasons why this process is nowadays used for manufacturing of highly customized small series parts. The process involves the usage of a hemispherical punch which gradually deforms the sheet metal blank fixed by two simple clamping rings, by following a path until the final shape of the product is obtained. The aim of this paper is to investigate and analyze the influence of the vertical step over forces involved in the process and obtained geometrical accuracy, which is one of the main drawbacks for large scale implementation of the process. A numerical analysis was carried out through finite element method with different step size for frustrum pyramid shaped parts made from the same material. In this way, the most appropriate vertical step can be chosen for further experimental research in order to obtain the most accurate parts and with as little stress as possible on the equipment involved in the process.

1. Introduction

Current trends in the automotive industry and not only aim to reduce costs and production time as much as possible, so new methods are needed to keep up with the demand. If until now, when a new car concept appeared, it was necessary to modify some components of the car body, the engineers tried to optimize a mold as well as possible, and later different defects of the part were highlighted and this whole process was resumed again. Now with the help of the incremental forming process they can produce prototypes at a much lower cost given that this process does not require a mold, but only two retaining rings for the sheet and a punch to deform it. This process receives a lot of attention among researchers to be as optimal as possible, and the deformed parts to have the best possible accuracy and as few defects as possible.

During the incremental forming process, a punch follows a predefined toolpath gradually deforming the sheet fixed with the help of two retaining rings. Figure 1 shows the principle of single point incremental forming process. For implementation of this process it is suitable to use numerically controlled machines due to the high stiffness and rigidity of the machines [1, 2] or industrial robots [3] due to the high flexibility of the shapes that can be made with them [4, 5]. Some researchers tried to build their own equipment for this process such as in paper [6]. The toolpath design can be done with the help of CAM software dedicated for milling applications or other new custom software proposed by researchers which generates toolpaths like in paper [7].

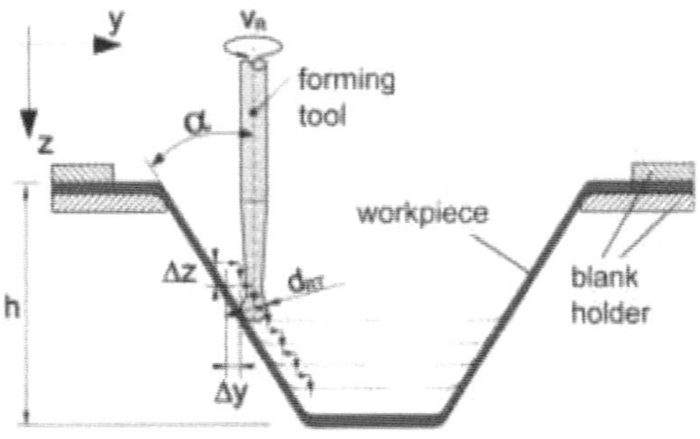

Figure 1. Single point incremental forming process [8].

The incremental forming process has some important advantages, such as: low production costs [9], high flexibility and customization of parts [10] and the material deformation is high [11] and suitable for prototype size production [12]. However, this process comes with some disadvantages that prevent its industrialization and implementation in large series production. One of these disadvantages and perhaps even the most important is given by the low accuracy of the parts manufactured by this process, regardless of the variant of forming procedure (with punch / total or partial die / multipoint / etc.).

Researchers have shown a concern for various factors that may influence the dimensional and shape accuracy of the part, but also on its roughness. The toolpath that the punch follows can have different influences on the deformation of the sheet as highlighted by Verbert with his experiments performed in paper [13]. Among the parameters studied by them is the effect of temperature on the deformability of the material, a problem that was investigated by Duflou in paper [14] where he explained how to perform the process of incremental forming with a laser in order to observe the behavior of the material in these conditions. The influence of lubrication for this process is also important to be studied. The use of lubricants is essential for the contact between two or more components of a mechanism and any deformation process because the life of the punch can be extended due to reduced friction and wear, and also improved heat distribution [15]. The vertical step size of the punch is another parameter which is influencing the accuracy of the part shape and can cause the phenomenon of waves that can be seen in figure 2.

The purpose of this paper is to study the influence of the vertical step size on the accuracy of the part and on the forces, which take place during single point incremental forming process. In the paper done by Lu and his collaborators are presented differences between the conical parts made by incremental forming process with the size of vertical steps of 0.1 mm, 0.6 mm and 1.4 mm. The material used by the researchers is A 7075-O, and the results obtained showed that the parts for which the 0.1 mm vertical step was used have a better roughness and are more accurate. [16].

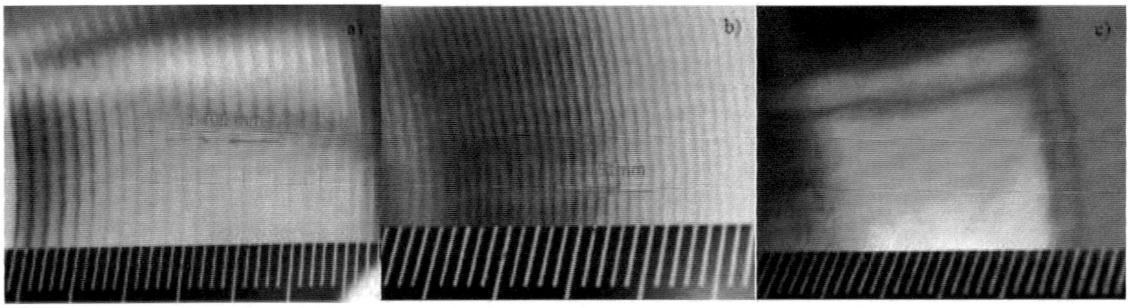

Figure 2. Waves effect on sheet metal according to vertical step size [16].

2. Numerical investigation

For the investigation of the influence of the vertical step size on the single point incremental forming process the sheet metal would be deformed in the form of a frustrum pyramid with the large base size 106mm, depth of 20 mm and the wall angle of 60°. In the first stage of the research that are presented in this paper four finite element analysis were performed with only one parameter varied that is the vertical step size. The FEM model is presented in figure 3 and consists of the following parts: two discrete rigid retaining rings, a analytical rigid tool and the deformable sheet metal which is the interest part. The undeformed shape of the sheet has the size of 200 x 200 mm and a thickness of 0.6 mm. The mesh of the steel sheet blank was performed using S4R elements (an 4-nodes, quadrilateral, stress / displacement shell element with reduced integration and a large-strain formulation) [17] with 6 points of integration. The blank used is made of structural steel S235 which according to the European standard EN10025 has the mechanical properties presented in Table 1. The material characteristics needed for finite element analysis were previously obtained by the tensile test, thus determining the plastic curve of the material.

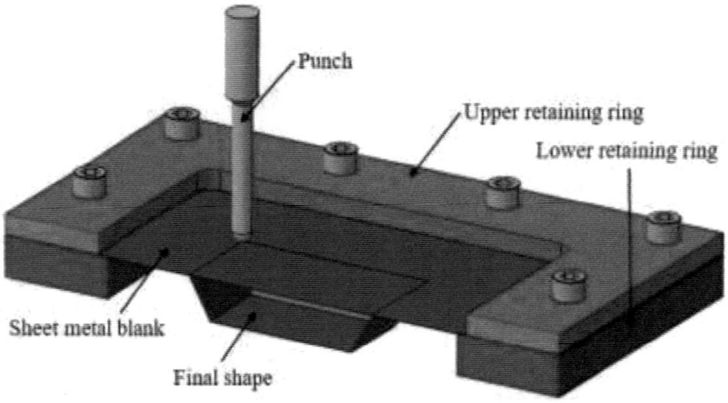

Figure 3. Analytical model with 60° wall angle frustrum pyramid.

Table 1. Mechanical properties of structural steel S235.

Material	Yield stress Rp0.2 [MPa]	Tensile strength Rm [MPa]	Elongation A [%]
S235	235	360 – 510	26

The two retaining rings are encastered so the sheet metal is held fixed and the tool has applied a displacement in all the directions, so it follows the trajectory necessary to create the frustrum pyramid.

3. Results and discussion

This section presents the results obtained from the finite element analyses that were performed. In figure 4 the profiles of the pieces are presented in comparison to the theoretical profile of the frustrum pyramid. It can be observed that in the case of the toolpath that uses the vertical step of 0.25 mm the profile is closer to the theoretical profile and that the sheet bending is much smaller compared to the other three variants.

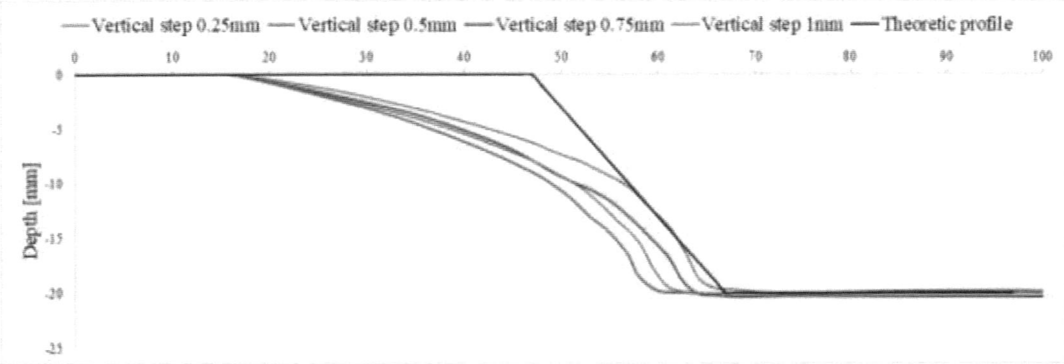

Figure 4. Comparison between analytical profiles and theoretical profile of frustrum pyramid.

Table 2 shows the differences in part accuracy obtained between the values of the vertical step and it can be seen that the most favorable case is when the step size of 0.25 mm is used.

Table 2. Springback, thickness reduction and maximum force on Z axis (Fz).

Vertical step [mm]	Springback [mm]	Thickness reduction [%]	Maximum value of the force measured on vertical axis Z [N]
0,25	0,87	55	1182
0,5	0,88	56.7	1287
0,75	0,83	56.7	1389
1	0.21	56.7	1600

Figure 5 presents the principal strains for all the vertical step size that were simulated and it can be observed that if the step is smaller the strains are bigger.

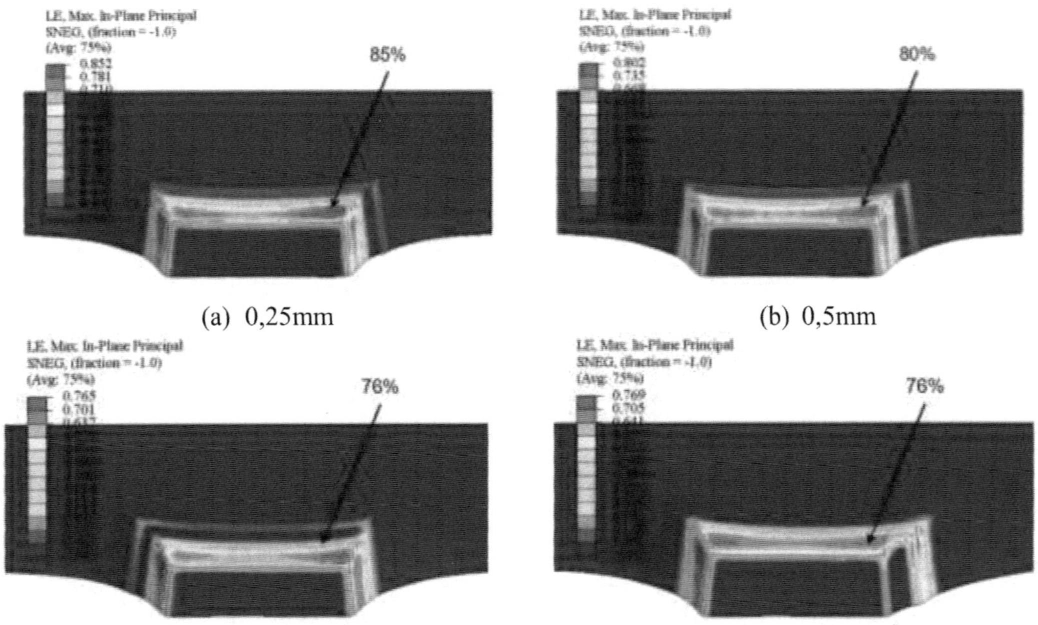

(a) 0,25mm (b) 0,5mm

(c) 0,75mm (d) 1mm

Figure 5. Principal strains for all the vertical step size.

IOP Conf. Series: Materials Science and Engineering **968** (2020) 012020 doi:10.1088/1757-899X/968/1/012020

The forces distribution in relation with the depth of the frustrum pyramid are shown in figure 6 and it can be observed that the smallest size of the vertical step presents the smallest F_z forces whereas the 1 mm step size presents the highest forces on Z axis. The results show that the forces are reduced with 26% in the case when the step size is 0.25mm as against of 1mm. This means that the forces for the minimum step size are reduced with a quarter of the forces of the maximum size step.

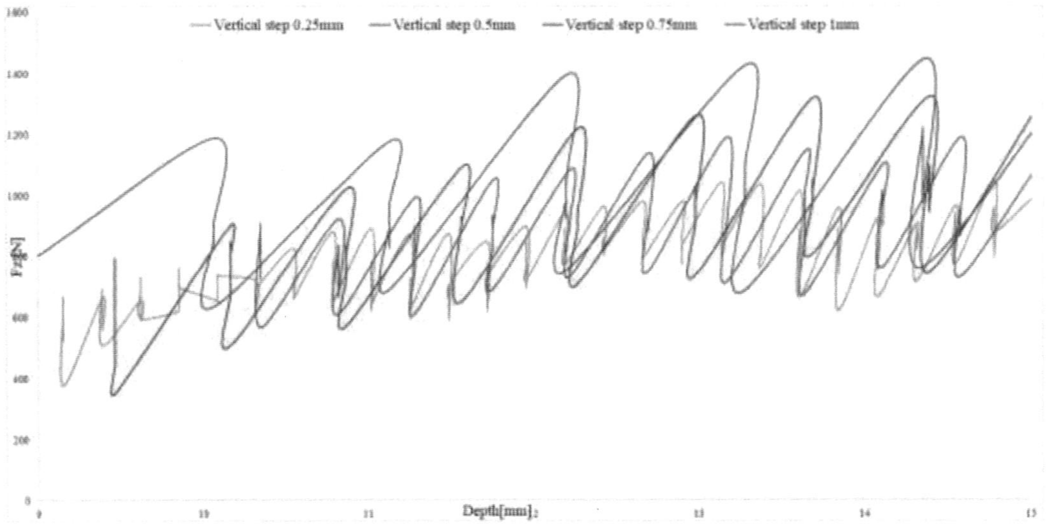

Figure 6. F_z distribution over depth of frustrum pyramid.

Thickness reduction is represented in figure 7 for the most favorable case of the vertical step, 0.25mm.

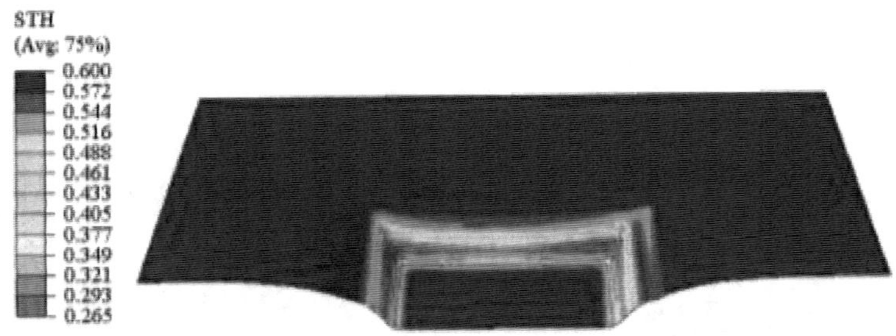

Figure 7. Thickness reduction for the vertical step size of 0,25mm.

4. Conclusions and further research

As can be seen from the results obtained from the numerical investigations performed using the ABAQUS / EXPLICIT finite element analysis for the 4 variations of the vertical step: 0.25 mm, 0.5 mm, 0.75 mm and 1 mm; the smallest vertical step brings the most advantages in terms of dimensional and shape accuracy, but also in the case of the forces obtained. Following these analyzes, it can be concluded that the smaller the vertical step is, the higher the precision of the part is and the forces can be reduced in the case of 0.25 mm to almost half of the values obtained in the case of 1 mm step size, and with the help of the research carried out in the field, it can be shown that a big vertical step size may affect the roughness of the part in such a way that it can be observed with the naked eye. In addition, it can be observed that by using different vertical steps there is no major influence on the thinning of the material.

Regarding future research, the authors of this article propose to verify the influence of various factors on the accuracy of the shape and size of the part using a KUKA KR210 industrial robot which is in the laboratories of the "Center for Studies and Research for Plastic Deformations" from University Lucian Blaga of Sibiu.

5. References

[1] Tera M, Breaz R, Racz S and Girjob C 2019 Processing strategies for single point incremental forming – a CAM approach *Int. J. Adv. Manuf. Technol.* **102** pp 1761-1777
[2] Bologa O, Breaz RE and Racz GS 2018 Using the Analytic Hierarchy Process (AHP) and fuzzy logic to evaluate the possibility of introducing single point incremental forming on industrial scale *Procedia Computer Science* **139** pp 408–416
[3] Breaz RE, Bologa O and Racz GS 2017 Selecting between CNC milling, robot milling and DMLS processes using a combined AHP and fuzzy approach *Procedia Computer Science* **122** pp 796–803
[4] Crenganis M and Csiszar A 2019 A Dynamic Model for KUKA KR6 *in SPIF Processes Material Science Forum* **957** pp 156 – 166
[5] Bârsan A 2019 A Brief Review of Robotic Machining *Acta Universitatis Cibiniensis* **71**
[6] Allwood J M, Houghton N E and Jackson K P 2005 The design of an incremental sheet forming machine *Advanced Materials Research* **6-8** pp 471–478
[7] Tisza M 2012 General overview of sheet incremental forming *Journal of Achievements in Materials and Manufacturing Engineering* **55** pp 113–120
[8] Nasulea D and Oancea G 2018 Integrating a New Software Tool Used for Tool Path Generation in the Numerical Simulation of Incremental Forming Processes *Journal of Mechanical Engineering* **64** pp 643-651
[9] Micari F, Ambrogio G and Filice L 2007 Shape and dimensional accuracy in Single Point Incremental Forming: State of the art and future trends *Journal of Materials Processing Technology* **191** pp 390–395
[10] Filice L, Ambrogio G and Gaudioso M 2013 Optimised tool-path design to reduce thinning in incremental sheet forming process *International Journal of Material Forming* **6** pp 173–178
[11] Oleksik M 2018 Comparative study about different experimental layouts used on single point incremental forming process *Acta Universitatis Cibiensis. Technical Series* **70**
[12] Breaz R E and Racz S G 2019 Considerations Regarding the Industrial Implementation of Incremental Forming Process *Materials Science Forum* **957** pp 111–119
[13] Verbert J, Duflou J R and Lauwers B 2007 Feature Based Approach for Process *Key Engineering Materials* **344** pp 527–534
[14] Duflou J R, Callebaut B, Verbert J and De Baerdemaeker H 2007 Laser Assisted Incremental Froming: Formability and Accuracy Improvement *Formability and Accuracy Improvement. CIRP Annals* **56** pp 273–276
[15] Tisza M and Fulop T 2001 A general overview of tribology of sheet metal forming *Journal for Technology of Plasticity* **6** pp 11–25
[16] Lu H, Li Y, Liu Z, Liu S and Meehan P A 2014 Study on step depth for part accuracy improvement in incremental sheet forming process *Advanced Materials Research* **939** pp 274-280
[17] *** Abaqus documentation

Acknowledgments
This research presented here was partially financed by a grant of the Romanian Ministry of Research and Innovation CCCDI-UEFISCDI, project number PN-III-P1-1.2-PCCDI-2017-0446/ nr. 82 PCCDI/2018, within PNCDI III, project title: "Intelligent manufacturing technologies for advanced production of parts from automotive and aeronautics industries".

IOP Conf. Series: Materials Science and Engineering **968** (2020) 012021 doi:10.1088/1757-899X/968/1/012021

Analysis of the influence of the hydrostatic ball burnishing pressure in the surface hardness and roughness of medium carbon steels

P Fernández-Lucio[1], H González-Barrio[1], G Gómez-Escudero[1], O Pereira[2], L N López de Lacalle[2] and A Rodríguez[2]

[1] Department of Mechanical Engineering, University of the Basque Country (UPV/EHU) Faculty of Engineering in Bilbao, Plaza Torres de Quevedo s/n, 48013 Bilbao, Spain.

[2] Aeronautics Advanced Manufacturing Center, University of the Basque Country (UPV/EHU), Parque Tecnológico de Bizkaia, 48170 Zamudio, Spain.

E-mail: pablo.fernandezd@ehu.eus

Abstract. Standing sectors in the industry such as railroad or plastic injection moulds present~~s~~ many challenges for manufacturing complex components in terms of finishing requirements and mechanical properties. Because of that, hydrostatic ball burnishing is considered an optimal solution since it reduces surface roughness and generates hardened surfaces and compressive residual stresses, increasing the performance and lifespan of this parts in terms of resistance and mechanical fatigue. Additionally, this technology could be integrated directly into the machining centres, what reduces times and dimensional errors arising from mooring changes. Therefore, lead times and production costs can be drastically reduced in comparison with other finishing techniques. The aim of this project is to analyse the use of the ball burnishing process to improve the final quality of medium carbon steel surfaces, minimizing surface roughness and improving hardness. In order to achieve significant results, different working pressures are analysed in terms of surface roughness and hardness in AISI 1045 and AISI P20 steels. The results showed a reduction in roughness parameters of more than 89% for both materials using a pressure of 10 MPa. Moreover, at a pressure of 30 MPa, the surface hardness has been increased by 15% and 34% respectively.

1. Introduction

With the aim of maintaining competitiveness in a globalized world, the different industrial sectors are forced to be in a permanent development, even the most traditional sectors such as the steel industry. In fact, the European Union has developed different plans in many different sectors to enhance this objective. This is the case of the railroad industry or the plastic injection moulds sector, among others.

Many of the components of these industries must withstand high cyclical working loads during their lifespan, for example railway axles (according to the UNE-EN 13103-1:2019). Indeed, high geometrical precision and roughness adjustments are required to ensure the adequate functions of the components. Therefore, many of them are considered critical from very high-added value parts.

Hence, to achieve these demanding requirements, different superfinishing techniques are used. Some of them are the hydrostatic ball burnishing ~~or~~ and the polishing. In fact, within the superfinishing technologies, hydrostatic ball burnishing offers greater advantages over the rest. Firstly, this technique can be applied directly on the same machine where the finishing process is performed. This implies a reduction in workpiece transportation times and, in a more important manner it removes positioning and fixing errors in multi-machine processes, something that is essential in components with restrictive geometrical requirements. Furthermore, it is a process that consists on material deformation, avoiding cutting. It is based on rolling a hard metal or ceramic ball under pressure on a surface. However, the main limitation of this process is geometrical. Due to the geometry of the tool, the angle between the normal direction of the burnished surface and the tool axis must be between ±28°. In figure 1 it can be seen a scheme of the transmitting force to the burnished surface. In this operation, the speed of the tool in relation to the surface can reach the maximum allowed by the machine.

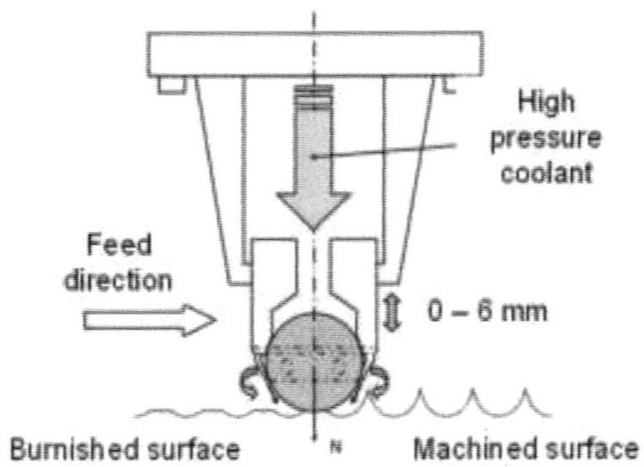

Figure 1. Scheme of the transmitting force [1].

By means of this plastic deformation, a displacement of the material in cold is produced, causing peaks of roughness to fill the hollows of the valleys, thus reducing the surface roughness [2]. The main characteristic of this technique lies on the fact that, in addition to improving the quality of the surface, it also generates compressive residual stresses, what increase performance and service life. This implies an improvement of the mechanical fatigue resistance, being very beneficial for components that are subjected to high loads during many work cycles [3].

In fact, *Travieso-Rodríguez et al.* [4] concluded that, when applying ball burnishing to AISI 1038 steel, the surface hardness of the workpiece is enhanced. Moreover, this improvement resulted in improving mechanical fatigue resistance to failure. In this sense, *Hua et al.* [5] claim that a 15% increase in the surface hardness in Inconel® 718 means an 83% improvement in mechanical fatigue failure.

The key to the correct implementation of hydrostatic ball burnishing lies in the influence of the process parameters on the workpiece and the appropriate choice according to the desired result. In line with this, *Saldaña-Robles et al.* [6] concluded that, using the right ball burnishing parameters, AISI 1045 steel surface roughness can be improved in more than an 80%. However, changing those burnishing parameters, the surface hardness can be optimized up to a 14%. In the literature, researchers agree that, in the case of obtaining an improvement in surface roughness and hardness, the most relevant parameters are burnishing pressure and force, ball diameter and burnishing speed.

The burnishing pressure corresponds to the pressure exerted by the fluid inside the tool and it is the responsible of controlling the burnishing force applied to the workpiece. An increase in the pressure means a significantly increase of the reached depth by the ball, although there is a limit beyond which there is no appreciable improvement [1].

Related to the roughness values, researchers establish different points of view. On the one hand, the study by *Luca et al.* [7] points out that the pressure is not excessively relevant for roughness values. On the other hand, *El-Taweel and El-Axir* [8] and *Swirad and Wdowik* [9] established that a higher burnishing force reduces the surface roughness, provided that a limit is not exceeded beyond which the roughness increases again. Since higher pressure also means higher burnishing force and vice versa, it is not clear the level of influence of this parameter on the final roughness. However, researchers agree that the greater the burnishing pressure is, the greater the surface hardness ups to a limit.

Furthermore, the diameter of the burnishing ball affects to the forces generated in burnishing, the larger the ball diameter is defined, the greater the area of ball-surface contact is obtained and, thus, the greater the force appears for the same pressure. This results in a higher surface hardness and, consequently, higher residual stresses, but with bigger ball trucks on the final surface. Hence, a compromise should be reached.

Finally, the increase in burnishing speed enhances surface hardness and minimizes the roughness up to a limit beyond which the surface finish deteriorates. Nonetheless, due to the fact that the optimal speed range is elevated, it is not as relevant as the burnishing pressure or force. Furthermore, it affects directly in the machining process [10].

Thereby, in this study, an analysis of the influence of pressure during the hydrostatic ball burnishing process on the surface roughness and hardness is performed for two different medium carbon steels, AISI 1045 and AISI P20, commonly used in railways axles and plastic injection moulds respectively. In the following table the main properties and the chemical composition of both materials are shown.

Table 1. Chemical composition and mechanical and physical properties.

Chemical Composition (%)								
Material	C	Mn	P	S	Cr	Mo	Si	Fe
AISI 1045	0.45	0.73	0.04	0.05	-	-	-	Balanced
AISI P20	0.40	1.50	0.01	0.01	1.90	0.20	0.30	Balanced

Mechanical and physical properties					
Material	Hardness	Young´s Modulus	Tensile Strength	Density	Poisson´s Component
AISI 1045	212 HB	200 GPa	413 MPa	7.85 g/cm^3	0.29
AISI P20	290 HB	207 GPa	862 MPa	7.86 g/cm^3	0.30

2. Experimental set up

The tests were carried out in a KONDIA® HS1000 machine centre with 18 kW of power in which both milling and hydrostatic ball burnishing were made on prismatic geometry test pieces of 20×20×10 millimetres.

In order to see the influence of the ball burnishing process, the surface of the workpieces was milled and used as a reference for the measurements. For this operation, a PPH 1600-CL1 quality 2003 was used and the cutting parameters can be seen in table 1. This cutting parameters are commonly used for finishing operations in medium carbon steels.

Table 2. Milling cutting parameters.

Diameter [mm]	**z [-]**	**v_c [m/min]**	**f_z [mm/(r*z)]**	**a_p [mm]**	**a_e [mm]**
16	2	250	0.05	0.20	0.20

For the hydrostatic ball burnishing process, a special burnishing tool was used with a 6 mm diameter ceramic ball (HG6-19E90-ZS20-X) as well as a hydraulic pump HGP 6.5 Ecoroll. Both, the tool and the hydraulic pump, were from the same manufacturer. The burnishing speed was stablished in 2000 mm/min for both steels and the feed direction was perpendicular to the milled direction in order to minimize the surface roughness. The tested burnishing pressures were 10 and 30 MPa.

IOP Conf. Series: Materials Science and Engineering **968** (2020) 012021 doi:10.1088/1757-899X/968/1/012021

3. Results and Discussion

Topographies of all the workpieces were obtained. Furthermore, surface roughness and hardness for each material was measured and compared with the reference milled ones.

3.1. Surface roughness

Workpiece surface topographies were obtained with a confocal optical microscope Leica DCM3D with a resolution of 0.1 nm. From those topographies, three random profiles were analysed obtaining, according with the ISO 4287:1997, R_a and R_z parameters.

In figure 2 the topographies of AISI 1045 steel are shown.

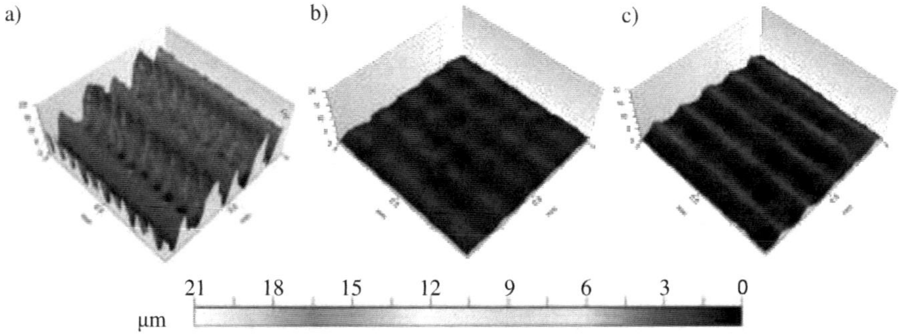

Figure 2. Topographies of the AISI 1045 surfaces. (a) Reference, (b) 10 MPa burnishing and (c) 30 MPa burnishing.

It can be seen that, with the application of hydrostatic ball burnishing, the peaks resulting from the milling operation are notably reduced with both pressures. However, between them the appreciation is not as clear. In fact, due to the higher waves that can be seen, it seems that the burnished surface with 30 MPa has exceeded the pressure threshold that optimizes the surface roughness.

In figure 3 it can be seen the surface topographies of AISI P20 steel. As in the case of AISI 1045, it can appreciate a great improvement in the surface quality.

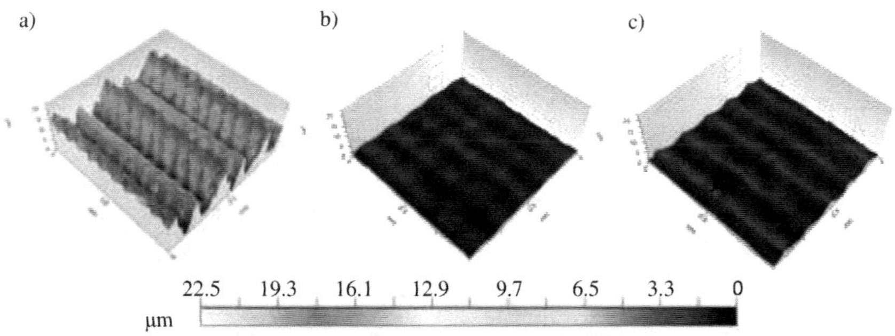

Figure 3. Topographies of the AISI P20 surfaces. (a) Reference, (b) 10 MPa burnishing and (c) 30 MPa burnishing.

The high peaks left by the milling operation almost disappeared after the hydrostatic ball burnishing process, leaving small waves in the tested surface. Moreover, as has been the case with AISI 1045, the smaller pressure left a smoother surface than the one left by burnishing with a pressure of 30 MPa. Figure 4 shows the comparison between the reference surface and the burnished surfaces at 10 and 30 MPa of the roughness parameters, R_a and R_z. It can be seen that the hydrostatic ball burnishing process decrease both parameters significantly in both AISI 1045 and AISI P20.

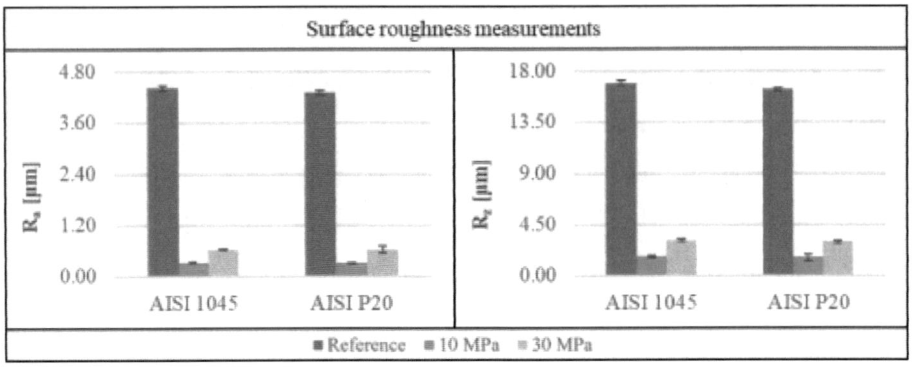

Figure 4. R_a and R_z roughness parameters for both steels.

In AISI 1045 steel, the hydrostatic ball burnishing at a pressure of 10 MPa reduce in more than a 90% and 85% the roughness parameters R_a and R_z respectively. However, when the pressure was of 30 MPa, that reduction was not as huge as in the previous case (84% and 80% respectively). Therefore, ball burnishing process in this steel at a pressure of 30 MPa has exceeded the threshold that optimizes the surface roughness.

Similar thing happened with AISI P20 steel. Nonetheless, in this case the reduction percentage for both pressures and parameters was a little bit higher. While R_a parameter was reduced in a 92% and 85% at pressures of 10 and 30 MPa respectively, R_z parameter decreased in a 90% and 81% respectively. Thus, for this material the pressure of 30 MPa also went over the roughness optimized threshold and in a mayor proportion than in AISI 1045 steel.

3.2. *Surface hardness*

The hardness measurements were obtained with a Future-Tech hardness tester. To make the comparison, five measures were carried out in each tested workpiece. In figure 5 can be seen the average of those measures.

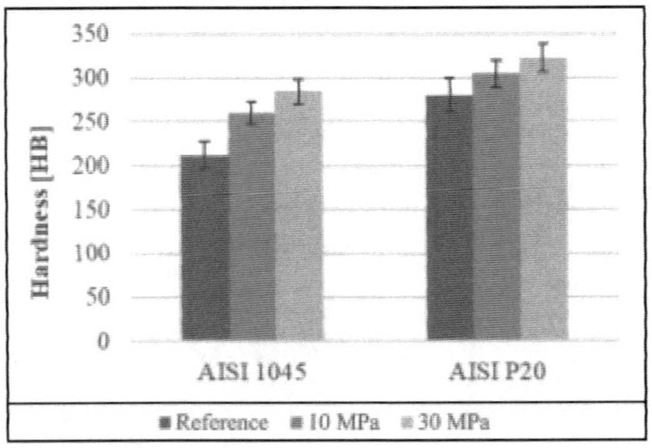

Figure 5. Comparison of hardness measurements for both steels.

As it can be seen, AISI P20 steel is harder than AISI 1045 from the milling stage. However, higher increases in hardness have been obtained in AISI 1045. While in AISI P20 the increase in hardness reached 8% and 15% using 10 and 30 MPa respectively, in AISI 1045 those percentages have improved a 22% and 34% respectively. It should be noticed that, in contrast with what occurred with the surface roughness, the more the pressure is applied the greater the hardness obtained in both steels. Hence, in the case of surface hardness it has not reached the threshold from which the hardness does not increase. Furthermore, this results agree with what *Saldaña-Robles et al.* [6] said: different combinations of hydrostatic ball burnishing parameters optimize different surface quality parameters.

4. Conclusions

In this work an analysis of the influence of hydrostatic ball burnishing pressure in different surface quality parameters has been made. After carrying out the tests and the different measurements, the following conclusions have been reached:

- In AISI 1045 steel there has been obtained an improvement in R_a and R_z roughness parameters of 92% and 89% respectively with a pressure of 10 MPa. In the case of AISI P20 steel, that improvement has overcome 90% in both parameters using a pressure of 10 MPa.
- In both materials, using a pressure of 30 MPa resulted in an exceeding of the surface roughness optimized threshold.
- Using a pressure of 30 MPa increases the surface hardness of both steels. The increase in AISI 1045 was 15% and in AISI P20 was 34%.
- Different combinations of the process parameters can optimize different surface quality parameters. Thus, the choice of hydrostatic ball burnishing parameters should be according to the desired result in the final component.

5. References

[1] Rodríguez Ezquerro A. 2015 Mejora funcional de superficies en componentes mecánicos mediante bruñido hidróstático con bola *University of the Basque Country* (Bilbao: Doctoral Thesis)

[2] Priyadarsini C, Ramana V S N V, Prabha K A and Swetha S 2019 A review on ball, roller, low plasticity burnishing process *Materials Today: Proceedings* Elsevier Ltd. pp 5087–99

[3] Avilés R, Albizuri J, Rodríguez A and López De Lacalle LN 2013 Influence of low-plasticity ball burnishing on the high-cycle fatigue strength of medium carbon AISI 1045 steel *Int J Fatigue* **55** pp 230–244

[4] Travieso-Rodríguez JA, Jerez-Mesa R, Gómez-Gras G, Llumà-Fuentes J, Casadesús-Farràs O and Madueño-Guerrero M. 2019 Hardening effect and fatigue behavior enhancement through ball burnishing on AISI 1038 *J Mater Res Technol.* **8**(6) pp 5639–5646

[5] Hua Y, Liu Z, Wang B and Hou X 2019 Surface modification through combination of finish turning with low plasticity burnishing and its effect on fatigue performance for Inconel 718 *Surf Coatings Technol.* **375** pp 508–517

[6] Saldaña-Robles A, Plascencia-Mora H, Aguilera-Gómez E, Saldaña-Robles A, Marquez-Herrera A and Diosdado-De la Peña JA 2018 Influence of ball-burnishing on roughness, hardness and corrosion resistance of AISI 1045 steel *Surf Coatings Technol.* **339** pp 191–198

[7] Luca L, Neagu-Ventzel S and Marinescu I 2005 Effects of working parameters on surface finish in ball-burnishing of hardened steels *Precis Eng.* **29**(2) pp 253–256.

[8] El-Taweel T A and El-Axir M H 2009 Analysis and optimization of the ball burnishing process through the Taguchi technique *Int J Adv Manuf Technol.* **41**(3–4) pp 301–10

[9] Swirad S and Wdowik R 2019 Determining the effect of ball burnishing parameters on surface roughness using the Taguchi method *Procedia Manufacturing* Elsevier B.V. pp 287–292

[10] López De Lacalle L N, Lamikiz A, Muñoa J and Sánchez J A 2005 Quality improvement of ball-end milled sculptured surfaces by ball burnishing *Int J Mach Tools Manuf.* **45**(15) pp 1659-1668.

Acknowledgements

This project has been carried out thanks to the collaboration with the PROCODA (High Value Processes Based on Knowledge and Data) Elkartek type 1 projects: KK-2019/00004, the IT1337-19 project of the Basque Government and the Mineco DPI2016-74845-R project. The authors would also like to thank the vice-rector of innovation, social commitment and cultural action of the UPV/EHU (Bizialab programme of the Basque Government) and the UPV/EHU itself for the financial aid for the pre-doctoral grants PIF 19/96 and PIF 19/161.

IOP Conf. Series: Materials Science and Engineering **968** (2020) 012022 doi:10.1088/1757-899X/968/1/012022

Creating an ethernet communication between a Simatic S7-1200 PLC and Arduino Mega for an omnidirectional mobile platform and industrial equipment

A I Marosan [1], G Constantin [2], A Barsan [1], M Crenganis[1] and C Girjob[1]
[1] Industrial Machinery and Equipment Department, "Lucian Blaga" University of Sibiu, Engineering Faculty, 4 Emil Cioran, 550025 Sibiu, Romania
[2] Robots and Manufacturing Systems Department, University "Politehnica" of Bucharest, Splaiul Independentei 313, Sector 6, 060042, Bucharest, Romania

E-mail: adrian.marosan@ulbsibiu.ro

Abstract. The degree of automation in the industry increases more and more every year, trying to reduce more and more or even eliminate the collaboration with the human operator. The basis of all automatic industrial equipment is PLCs. This paper presents a method of extending the number of inputs and outputs of a Siemens Simatic S7-1200 PLC, using an Arduino Mega development board and an Ethernet communication. Following the realization of this communication, the number of inputs and outputs will increase considerably, being able to connect various sensors used for the construction of an omnidirectional mobile platform, but also for equipment for industry. The advantages of using this communication are multiple, both from a financial point of view and from the point of view of the flexibility and integration of robotic systems in industrial equipment.

1. Introduction
Mechatronics and robotics as rather new branches of technical sciences have created a big demand of communication protocols between the devices within industrial systems [1]. The digital industry is undergoing a major transformation due to recent developments in information technology and communication. They have a massive impact on all known technological aspects, having a connectivity with a speed impossible to achieve in the past. As a result, the barrier between the two physical and digital worlds is getting smaller. This leads to rapid progress, which generates new trends in the industrial world, such as: Cyber-Physical Systems (CPS), Internet of Things (IoT), cloud computing, Industry 4.0 and Smart Grids (SG) [2, 3].

The Industry 4.0 concept will create new skills requirements in the future with the digitization of future production. The demands on the use of emerging technologies will put pressure on the type of qualification and finding new employees with superior cognitive skills and abilities. It also requires interdisciplinary thinking, a high dissemination capacity and the ability to put into practice, but also to have a tactical and highly operational understanding in order to be able to easily develop and implement all Industry 4.0 solutions [4].

For automation and monitoring of industrial equipment, PLCs are used due to their proven reliability and flexibility over time. In the traditional scheme, it is specified that production systems must operate for a long time with high reliability. Of these consonants, programmable logic controllers (PLCs) are the most widely used automation devices in industrial processes. As for open source hardware, Raspberry Pi, Arduino, Beagle-Bone Boards, Intel Edison or Phidget, are devices that were not created for an industrial purpose. The best-known example is Arduino, a cheap development board based on a powerful Atmega microcontroller. In terms of use in research and development but also academia, Arduino is a major trend that has become a powerful tool to develop and create various applications in the fields of automation or data acquisition. It is well known that programming and configuring Arduino chips is easy and flexible [5].

From the point of view of Arduino software, it has the same IDE interface (Integrated Development Environment) which is used for all types of development boards and is compatible for several types of operating systems. This IDE is open source with a friendly interface that is easy to use. C / C ++ are the programming languages used, which allow the user to create from scratch a simple program based on procedures in a single file, to a more complex program oriented on objects from several types of files. Other advantages of the Arduino development board are the information that is easy to find and in large quantities about it, information being available on the manufacturer's website, but also in books in the form of tutorials with applicability in various fields. Another reason why the Arduino development board is so popular is the number of tutorials and courses that are often found for free. These courses are aimed at people specializing in electronics and robotics, but also teachers or students who want to use the platform for education and research [6].

2. Design of the communication hardware system

2.1. Industrial communication in the perspective of Industry 4.0

For future Industry 4.0 technologies, engineers need to improve all their professional, social and personal skills. They need an interdisciplinary knowledge of production systems and processes, automation technology, information technology, ergonomic principles, etc. In addition, communication and cooperation skills in interdisciplinary groups are also important. The future modern engineer who is part of Industry 4.0 must have a holistic view of complex production systems [7,8].

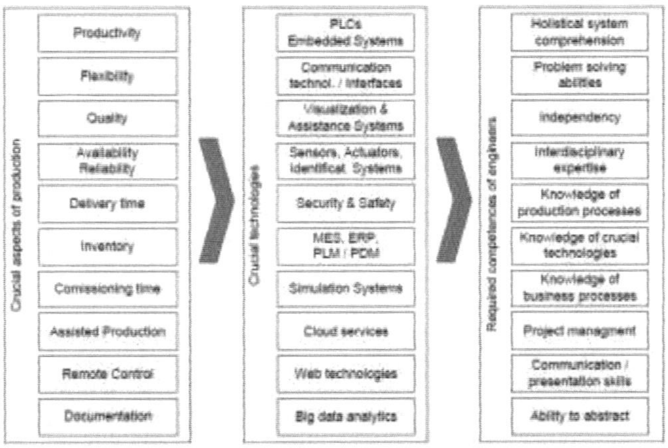

Figure 1. Critical aspects related to production, technology, and personnel [7].

Figure 1 illustrates this overview showing the need for aspects related to production, technology, and the qualification of engineers. This paper will refer to the technology for the development of communication and industrial interfaces, integrated into Industry 4.0. Nowadays, equipment will experience a massive development, and the elimination of the human factor from production processes is becoming increasingly evident. Thus, the integration of robots on production lines is becoming an increasingly intense trend in the industrial world. Autonomous mobile robots also have a new perspective, which know a rapid development in terms of their integration in the industry [7].

IOP Conf. Series: Materials Science and Engineering **968** (2020) 012022 doi:10.1088/1757-899X/968/1/012022

2.2. Communication block diagram between Simatic S7-1200 PLC, Raspberry pi-B and Arduino Mega
Figure 2 below shows a demonstration communication model between the control system of an omnidirectional mobile platform consisting of a Raspberry pi mini-computer and an Arduino Mega development board, with a Siemens S7-1200 PLC found on most industry equipment. The need for such communication becomes necessary when you want a connection between an industrial equipment and a mobile robot or when you want to use the PLC to order actuators for robotics, or to monitor special sensors also in robotics which have special communications such as a gyroscope sensor with I2C communication. In these block diagrams we have a 22V LiPO battery, to which are connected two voltage regulators, one step-up and one step-down. The step-up controller is used to power the PLC at a voltage of 24V, and the step-down controller is used to power the Arduino development board with 5V and the Raspberry pi B mini-computer.

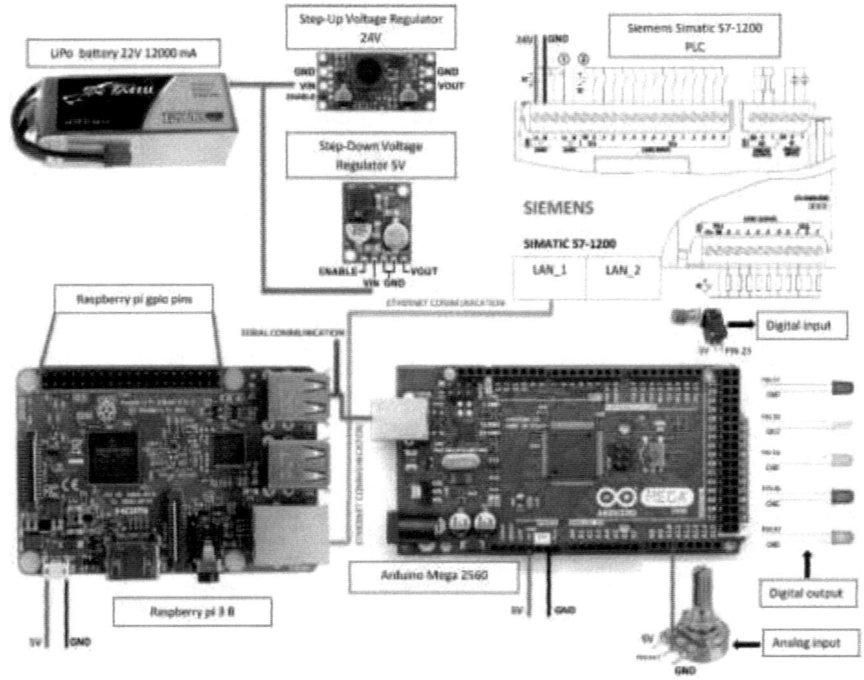

Figure 2. Communication block diagram for an omnidirectional mobile platform with a Simatic S7-1200 industrial PLC.

3. Ethernet communication between Raspberry Pi B and Simatic S7-1200

This paper presents the communication between Arduino and Simatic S71200, which is done through Raspberry Pi because the data transmission between Arduino and S7-1200 can be done only through a database created in TIA Portal. The database is updated all the time with new values. Because the computing power of the Arduino board is limited, there is a fairly significant delay. This delay can lead to many errors in dynamic processes. Using Raspberry Pi we greatly reduce this delay but due to the fact that Node-Red uses java the processor will be quite demanding and that is why it is recommended to use a Raspberry Pi with higher performance than the latest generation. The Raspberry Pi 3 Model B is a mini-computer based on a 64-bit Broadcom BCM2837 64-bit Core CPU and has 1GB of RAM. It also offers Wi-Fi and Bluetooth Low Energy features to improve functionality and the ability to power more powerful devices through USB ports. Also, this model has 4 × USB 2 ports and extended GPIO with 40 pins.

The Siemens Simatic S7-1200 PLC is an industrial programmable controller with a 1215C DC / DC / DC CPU −40 ... + 70 °C with conformal coating based on 6ES7215-1AG40-0XB0. It has 2

IOP Conf. Series: Materials Science and Engineering **968** (2020) 012022 doi:10.1088/1757-899X/968/1/012022

ethernet ports with PROFINET. This PLC has 14 digital inputs, 10 digital outputs, two 0–10 V analog inputs and two 0–20 mA analog outputs. The programming memory capacity is 125 KB.

The first step to achieve communication between the two devices was to install the Raspbian operating system on the Raspberry Pi. After installing this OS, an interface program between PLC and Raspberry is necessary (this is Node-RED).

Node-RED is a programming tool for connecting hardware devices, APIs and online services in a simple and fast way. It has a friendly interface and offers an editor, which makes it easy to connect streams along with a wide range of nodes in the palette that can be implemented at run time.

After installing the Node-RED program, the next necessary step to be able to communicate with the Siemens Simatic S7-1200 PLC is to install a library specially created for this type of PLC. To install this library, the following steps should be performed:

- Open the Raspbian programming console;
- Enter the Node-RED directory using the command: *cd $ HOME./node-red*;
- Then install the library using the command: *npm install node-red-contrib-s7*.

After the installation is complete, a Raspberry Pi reset is required. In order to be able to communicate with the PLC, it is necessary to configure the mini-computer with a fixed IP, which is of the same IP class as the PLC with which to communicate. This operation is shown in Figure 3a. Immediately after setup, a Raspberry reset is required to take over the new settings. You can also use a router that can be configured via DHCP to always assign a type of IP to both the PLC and the mini computer. The advantage of using a router is that there is also internet access, if it is necessary to install a new library for Node-RED. Figure 3b shows how to define GPIO pins in Raspberry Pi, how they can be defined as input or output pins depending on what type of block are selected from the Node-RED.

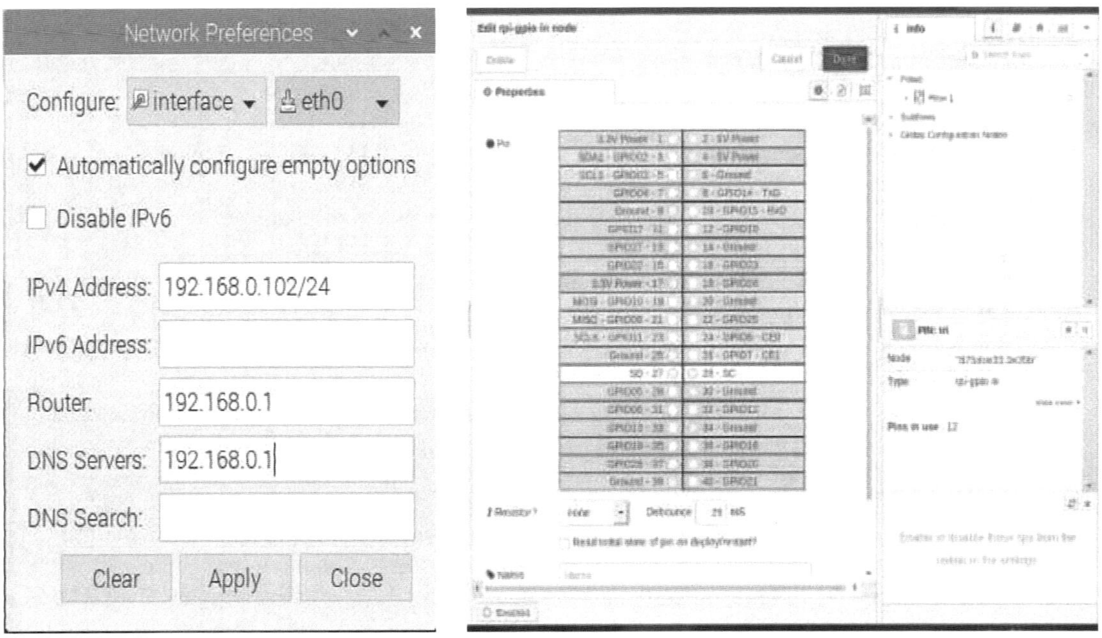

(a) (b)

Figure 3. Initial Raspberry Pi settings: a) Static IP setting, b) GPIO pin setting.

(a) (b)

Figure 4. Communication settings between Raspberry and PLC: a) IP PLC setting, b) Defining the variable in the PLC.

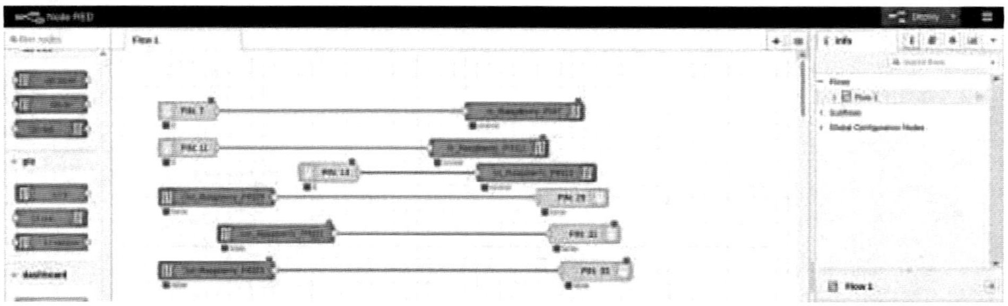

Figure 5. Making connections between GPIO pins and PLC variables in node-red.

The necessary settings for the s7 in and s7 out block of the Node-RED are presented in Figure 4a, where the IP of the PLC is defined as well as the type of communication that can be serial via the USB port or ethernet via TCP-IP. After this configuration it is necessary to define the place where the information is transmitted, more precisely the variable from PLC. This step is shown in Figure 4b. The connections made between the input and output pins of the mini-computer and the PLC variables are illustrated in Figure 5. Here one can see that Node-RED makes this interface and gives real-time information about the connection status but also about the input and output status from raspberry pi.

4. Communication between Simatic S7-1200 and Arduino Mega

Due to the fact that the Raspberry Pi does not have pins dedicated to analog inputs and outputs, it is necessary to use more expensive shields or analog-to-digital converters, for these reasons the use of an Arduino board is advantageous to use for that it has a large number of analog inputs and outputs and the acquisition cost is low, but the software part is easier to use and specialized libraries are made for sensors used in robotics. This type of communication between PLC and Arduino Mega will be done through the Raspberry Pi mini-computer using the whole Node-RED interface program. The reason or need for such a connection arises due to the need to integrate in the industry different types of robots that use such control boards for their control. Arduino Mega is a development board based on an Atmega microcontroller with a large number of inputs and outputs, but the main advantage is the price which is considerably lower than an output input module for a PLC. The reliability of this component can be considerably increased, if it is provided with a stabilized power supply and if an interface with optocouplers is made on the input and output side. To achieve this communication, a new library that must be installed in Node-RED is needed. Following the instructions above, the command to install this library is: *npm install node-red-node-serial port*. It is important to note that when this command is run, an internet connection is necessary. Once the library is installed, the Arduino board must be connected to the Raspberry Pi via USB, as shown in Figure 2. Then, in Node-RED the block connections between the serial input or output port and the input and output blocks of the PLC should be made, as exemplified in Figure 6.

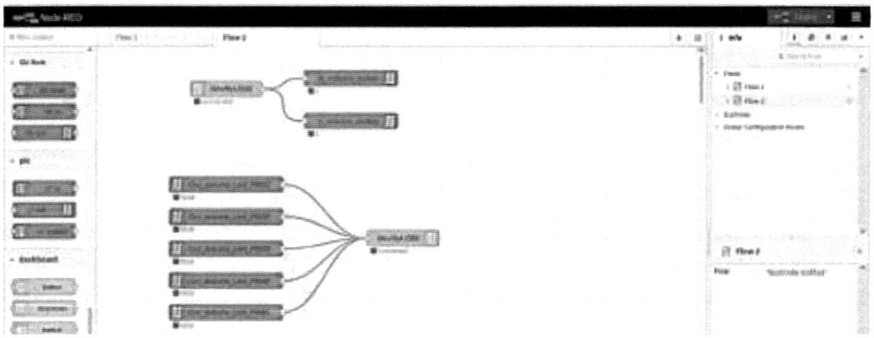

Figure 6. The connection between the serial blocks and the s7 blocks.

A program must be written on the Arduino development board that will define the pins on the board as input or output and also transmit or receive on the serial port the information from the PLC. In this case, according to Figure 2, pins from the Arduino were defined as output, pin 23 being digital and pin A13 being analog. In PLC, these will be input pins that will be stored in variables IW6 and IW8. Also, the inputs from PLC M0.0, M0.1, M0.2, M0.3, M0.4 will control the outputs on the Arduino, in this case 5 LEDs that will be connected to pins 22, 30, 38, 46, and 52. These types of variables are defined in PLC tags according to Figure 7 and their status can be tracked in real time as shown in Figure 8.

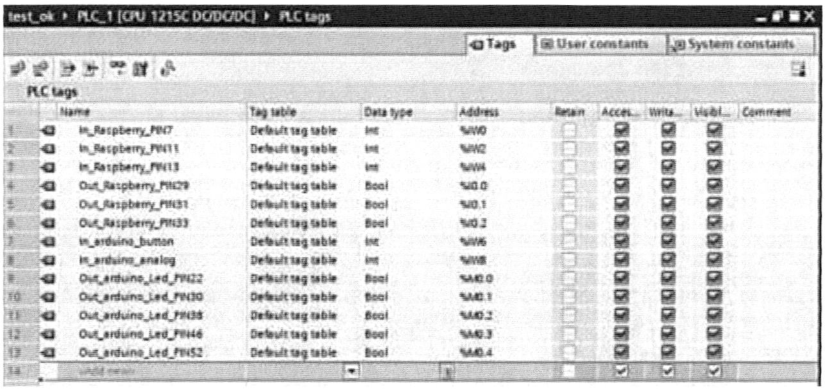

test_ok ▸ PLC_1 [CPU 1215C DC/DC/DC] ▸ PLC tags

Tags | User constants | System constants

PLC tags

	Name	Tag table	Data type	Address	Retain	Acces...	Writa...	Visibl...	Comment
1	In_Raspberry_PIN7	Default tag table	Int	%IW0	☐	☑	☑	☑	
2	In_Raspberry_PIN11	Default tag table	Int	%IW2	☐	☑	☑	☑	
3	In_Raspberry_PIN13	Default tag table	Int	%IW4	☐	☑	☑	☑	
4	Out_Raspberry_PIN29	Default tag table	Bool	%Q0.0	☐	☑	☑	☑	
5	Out_Raspberry_PIN31	Default tag table	Bool	%Q0.1	☐	☑	☑	☑	
6	Out_Raspberry_PIN33	Default tag table	Bool	%Q0.2	☐	☑	☑	☑	
7	In_arduino_button	Default tag table	Int	%IW6	☐	☑	☑	☑	
8	In_arduino_analog	Default tag table	Int	%IW8	☐	☑	☑	☑	
9	Out_arduino_Led_PIN22	Default tag table	Bool	%M0.0	☐	☑	☑	☑	
10	Out_arduino_Led_PIN30	Default tag table	Bool	%M0.1	☐	☑	☑	☑	
11	Out_arduino_Led_PIN38	Default tag table	Bool	%M0.2	☐	☑	☑	☑	
12	Out_arduino_Led_PIN46	Default tag table	Bool	%M0.3	☐	☑	☑	☑	
13	Out_arduino_Led_PIN52	Default tag table	Bool	%M0.4	☐	☑	☑	☑	
14	<Add new>								

Figure 7. The connection between the serial blocks and the s7 blocks.

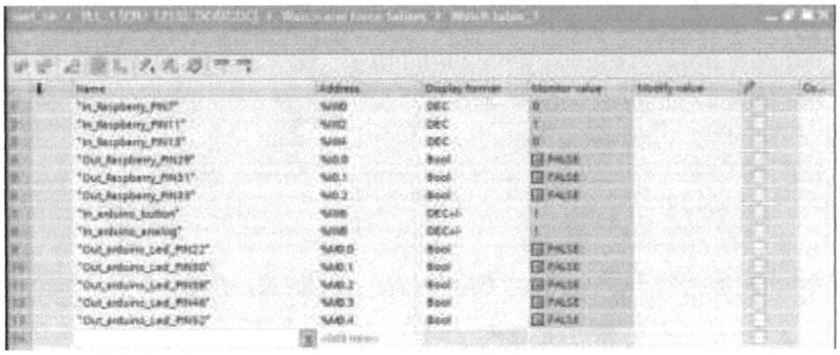

Figure 8. The connection between the serial blocks and the s7 blocks.

5. Conclusions

This paper presents a communication model between a Siemens Simatic S7-1200 PLC and an Arduino Mega development board via a Raspberry Pi mini-computer model B. This communication time helps to facilitate the integration of robots that were not created for industrial purposes in the industry, and to further develop the new concept of Industry 4.0. Both Arduino and Raspbery Pi are low-cost components that use open source software. This is also ideal for education and research. Also, this type of communication offers PLCs the possibility to expand to a large number of inputs and outputs at considerably lower costs than the use of special modules for PLCs. This work was carried out following a research on an omnidirectional mobile platform on which an industrial PLC was integrated having as main advantages the following:

- wired communication with other industrial equipment through the PROFINET protocol;
- connection of special industrial sensors that use the IO-LINK protocol;
- using special functions in the TIA Portal, such as Auto-tuner PID, to automatically find the parameters needed for a high-performance PID controller;

The most important advantage is that they can connect a wide variety of sensors and actuators without the need for special modules. The obtained results will be used to achieve a communication between an omnidirectional mobile platform that is controlled by a PLC and an industrial equipment that also uses a PLC. Further, starting from this work, a Wi-Fi communication will be realized, in which the omnidirectional mobile platform can communicate in real time at the same time with several industrial equipment's that are based on PLCs.

6. References

[1] Breaz R E, Bologa O, Racz G S and Chicea, A L 2015 *INTED Proceedings, 9TH International Technology, Education and Development Conference* (Madrid) pp 3349-3358
[2] Gonzalez I and Calderon A J 2019 *Sustainable Energy Technologies and Assessments* **36** 100557
[3] Breaz R E, Bologa O and Racz G S 2017 *Procedia Computer Science* **122** pp 796–803
[4] Gron G H, Lindgren K and Nielsen H I 2020 *Procedia Manufacturing* **45** pp 196–201
[5] Faugel H and Bobkov V 2013 *Fusion Engineering and Design* **88** pp 1276– 1279
[6] Candelas F A, Garcia G J, Puente S, Pomares J, Jara C A, Perez J, Mira D and Torres F 2015 *IFAC-PapersOnLine* **48-29** pp 105–110
[7] Simons S, Abe P and Neser S 2017 *Procedia Manufacturing* **9** pp 81–88
[8] Bârsan A 2019 *Acta universitatis cibiniensis –technical series* **71**

Acknowledgments

The experimental test unfolded within this research work were supported by a grant of the Romanian Ministry of Research and Innovation CCCDI-UEFISCDI, project number PN-III - P1-1.2 - PCCDI-2017-0446 /nr. 82PCCDI/2018, within PNCDI III, project title: "Intelligent manufacturing technologies for advanced production of parts".

IOP Conf. Series: Materials Science and Engineering **968** (2020) 012023 doi:10.1088/1757-899X/968/1/012023

Tool-holder working unit used for robot-based incremental sheet forming

A Barsan, M Crenganis, A I Marosan and A L Chicea
Lucian Blaga University of Sibiu, Engineering Faculty, 4 Emil Cioran, 550025 Sibiu, Romania

E-mail: alexandru.barsan@ulbsibiu.ro

Abstract. The diversity of industrial robot applications is constantly growing, and their use in manufacturing processes is increasing year by year. One of the most flexible sheet metal forming process, used mainly for rapid prototyping or small series production, is represented by the single point incremental forming (SPIF). Usually, incremental sheet forming processes are performed by means of CNC milling machines or industrial robots, both having advantages and disadvantages. Due to the superior number of axes, especially when compared with 3-axis CNC milling machines, one of the most obvious advantage of the industrial robots rely upon their superior kinematic. The approach of this paper tackles the problem of designing a tool-holder working unit for SPIF process performed through a KUKA KR 210-2 industrial robot. After designing of the working unit, and simulating the tool path, the generated program code was used afterwards to successfully control the robot to obtain a truncated cone-shape part.

1. Introduction
In an obvious global competition, industry requirements suggest the use of small series and large variety production. Due to their need to develop very complicated forming tools, conventional sheet metal forming processes, like deep drawing or stamping, are not able to meet the growing demand for shortening product development cycles. Recently, SPIF process has attracted a continuous growing interest in the manufacturing of prototype parts or small series production due to the customer demand custom products [1-3].

Compared with conventional sheet metal forming processes, the SPIF process present some unique advantages, such as low cost of tools, increased formability, and process flexibility. SPIF is a forming process where a sheet metal is incrementally formed into a designed shaped, usually, by a round tipped tool, controlled by a CNC machine or an industrial robot, that moves along a designated tool path.

This paper describes the steps taken for development of a proper tool-holder working unit for SPIF process performed through a KUKA KR 210-2 industrial robot.

2. Single point incremental forming
The sheet incremental forming process is one of the most flexible metal forming processes. This relatively new process is used in the forming of sheet metal part, resulting in the production of small series or prototypes metal parts, and beyond rapid prototyping and unique parts.

Figure 1 present the principle of the SPIF process, where the metal sheet (2) is fixed between a retaining frame (3) and a support plate (1). With no need for high-cost dedicated dies or tooling, based on a simple hemispherical punch, noted with (4), that executes a series of combined movements following a predefined path, the metal sheet is incrementally formed into its final shape.

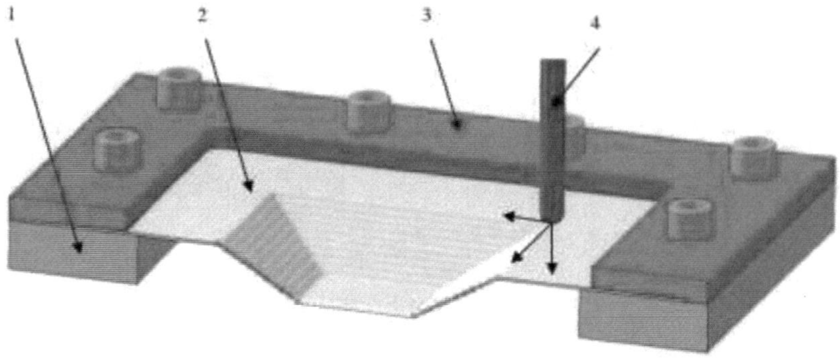

Figure 1. The principle of the SPIF manufacturing process.

In contrast with conventional metal forming processes, the SPIF process is relatively slower, but at the same time does not require an expensive technical equipment. The available power of the technical equipment used, the speed of the punch and the length of the tool-path necessary to achieve the desired profile are directly linked to the manufacturing time of the parts. With such a great process flexibility, various shapes and sizes of parts can be manufacture with the same equipment.

Given the fact that the incremental forming process has a high degree of novelty, it is not possible at this time to discuss a certain type of technological equipment dedicated to this process and to be widely accepted. Thus, it was found mainly four approaches to the technological equipment used for processing by the incremental deformation process: specialized forming machines [4], CNC milling machines [3], parallel robotic structures [5] or articulated industrial robots [6].

The high flexibility and high travel speeds of the robots are favorable in the incremental forming process to produce parts at a lower deformation force than one obtained with the CNC milling machine [7-9]. Therefore, industrial robots consume less energy compared to CNC milling machine and make the incremental forming process economical. However, the robot forming process has not yet fully optimized and implemented in industry. But with major automotive companies, such as Ford, interested in the SPIF technology, and in order to develop the process for industry adaptation, different studies regarding the usage of industrial robots can be found [1, 10].

With multi-parameters that are affecting the process, to validate the complex methodology for system design and prototyping, Callegari et al. [11] performed several experiments. It was concluded that most the important parameters are number of axes of the task, the size and the toolpath of the punch, types of materials, the kind of interpolation between the points. Ismail et al. [12] conducted experimentally investigation of robot based SPIF on AA3003 aluminum alloy sheet. Using Taguchi orthogonal array and ANOVA is was showed that a better surface roughness of the part can be achieved at 150 mm/s of robot speed, with a 0.3 mm of step size and 45 degree of wall angle. Ingarao et al. [13] reports an analysis of energy consumption, including a study of the power of single-point incremental forming processes, developed on a 6-axis robot. By analyzing the production mode time share and the energy demand for four different parameter combinations, it was concluded that the forming time is the most dominant factor in the energy demand of SPIF processes. Belchior et al. [14] studied the ability of Fanuc S420iF industrial serial robot to form a part with SPIF process. The tool center point pose errors are calculated with the elasto-geometrical model and the corrected tool path is generated. Based on this elastic model of the robot and the force prediction, the experimental result showed a difference of 1 mm between the finals shape made by milling machine and the one made by the robot.

In all research studies presented so far, researchers mention the presence of forces on three directions during SPIF process. Knowledge about the size of the deformation force and the deformations distribution in the incremental forming process is very important, especially in determining the adequate equipment and optimal parameters in the process for sheet forming. A preliminary step before determining the magnitude of forces and hence if the technical equipment is suitable for incremental forming process, is to use a proper tool-holder working unit for robot-forming.

3. Tool-holder working unit for robot-based incremental sheet forming

Designing a proper tool-holder working unit for SPIF process represent an important step ahead experimental research. Given that punches with different sizes are usually used for incremental forming study, to be able to change them very quickly during the tests and to ensure their secure fixation, a suitable punch fastening system must be designed.

The experimental layout, presented in figure 2, used for SPIF consists of the KUKA robotic system (1), the tool-holder working unit assembly for fixing the punch (2), the punch (3) and the fastening system (4).

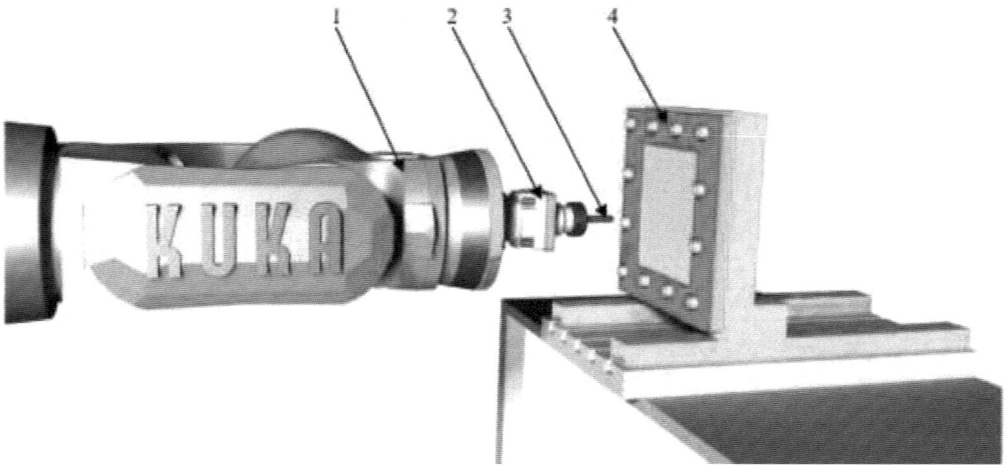

Figure 2. Experimental layout.

The tool-holder working unit assembly is made of a connecting flange attached to the last element of the robotic arm, a force sensor, a tool-holder, and a punch. The assembly was fastened with four socket countersunk head screws UNF 1936 series ASME B18.3 3/8-24x1 Inch ft.

The connecting flange allows the link between the last element of the robotic arm and the force sensor. The force sensor is a 3-component charge output force sensor, model 261A13, made by PCB Piezotronics.

The punch tool-holder unit was designed starting from a milling arbor chuck holder CNC tool. Thus, an ISO30 ER32 tool-holder for ER32 collets has been reconfigured to be used for the incremental forming with the robot. An ER-32 collet chuck with an inside diameter of 10 mm was used to ensure the secure fixing of the punch in the tool-holder.

The tool-holder and the force sensor are fastened through a square flange. The link between the square flange and the sensor was made with four socket countersunk head screws UNF 1936 series ASME B18.3 3/8-24x1 Inch ft.

The sheet fastening system has been designed and produced for orientation and fixing the metal sheet, with assuring their free deformation during the forming process. The fastening system consist of a retaining frame fixed on support frame, a support plate, and a metal sheet. To improve the geometrical accuracy of the parts produced by SPIF, Popp et al. [15] proposed the used of a pressurize fluid under as a support die instead of the fixed backing plate. Even if the simulation results showed that the profile obtained was closer to the theoretical profile, further research must made to validate the proposed model.

IOP Conf. Series: Materials Science and Engineering **968** (2020) 012023 doi:10.1088/1757-899X/968/1/012023

4. The robot-forming system

4.1. The used equipment

The custom layout used for SPIF study consists of a KUKA KR 210-2 2000 industrial robot, a custom fastening system, a custom tool-holder unit, and a punch. The overall dimension for the KR 210-2 2000 robot are presented in figure 3 and table 1.

To simulate the trajectory of the punch for experiments, a 3D model of the desired part was designed in CATIA V5. Due to the development of CAD/CAM software solutions, as well as the capability of robotic controllers, processing operations that require continuous path control can now be performed with the help of industrial robots. Thereby, the commercial SprutCAM software program, was used to simulate machining process. SprutCAM includes a module for robotic processing, where the trajectories of the robot's final effector are generated as well as cutter location data files, based on the 3D model of the formed part and the kinematics of the robotic structure. Specific robotic process problems related to the singularities, collisions and/or reach limitation are checked and solved through this module.

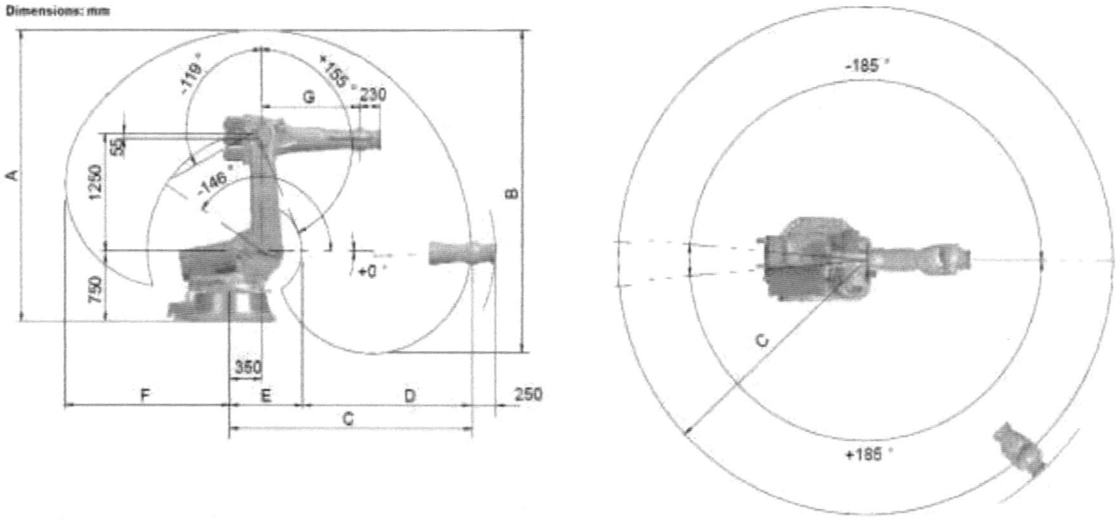

Figure 3. Overall dimensions and workspace of KUKA KR 210-2 2000 [16].

Table 1. Dimensions of KUKA KR 210-2 2000 industrial robot [16].

A [mm]	B [mm]	C [mm]	D [mm]	E [mm]	F [mm]	G [mm]	Volume [m^3]
3100	3450	2700	1875	825	1788	1100	55

4.2. The part geometry and working paramenters

A truncated cone-shape whose geometry is presented in figure 4.a was chosen to perform the experimental study. As conceptually illustrated in figure 4.b, the used contouring trajectory needed to shape the truncated cone-shape was a spiral tool path. The metal sheet is made from DC04 non-alloy steel, with very good deformability properties.

Table 2 include the geometrical data for this experiment. By using these working parameters, it was possible to obtain the truncated cone-shape without failure occurrence, as can see in figure 5.

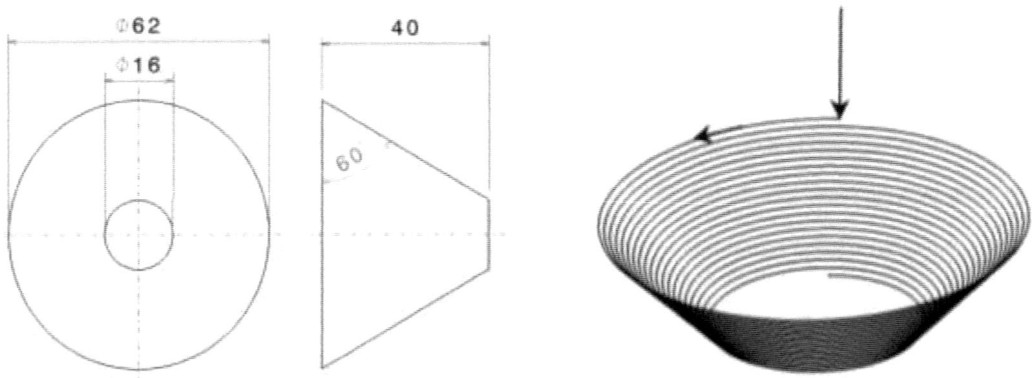

a) b)

Figure 4. Truncated cone-shape: a) dimensions, b) circular toolpath.

Table 2. Geometrical data of SPIF process.

Geometrical data	Dimensions
Metal sheet size	250 x 250 [mm]
Metal sheet thickness	0.6 [mm]
Punch diameter	8 [mm]
Vertical step size	0.5 [mm]
Height of the truncated cone-shape	40 [mm]
Wall-angle of the part	60 [°]

Figure 5. Obtained part.

5. Conclusions

The first part of the paper presents the SPIF process, as well as the technical equipment on which the metal sheet can be successfully incrementally formed. The use of the KUKA KR 210-2 industrial robot as an alternative to CNC machines can meet the SPIF process requirements.

The second part of the paper presents a newly developed tool-holder working unit for SPIF process, as well as the fastening system, which increases the flexibility of the SPIF process.

SprutCAM, a commercially available CAM software package, was used for building the kinematic models and running the simulation. Finally, a truncated cone-shape made from DC04 non-alloy steel was successfully formed.

Further research will be performed on three main directions:

- the influence of process parameters on the part accuracy;
- study of the forces within the SPIF process;
- the effect of keeping the tool always perpendicular to the processed surfaces.

6. References

[1] Behera A K, de Sousa R A, Ingarao G and Oleksik V 2017 *J. Manuf Process.* **27** 37-62
[2] Racz G S et al. 2019 *IOP Conf. Ser.: Mater. Sci. Eng.* **564** 012023
[3] Tera M et al. 2019 *Int J Adv Manuf Technol* **102** 1761–1777
[4] Amino M et al. 2014 *Procedia Engineering* **81** 54-62
[5] de Sousa R A et al. 2014 *Structural Engineering and Mechanics* **49(5)** 645-660
[6] Chera I, Bologa O et al. 2013 *Ann Oradea Univ Fascicle Manag Technol Eng* **1** 53–58
[7] Belchior J, Leotoing L, Guines D et al. 2014 *J Mater Process Technol* **214(8)** 1605–1616
[8] Breaz R E, Bologa O and Racz S G 2017 *Procedia computer science* **122** 346-353
[9] Chen Y and Dong F 2013 *Int J Adv Manuf Technol* **66** 1489-1497
[10] Bologa O, Breaz R E and Racz S G 2018 *Procedia computer science* **139** 408-416
[11] Callegari M, Amodio D, Ceretti E and Diardini C 2006 *Intech* **702** 493-514
[12] Ismail N A et al. 2019 *Int J of Int Eng* **11(1)** 217-224
[13] Ingarao G et al. 2013 *Global Conf on Sustainable Manuf* **1** 131-136
[14] Belchior J, Guines D, Leotoing L and Ragneau E 2013 *Trans Tech Publications Ltd.* **554** 1282-1289
[15] Popp M, Oleksik M, Racz S G and Rusu G P 2019 *ACTA Universitatis Cibiniensis* **71(1)** 62-66
[16] KUKA AG, company website, https://www.kuka.com/en-us/services/downloads/, Accessed: 2020-07-15.

Acknowledgments

The experimental test unfolded within this research work were supported by a grant of the Romanian Ministry of Research and Innovation CCCDI-UEFISCDI, project number PN-III - P1-1.2 - PCCDI-2017-0446 /nr. 82PCCDI/2018, within PNCDI III, project title: “Intelligent manufacturing technologies for advanced production of parts”.

Using open source software CNC controllers and modular multi-axis mechanical structure as integrated teaching environment for CAD/CAM/CAE training

R E Breaz[1], S G Racz[1], C E Girjob[1], M Tera[1] and C Biriş[1]

[1] Lucian Blaga University of Sibiu, Faculty of Engineering, Victoriei, 10, Sibiu, 550024, Romania

Email: radu.breaz@ulbsibiu.ro

Abstract. Open source CNC controllers, such as Linux CNC are used more and more because they are open source projects and consequently are available for free. Moreover, these controllers are highly customizable, and their capabilities are close to the commercial solutions. The paper presents how using Linux CNC controller and a modular structure of a multi-axis CNC machine-tools, built by a commercial company, an integrated environment for training students in implementing CAM techniques was developed. Aside the CNC controller and the machine-tool, a CAM software package was used for programming complex machining operations. The kinematic model of the machine was built, for the user to be able to simulate, in a realistic way the multi-axes machining operations. The integrated teaching environment enable the users to be trained in CNC programming for three to five axes machining operations. Advanced programming strategies, such as tool center point management (TCPM) can also be tested.

1. Introduction

CNC machine-tools are the backbone of the modern manufacturing industry, as the most used technological equipment. Even if additive manufacturing is spreading fast, subtractive manufacturing will keep its ground for a long time to come. Moreover, hybrid solutions of CNC machine-tools are launched on the market, which encompass both additive and subtractive manufacturing capabilities.

Being able to implement CNC machine-tools in modern manufacturing systems, in conjunction with the use of CAD/CAM/CAE techniques is a competence of the greatest importance for mechanical, manufacturing, and industrial engineers. Thus, CNC training is provided for students in technical universities all over the world as core activity within many subjects form the curriculum.

CNC training, from a state-of-the-art point of view should be regarded as the integration of CAD/CAM/CAE techniques in a machining environment based upon CNC machine-tools. 3-axis and 5-axis milling machining centers are the core of the modern CNC machine-tools [1, 2].

However, commercially available CNC machine-tools, particularly 5-axis machining centers are very expensive and their use for training purposes in technical universities could be problematic due to cost-related issues.

The approach presented in this paper is based upon the use of a modular, low-cost 5-axis machining center, specially developed for training, and controlled by means of an open-source CNC software controller.

IOP Conf. Series: Materials Science and Engineering **968** (2020) 012024 doi:10.1088/1757-899X/968/1/012024

2. Hardware CNC controllers vs. software CNC controllers

Commercially available CNC multi-axes CNC machining centers are controlled by means of hardware CNC controllers, specifically developed for a certain class of machines. Hardware CNC controllers are based upon the specific hardware embedded on the machine and consequently are not configurable. In opposition, as presented in figure 1, open source software controllers [3-5] are designed to be highly configurable and to be integrated with any kind of hardware modules existing on a CNC machine-tool.

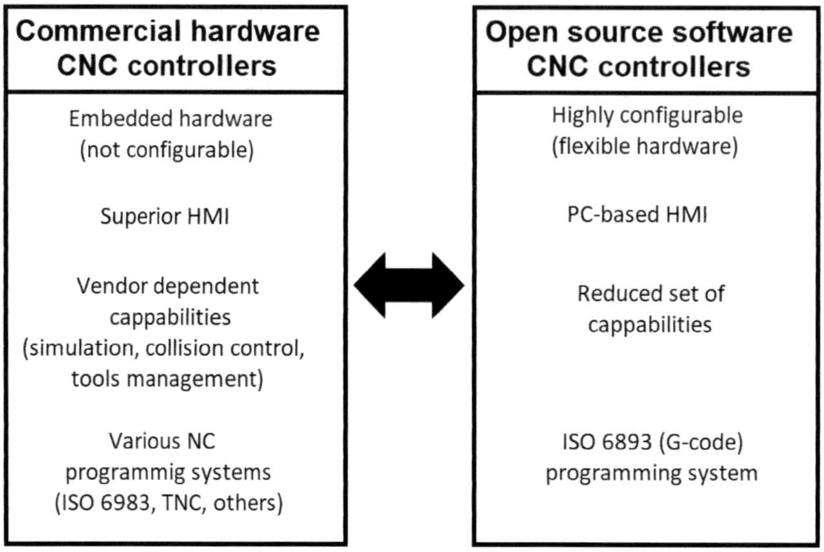

Figure 1. Differences between hardware and software CNC controllers

Hardware CNC controllers are fitted with specifically designed Human-Machine Interface (HMI). Usually in the form an operator panel, the HMI has hardware buttons and knobs for each specific function. Designed to provide the user with ease of access and ergonomics, these HMI's are far superior to the ones provided by the open source software controllers which are usually PC-based. It means that usually no buttons and knobs are available for the user and every function is implemented at the level the computer keyboard and mouse.

Moreover, CNC hardware controllers are fitted with a lot of supplementary capabilities, such as machining simulation, collision control, tools management, depending on the vendor and on the price the user is willing to pay for them. The software CNC controllers are fitted with a reduced set of capabilities, strictly necessary for controlling the machining process on the CNC machine-tool. However, modern CAM software packages can compensate the reduced set of capabilities of the software CNC controllers by providing them off-line.

Hardware CNC controllers are supporting various NC programming systems (such as ISO 6983 G-code, Heidenhain TNC and many others) depending on the manufacturer, while software CNC controllers are usually supporting only ISO 6893 G-code programming system.

3. CNC machine-tool and CNC controller

The main objective of the research was to develop a low-cost, yet effective integrated layout for CNC training, in the form of a CNC milling machining center.

A 3-axis CNC industrial machining center, figure 2a, developed by General Numeric company from Brașov was fitted with two supplementary rotary axes [6,7], to increase its technological capabilities. The rotary axes unit (figure 2b) consists of two ISEL specialized modules, DSH-S (tilting) and RDH-S (rotary). The unit is based upon harmonic drive manufactured by ISEL. Linux CNC [8] software controller was used to control the machine. Being a software CNC controller made the operation of adding two supplementary axes a simple one, due to the high configurability of the controller. The operating panel of the machine, PC-based HMI, (figure 2c) consists of a PC using Linux as operating system.

All machine axes are fitted with servomotors and consequently all motion control systems on each axis are closed-loop ones, a fact which guarantees high machining accuracies.

The machine was designed by General Numeric company in a modular, reconfigurable way, a fact which made the task of adding supplementary modules an easy one.

The machine frame and some other structural elements were designed and manufactured in house by the company, while the servomotors, transducers, transmission systems, other actuating and control modules were purchased from specialized vendors (especially ISEL company).

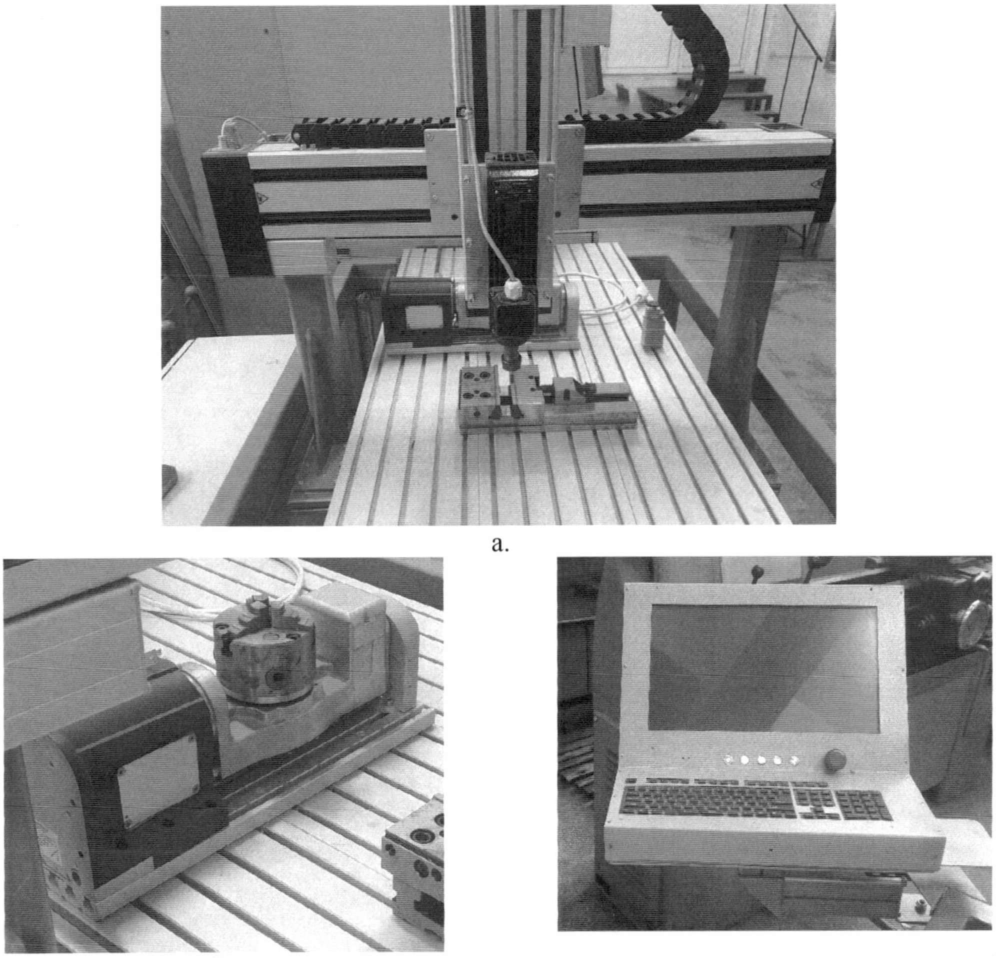

a.

b. c.

Figure 2. The modular 5-axes CNC milling machining center (a); the rotary axes (B, C) unit (b); the operator panel (c).

4. Integrated teaching environment

The CNC milling machining center was designed as an integrated teaching environment. The main objective was to provide the students the necessary training to implement CAD/CAM/CAE techniques in machining systems, figure 3.

Training in the field of CAD techniques is facilitated by making the 3D model of the machine, which was necessary for developing its kinematic model [9]. Moreover, for each part machined on it, the students must build the 3D model of the part, which furthermore will develop their abilities in using CAD techniques.

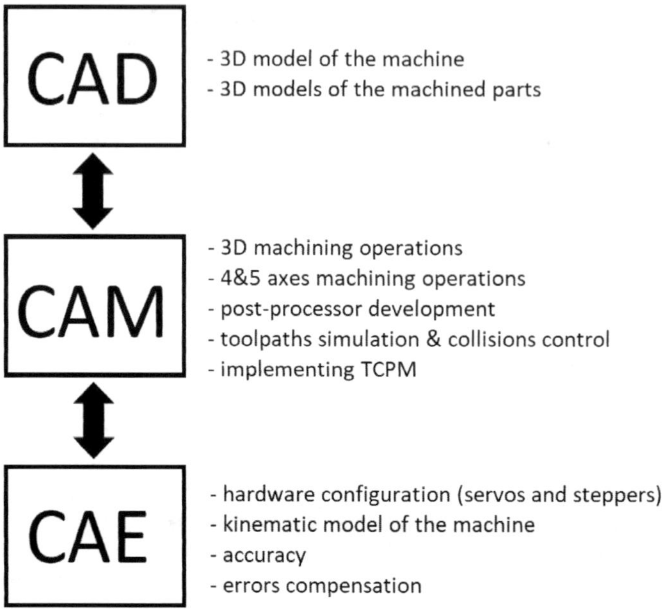

Figure 3. Training in the fields of CAD/CAM/CAE techniques

The core of the integrated teaching environment is intended to be the training in the field of CAM techniques. The students will program the machine, using dedicated CAM software packages to machine parts by means of 3D operations (which require only the use of X, Y, Z linear axes) and by means of 4&5 axis machining operations (which requires the use of both linear and rotary axes). The students will be also trained to develop specific prost-processors for Linux CNC software controller in every CAM software package used to program the machine, Is it here to mention that there are no limitation for the CAD/CAM software packages to be used in conjunction with the machine.

Training in the field of CAE techniques will be ensured be configuring the hardware modules of the machine using Linux CNC and studying and compensating the machining accuracy. It is her to be noticed the fact that the students will also be trained to build the kinematic model of the machine, which is required for realistic simulation of 4&5 axes machining operations, figure 4.

Figure 4. Kinematic model of the machine

Advanced CAM techniques such as realistic toolpath simulation, TCPM (tool center point management) and collision control (implemented by developing the kinematic model of the machine [9] could also be studied and implemented by using the integrated teaching environment.

The part presented in figure 5a was subject to an application for the post-graduate students (at master studies. It is here noticeable the fact that the part requires 4&5 axes operations to be machined (figure 5b).

The model of the part was provided by Sonnek Engineering company, form Şura Mică, Sibiu county.

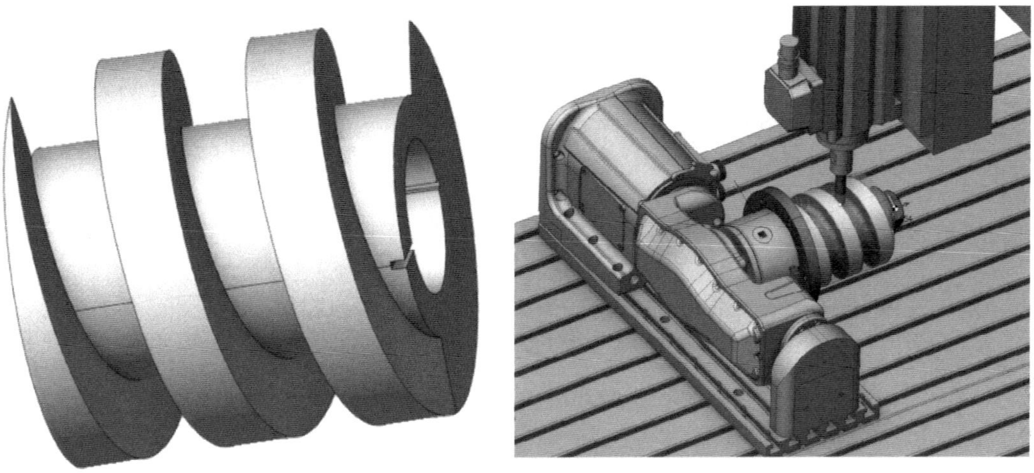

a. b.

Figure 5. Model of the machined part(a); 4&5 axes machining simulation

An example of the machined part is presented in figure 6.

Figure 6. Example of machined part

5. Conclusion

A complex research and teaching program which involved the development of an integrated teaching environment for CAD/CAM/CAE techniques was presented in this work. The main steps of building a 5-axes CNC milling machining center in a modular manner, in co-operation with an industrial company as the core of the teaching environment was introduced.

By using the integrated teaching environment presented in this paper, students are trained in designing and modeling individual parts and assemblies (CAD techniques), in programming CNC machine-tools using specialized software packages (CAM techniques) and in analyzing the complex kinematic of the machine-tool by building its kinematic model (CAE techniques). Consequently, the main objective enounced in the beginning of this paper, training the students to be able to implement CNC machine-tools in modern manufacturing systems, in conjunction with the use of CAD/CAM/CAE techniques, can be fulfilled by means of the proposed integrated teaching environment.

The main training directions and actions facilitated using the proposed integrated teaching environment to train the engineering students were also highlighted. An example of an application which involves the machining of a complex part on the proposed CNC machine-tools is also presented in the paper.

6. References

[1] Bologa O, Breaz R E, Racz S G and Crenganiş M 2016 *Procedia Comput. Sci.* **91** 184–92
[2] Bologa O, Breaz R E, Racz S G and Crenganiş M 2016 *Procedia Comput. Sci.* **91** 683–89
[3] Tianliang H, Chengrui Z, Riliang L and Peng L 2009 *Int. J. Adv. Manuf. Technol.* **40(5)** 541–552
[4] Park S, Kim S.-H and Cho H 2006 *Int. J. Adv. Manuf. Technol.* **27**, 788-796
[5] Sperling W and Lutz P 1996 *Robot Manuf.* **6** 613–620
[6] Suh S-H and Lee J-J 1998 *J. Manuf. Sci. Eng.* **120(1)** 120-28
[7] Chen Z C, Dong Z and Vickers G W 2003 *Comput. Ind.* **50** 319–31
[8] *** http://www.linuxcnc.org/
[9] Chicea A L, Breaz R E and Bologa O 2015 *Applied Mechanics and Materials* 809-810 1004-09

Acknowledgments

The experimental test unfolded within this research work were supported by a grant of the Romanian Ministry of Research and Innovation CCCDI-UEFISCDI, project number PN-III - P1-1.2 - PCCDI-2017-0446 /nr. 82PCCDI/2018, within PNCDI III, project title: "Intelligent manufacturing technologies for advanced production of parts".

Conceptual solutions for digital hydraulic pumps

P Drumea, I C Dumitrescu and M Al Hristea
INOE 2000 - Subsidiary Hydraulics and Pneumatics Research Institute, 14 Cutitul de Argint, 040558, Bucharest, Romania

E-mail: dumitrescu.ihp@fluidas.ro

Abstract. One of the methods by which specialists in the field of hydraulic drives try to increase the efficiency of the systems is the use of digital hydraulics. Digital hydraulics is a system with one or more digital components that actively controls the output of the system. One of the basic elements of such a system is the digital hydraulic pump. The first proposed and recognized variant was the one with digital pumping element; from this solution, it was possible to produce digital hydraulic pumps on an industrial scale. In this article we present some solutions regarding the creation of digital hydraulic pumps starting from several fixed flow hydraulic pumps that can ensure the variation of the flow through a special selection of them. The displacement of each pump is different and chosen taking into account the binary criterion. The article gives an overview of several such systems developed within the institute.

1. Introduction

1.1. Introduction to hydraulic drives

Hydraulics have existed for a very long time, since antiquity, when the first technical books were written, with chapters related to water or air-based drive, such as Archimedes' work "On floating bodies". And yet there was a long, blurry period, in which almost 1000 years seem to have been without important, known technical-scientific achievements. However, interesting ideas reappeared in the early Middle Ages, so that Leonardo da Vinci's work "On the movement and measurement of water" is a benchmark. A real technical boom occurred in the 16th and 17th centuries when in countries (or regions of these countries) such as Italy, France, Germany, England or Russia, research has been developed, books have been written and many equipment and machines have been developed, which over time have entered the history of the field of hydraulic drives, for example the writings of Galileo and Newton. Starting with the 18th century, the basic technical-scientific elements of hydraulics were created by personalities such as Leonhard Euler, Daniel Bernoulli, Chezy, Darcy, Stokes, Reynolds, Jukovski, Prandtl, and so on.

In Romania, the first achievements in hydraulics belonged to professors such as D. Pavel, A. Barglazan, George Constantinescu, Dionisie Ghermani, Caius Iacob, D. Dumitrescu, E. Carafoli, C. Mateescu in the mid-20th century. However, one could not talk still about achievements in the field of hydrostatic or pneumatic drives. The first specialists in the field appeared in the 60s and 70s of the last century, when engineers such as Marin Virgil, Theodor Demetrescu, N. Oprescu, S. Dobrescu, I. Mazilu, A. Popov, G. Radulescu, G. Tarlescu, R. Visan, D. Teneslav, D. Lazar, R. Moscovici, N. Arama, D.C. Popescu, M. Popescu, S. Medar designed and put into production the first hydraulic and pneumatic equipment. During that period, the first specialized factories appeared, such as those in Sibiu, Plopeni, Bucharest, Focsani, Ramnicu Valcea, Bistrita, Sinaia, Cluj-Napoca, Stei, Braila, Galati.

IOP Conf. Series: Materials Science and Engineering **968** (2020) 012025 doi:10.1088/1757-899X/968/1/012025

Nowadays the field is preserved by several production units in Bucharest, Iasi, Ramnicu-Valcea, Sibiu and several research centers, such as INOE 2000-IHP Bucharest and research centers of technical universities in Iasi, Cluj-Napoca, Bucharest, Timisoara, Galati and Brasov.

1.2. The concept of digital hydraulics

Hydraulic drives have become, since the interwar years of the last century, some of the basic drives in technology, along with mechanical and electrical drives. The main steps in modernizing hydraulics, over time, were those related to the development of high-quality pumps, advances in servotechnics and computerization of hydraulic systems and equipment. One of the great novelties of the last years, starting with 1990, is the emergence of digital hydraulics, which is intended to be a replacement (probably a partner) of hydraulic servosystems. The idea is part of the intention and mostly of the desire of specialists that the energy (be it electrical or mechanical) consumed by the hydraulic system be as close as possible to the value of useful energy, conveyed to the machine by the hydraulic installation. A solution prior to the novelties of the last years is the one in which the pumping group consists of 3 pumps that are introduced into the hydraulic circuit depending on the flow required at a given time; Dr. Heinrich Theissen of Aachen calls them digital pumps [1]. The selection of pumps with the help of simple on / off distributors can be done by several methods, of which the German specialist considered that it would be of interest to use the binary system or even better the ternary system. If a brief analysis is made, it is possible that this variant fall, roughly speaking, in the field of digital hydraulics with parallel-connected components.

The solution, although old, is still used in many applications, significantly improved, but there are specialists who do not think of it as modern digital hydraulics. What is being blamed on this solution is the low response time and the complexity of the pumping group, which we should probably discuss more, considering that in fact these elements do not give the quality of being "digital".

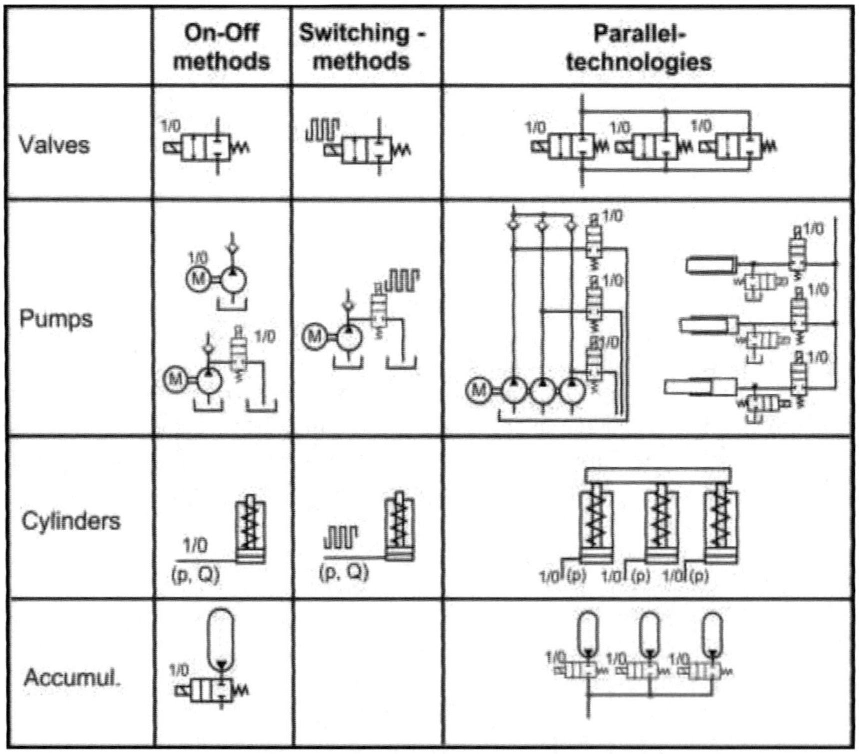

Figure 1. Graphical synthesis of traditional hydraulics vs. digital (switching technology and parallel connection) hydraulics [2].

It should be noted that digital hydraulics does not mean traditional, classic or modern hydraulics with analog type elements that we drive with digital electronics. As a result, a computer-driven servo-hydraulic system is not digital hydraulics, although in the opinion of the authors when it comes to an electro-hydraulic device, hence mechatronic device, the optics may change. Digital hydraulics means hydraulic systems that have in their structure at least one hydraulic element able to ensure precisely and actively controlled discrete values at the output. This involves, in addition to digital hydraulic elements, proper electronic control and therefore a good computerization of the system. The main advantages of digital hydraulics, more hoped for than achieved, are related to the speed, robustness, simplicity, reliability and energy efficiency of the products. Although worldwide there is an important development, unfortunately chaotic, often specific to a piece of equipment, it seems that the idea that digital hydraulic systems can be grouped on two main directions has been accepted quite widely, namely the first direction is based on **parallel connection** and the second is based on **switching technology**. Both directions are in their early stages, even if some achievements from the production of important companies are attached to them. An interesting synthesis is made by Prof. Scheidl starting from an article written by Prof. Linjama, and it is shown in this material in the table in figure 1 [2, 3, 4].

2. Digital hydraulic rotary pumps and motors

The arguments that make digital hydraulics a technology of great interest seem to be best supported by the obvious advantages of digital hydraulic pumps and motors, which are high-impact innovative equipment. The major differences between conventional and digital pumps are, according to Luke Wadsley at Sauer Danfoss, related to the way in which the displacement is achieved and the cylinder piston makes the switch from suction to discharge [5]. The permanently variable stroke of the classic pumps made by a special plate is replaced by a cylinder which has a fixed stroke and two valves, one inlet and one discharge valve, electronically controlled. The simplified diagram in figure 2 shows the basic functional element, which is actually equivalent to that of classic in-line piston pumps. In this way, both the suction-to-discharge switching and the adjustment of the displacement are performed electronically.

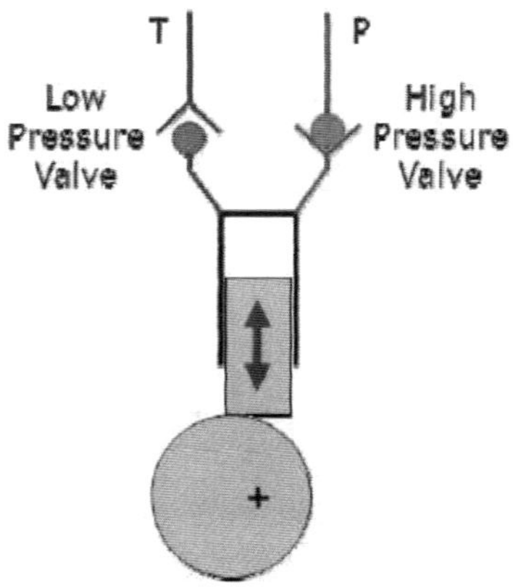

	Low Pressure Valve	High Pressure Valve
Pump (DDP)	Actively Controlled	Passive Component
Pump/ Motor (DDPM)	Actively Controlled	Actively Controlled
Motor (DDM)	Actively Controlled	Actively Controlled

Figure 2. Element of a digital hydraulic pump [5].

Several very interesting solutions for digital hydraulic pumps and motors belong to the ARTEMIS company; they are controlled directly by means of fast directional control valves driven with embedded microprocessor.

It should be noted that most of the vehicles and mobile machines that exist on the market are based on very old technologies, which control the stroke of the pump piston with the help of the distribution plate (swash plate), a mechanism patented since 1893 and introduced industrially since the 30s of last century. The mechanism has several disadvantages (friction, leakage, etc.), highlighted in recent years when energy efficiency has become extremely important.

The fact that in the Artemis project the cylinders are individually controlled led to the reduction of energy losses, to the increase of the response time and to the reduction of the noise level.

The specialist Dan Helgerson considers it extremely important that Stephen Salter (co-founder of Artemis) patented the Digital Displacement Pumps (DDP) and Motors (DDPM) in 1994. The implementation of this equipment on the transmission of excavators has led to a reduction in fuel consumption by 20% and especially an increase in productivity by 30%.

The industrial variant is the one shown in figure 3, which clearly shows that the active pistons are radial and the fact that the distribution and control equipment is compactly placed in the system; practically one can reckon that there is a technological grouping of several piston pumps under a central logical control.

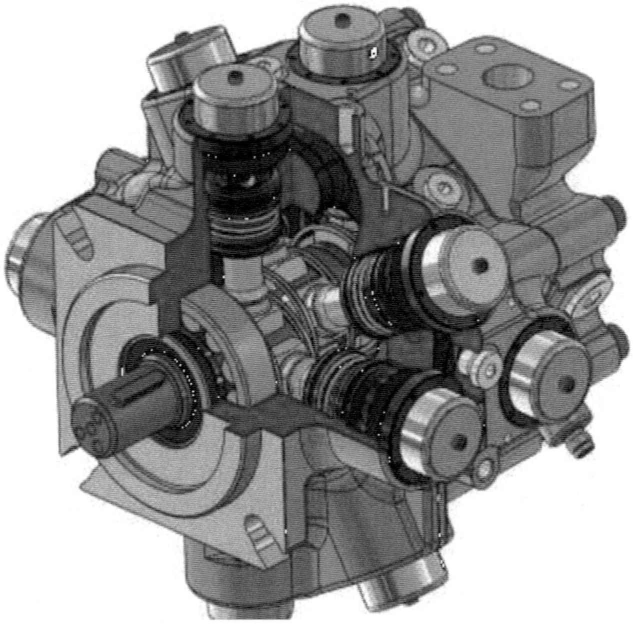

Figure 3. Digital hydraulic pump - industrial variant [6].

3. The situation of digital hydraulics in Romania

In Romania, there are few researches in the field of digital hydraulics, and the great achievements are even fewer. The only theoretical attempts and achievement of some components belong to the research group at INOE 2000 – IHP, which has developed a hydraulic directional control valve, currently at the stage of a functional model, for which a patent application was filed, 2 variants of digital hydraulic cylinders and a variant of digital hydraulic pump [7, 8, 9]. It is worth noting the existence of two stands (shown in figure 4 and figure 5) specialized for digital hydraulic equipment, the one for cylinders being extremely interesting and useful.

A small group led by the professor I. Bordeasu from Politehnica University in Timisoara and another group led by the professors Vaida and Opruta from Technical University in Cluj-Napoca joined the first group of researchers from INOE 2000 – IHP. The version addressed by the specialists from IHP Bucharest is that of digital switched hydraulics, represented on an international scale by the professor Scheidl from the University of Leinz, Austria. The group of specialists from IHP, consisting of PhD. eng. Petrin Drumea, PhD. eng. M. Blejan, PhD. eng. R. Radoi, and four PhD. eng. students – M. Al. Hristea, B. Tudor, I. Pavel and St. M. Sefu - have research interests in digital hydraulic valves, digital hydraulic cylinders and digital hydraulic pumps [10].

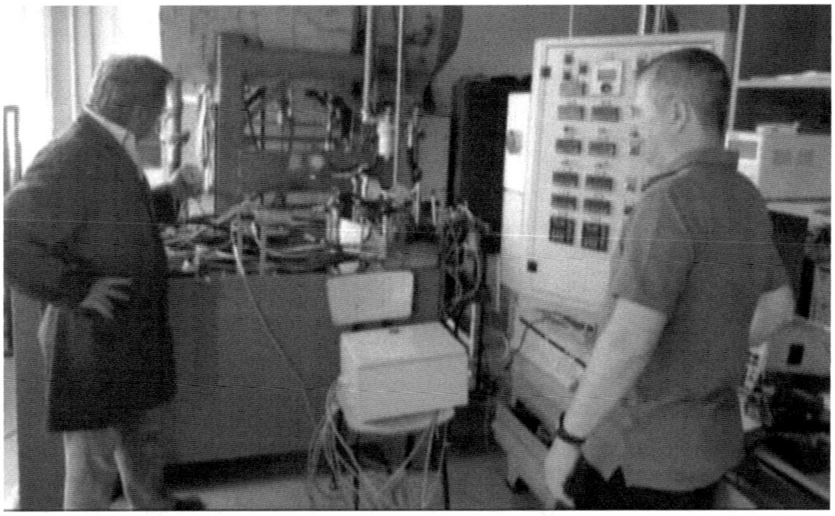

Figure 4. Servo equipment test stand and digital switching distribution.

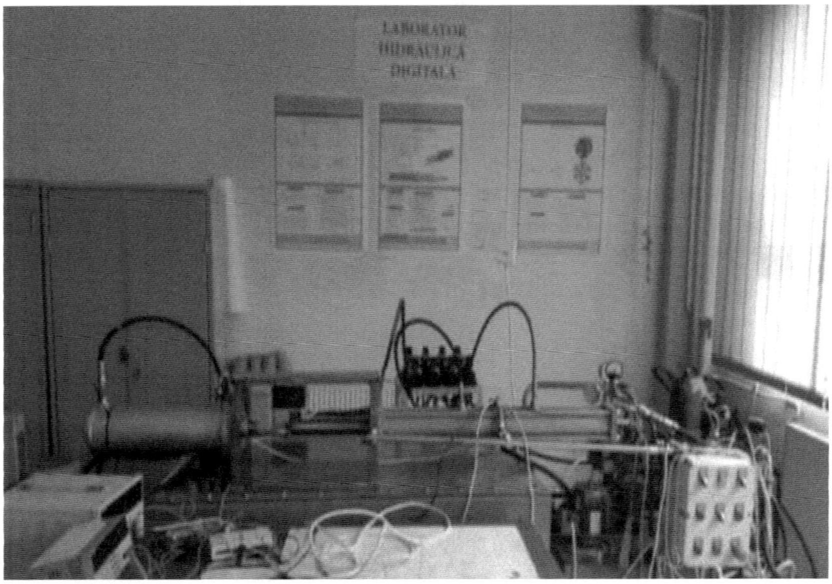

Figure 5. Digital hydraulic cylinder test stand and parallel distribution.

4. The digital pumping unit

Most variants of digital hydraulic pumps are similar to in-line (triplex) pumps, to which selection, intake and discharge blocks have been added and which, through a suitable construction, can deliver different flow rates depending on the number of cylinders which are part of its structure, such as the solutions of American specialists Luke Wadsley and Elton Bishop [11]. Also, the solutions of pumping groups with fixed gear pumps or water transfer pumps have been known for a long time; they allow a primary adjustment of the flow in only a few steps. All these solutions have the disadvantage of the small number of flow steps, the complexity of the flow selection system and most of the time the large number of drive motors, usually equal to the number of driven pumps. Such groups are found in hydraulic drives for presses, for hydraulic installations in metallurgy and for mobile equipment in which multi-stage speed hence multi-stage flow is required, as shown by the German specialist H. Theissen at the HERVEX 2011 conference [1]. The complexity of these types of pumping units has led to the increase in the price of hydraulic installations and, as a result, of the entire industrial equipment that includes them.

The solution presented by the authors, which, among other elements, is the subject of a patent application, eliminates all these disadvantages having in its structure a single drive motor, only 2 double pumps with different displacement observing the increase of the size in the terms of the binary numbers, a simple pump selection system by introducing on/off valves, controlled by the digital electronic block, which selects according to the indications in table 1which pumps are connected to the mains and which pumps remain directed to the tank.

The advantages of the digital pumping system, according to the elements presented in figure 6 and table 1, are as follows:

- The mechanical block is simplified.
- With a number of 4 fixed pumps, a range of 15 flow rates is easily accessible through an extremely simple digital electronic control. The 15 flow rates finally ensure a flow adjustment accuracy of approximately 7%, which is sufficient in the vast majority of industrial applications.
- The fixed pumps not involved in achieving the chosen flow are driven, but they are discharged to the tank by means of on/off valves, and as a result the energy consumption is reduced to the level of the useful one and also the wear is reduced.
- The efficiency of the group (over 80%) is quite high, all components being simple, with good efficiency and easy (without additional consumption) to operate.
- This solution eliminates the disadvantage of starting and stopping the pumps from time to time, in connection with the selection according to table 1.

Returning to the diagram in figure 6 and table 1, we mention that they are a hydraulic diagram of the digital group, and respectively a table for selecting the size of the pumps and for choosing the actuation of the on/off valves by the electronic control block (*8*). The digital pumping group, as shown in figure 6 and table 1, is based on a group of 4 fixed gear pumps (*2*, *3*, *4* and *5*), grouped two by two in double pumps, which are driven by an electric drive motor (*1*) through the two output ends. The speed of the electric motor and consequently of the pumps (*2*, *3*, *4* and *5*) is 1500 rpm. In accordance with the information displayed in table 1, 15 combinations of pumps hence 15 flow sizes can be selected. The pumps included in the system are differentiated by their displacement volume (Vg), as shown in table 1. The numbering of the pumps (displacement volumes) is also kept for the on/off valves (*2*, *3*, *4* and *5*). The pressure level taken into account is 210 bar, provided by a drive motor of about 30 kW.

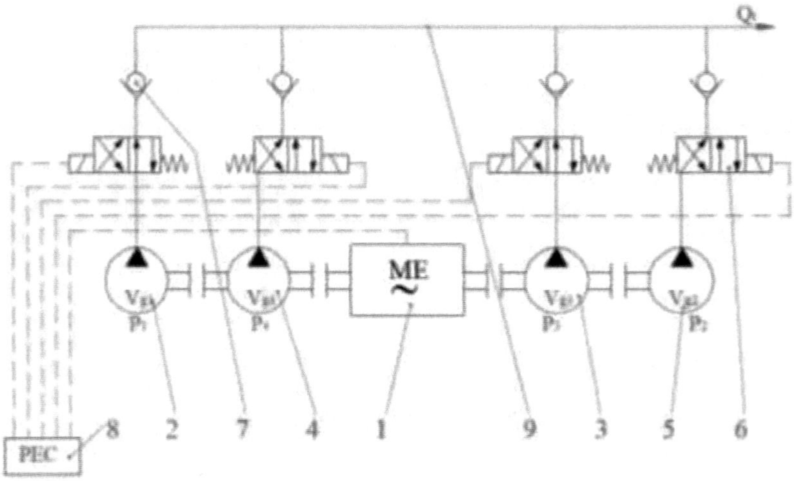

Figure 6. Digital pumping system with selection of fixed pumps by means of on/off hydraulic directional control valves.

Table 1. Possible combinations of pumps.

Item no.	Combination of pumps	Displacement [cm³/rev]				Total displacement [cm³/rev]	System flow [l/min]	Power required [kW]	Pump torque [daNm]
		Vg4	Vg3	Vg2	Vg1				
1	Vg1				4	**4**	6	2	**1.27**
2	Vg2			8		**8**	12	4	**2.54**
3	Vg1 + Vg2			8	4	**12**	18	6	**3.81**
4	Vg3		16			**16**	24	8	**5.08**
5	Vg3 + Vg1		16		4	**20**	30	10	**6.34**
6	Vg3 + Vg2		16	8		**24**	36	12	**7.61**
7	Vg3 + Vg2 + Vg1		16	8	4	**28**	42	14	**8.88**
8	Vg4	32				**32**	48	16	**10.15**
9	Vg4 + Vg1	32			4	**36**	54	18	**11.42**
10	Vg4 + Vg2	32		8		**40**	60	20	**12.69**
11	Vg4 + Vg2 + Vg1	32		8	4	**44**	66	22	**13.96**
12	Vg4 + Vg3	32	16			**48**	72	24	**15.23**
13	Vg4 + Vg3 + Vg1	32	16		4	**52**	78	26	**16.49**
14	Vg4 + Vg3 + Vg2	32	16	8		**56**	84	28	**17.76**
15	Vg4 + Vg3 + Vg2 + Vg1	32	16	8	4	**60**	90	30	**19.03**

5. Conclusions

5.1. Digital hydraulics is one of the ways to modernize hydraulic drives in order to improve the energy efficiency of systems.

5.2. The main element of the hydraulic system, the pump, is also the most important in the technical process of lining up the level of energy consumed to that of the energy actually required. Therefore, specialists need to focus mainly on modernizing the pumping units.

5.3. The solution presented in this article is known in principle, but it is new in terms of technology and structure; the authors join those specialists who consider this type of aggregate as being a digital hydraulic pump.

5.4. Digital hydraulics can be a high-efficiency research niche for specialists in our country.

6. References

[1] Theissen H 2011 Fluid Power for sustainability *Proc. Int. Salon of Hydraulics and Pneumatics – HERVEX (Calimanesti – Caciulata)* pp 137–57

[2] Scheidl R, Winkler B, Kogler H, Ladner P, Haas R and Lukachev E 2016 Digital Fluid Technologies for the Steel Industry *BHM Berg- und Hüttenmännische Monatshefte* **161** (11) 504–9

[3] Scheidl R 2008 Hydraulic Switching Control - Principles and State of the Art *Proc. The First Workshop on Digital Fluid Power (Tampere)* ed M Linjama & A Laamanen pp 31-49

[4] Linjama M 2011 Digital Fluid Power - State of the Art *Proc. The Twelfth Scandinavian Int. Conf. on Fluid Power SICFP'11 (Tampere)* **3** pp 331-54

[5] Wadsley L 2011 Optimal System Solutions Enabled by Digital Pumps *Proc. The 52nd National Conference on Fluid Power (Las Vegas)* pp 7-12

[6] Helgerson D 2018 Digital Displacement Pumps *Fluid Power Journal* (3) 40-41

[7] Drumea P, Pavel I, Matache G and Bălan I 2017 Digital linear hydraulic motors *Proc. Int. Conf. on Hydraulics and Pneumatics – HERVEX (Baile Govora)* pp 153-62

[8] Drumea P, Rădoi R, Tudor B and Bordeașu I 2016 Digital hydraulics solutions *Proc. Int. Conf. on Hydraulics and Pneumatics – HERVEX (Baile Govora)* pp 73-79

IOP Conf. Series: Materials Science and Engineering **968** (2020) 012025 doi:10.1088/1757-899X/968/1/012025

[9] Drumea P, Rădoi R, Tudor B, Hristea Al and Bălan I 2016 Normally closed switching valve for high frequency *Proc. Int. Conf. on Hydraulics and Pneumatics – HERVEX (Baile Govora)* pp 92-97
[10] Drumea P, Stoila M Al and Simionescu St M 2013 Digital hydraulics reduces energy consumption in Fluid Power drives *Proc. 2nd Int. Conf. of Thermal Equipment, Renewable Energies and Rural Development TE-RE-RD (Baile Olanesti)* pp 307-15
[11] Bishop E 2009 Digital hydraulic transformer – approaching theoretical perfection in hydraulic drive efficiency *Proc. The 11th Scandinavian Int. Conf. on Fluid Power SICFP'09 (Linköping)* p 19

Acknowledgements
This paper has been funded by the Romanian Ministry of Education and Research under Programme I-Development of national R&D system, Subprogramme 1.2–Institutional performance–Projects financing excellence in R&D&I, Project acronym: PRO-INSTITUTIO, Financial Agreement no. 19PFE/17.10.2018, Phase 5.

IOP Conf. Series: Materials Science and Engineering **968** (2020) 012026 doi:10.1088/1757-899X/968/1/012026

Static Structural Analysis Analytical and Numerical of Ball Bearings

S. Belabend[1], V. Paunoiu[2], N. Baroiu[2], R. Khelif[1] and I. Iacob[2]
[1] "Badji Mokhtar Annaba" University of Annaba, Faculty of Engineering Science, BP 12, 23000 - Annaba, Algeria
[2] "Dunarea de Jos" University of Galati, Faculty of Engineering, Department of Manufacturing Engineering, 47 Domneasca St., 800008 - Galati, Romania

E-mail: b.belabend@gmail.com

Abstract. FEM analysis is a very efficient method for achieving results of stresses at different loading conditions according to forces and boundary conditions applied to the component the static analysis. The purpose of the study is to collect data's using two different softwares and then to compare them with analytical results. This work aims at analysing the behaviour of the ball bearings under a static load, using Solidworks, ANSYS and MESYS software. The comparison is done between the analytical results using the Hertzian theory, ANSYS and MESYS, for two different cases of loading.

1. Introduction

Bearings are important mechanical components that help to maintain both linear and rotational movement of a machine. The loads which the bearings are supposed to carry are crucial if we seek to ensure a long operational life to them. Loading is considered to be static in several cases such as for stationary bearings, low oscillations when working at low speeds or taking large short shocks while rotating [1-4].

Hertz's theory [1] is generally used to calculate the contact pressure and radial stiffness of ball bearings. The limitation of the hertz theory is transcribed in the fact that it can only consider a small portion of the contact surface of the ball bearings and it cannot include the total structural deformation of the ball, the outer ring, and the inner ring, as well as housing, this limit the engineering calculations when the total structure of the aforementioned deformation is considered. The finite element method remains very practical for structural analysis and gives convincing results in many types of engineering calculations. The contact condition between the rollers and the rings is complex and the use of FEM analysis a delicate task. These contact problems are too difficult. First, before the problem is solved the specific contact area is unrecognized. With the change in load, material, boundary condition, or other external factors, contact or separation will occur between surfaces. When the load is 0, the contact zone is reduces to a point, i.e. the point contact. As the load increases during operation, the inner ring of the bearing, the outer ring and the rolling elements cause deformation in the contact area so that the point contact becomes a surface contact. Besides, the contact area gradually becomes elliptical and generates residual stress. This phenomenon is not easy to predict. Furthermore, the effect of friction which as for the contact is not easy to take into account. They may be variable with loads of change.

This paper, presents a simplified analysis of the ball bearing static behaviour using the finite element method for determining the values of the contact pressure. The research methodology is presented in figure 1. The obtained FEM results were compared with the results obtained through the methodology of Hertz theory based on specific analytical equations and also with those obtained with dedicated MESYS software.

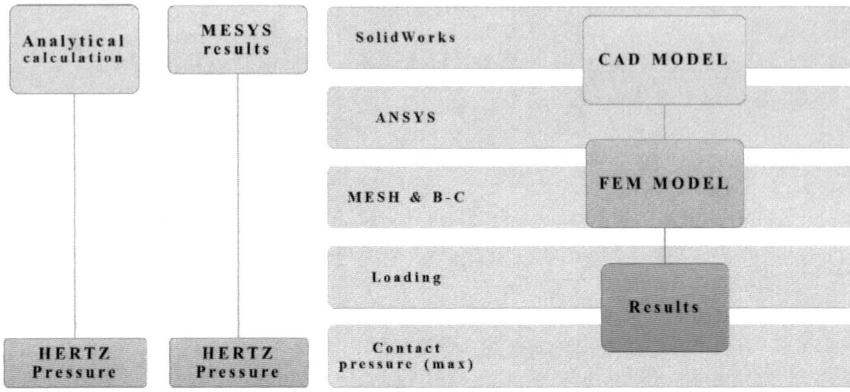

Figure 1. Research methodology

2. Theoretical analysis of ball bearing static behaviour

In the elastic area, it is possible to calculate the contact pressure and the resulting deformation at the contact points of the rolling elements and orbits by means of Hertz theory, concerning the contact of the volume elastic bodies.

This theory is based on these following assumptions, [8, 9].

1. The material of the bodies in contact is considered homogeneous and isotropic.
2. The tangential forces are not considered in the contact area.
3. Contact stresses and deformations satisfy the differential equations for stress and strain of homogeneous, isotropic, and elastic bodies in equilibrium.
4. Contact is limited to a small portion of the surface (the stress disappears at a great distance from the contact zone).
5. In the absence of an external force, the contact zone degenerates to a point.

When two elastic solids are in contact under load F, figure 2, a contact area develops. For a point contact, the area, in general is assumes to be an elliptical shape and has a semi-major (a) in one direction and a semi-minor (b) in the perpendicular direction.

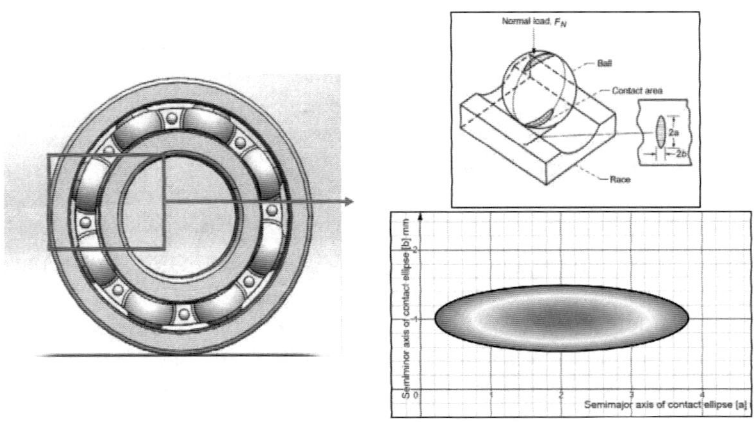

Figure 2. Contact between two spheres [2].

When the ball bearing is operating, there are usually more than one axis ball that is supporting the load axes. The condition is more complex between the rollers and the rings.

When the load increases in running, the bearing inner ring, outer ring, and rolling elements bring forth plastic deformation in the contact area, so the point contact becomes face-contact. Furthermore, the contact area gradually becomes an ellipse and generates residual stress [5, 9, 10].

The equations used in the theoretical analysis of a ball bearing are presented in the table 1:

Table 1. Equations used in the static analysis of contact pressure [10,11].

Contact radius a, b	a and b are the dimensions of the contact ellipse, see the figure 1	$a = 1{,}145.n_a.(F.K.\gamma)^{1/3}$ $b = 1{,}145.n_b.(F.K.\gamma)^{1/3}$	(1) (2)
n and e are used to simplify the equations		$n_a = \frac{1}{k}(\frac{2.k.E(e)}{\pi})^{1/3}$ $n_b = (\frac{2.k.E(e)}{\pi})^{1/3}$ $n_d = \frac{1}{E(e)}.(\frac{\pi^2.k.E(e)}{4})^{1/3}$	(3) (4) (5)
		$e = \sqrt{1 - \left(\frac{a}{b}\right)^2}$	(6)
E(e) Elliptic Integral		$E(e) = \int_0^{\frac{\pi}{2}} \sqrt{1 - e^2.sin^2(\varphi)d\varphi}$	(7)
Coefficients in equation for locus of contacting points		$K = \frac{1}{\frac{2}{R_1} - \frac{1}{R_2} + \frac{1}{R_3}}$	(8)
Relation between the elastic parameters of the two solids in contact		$\gamma = \frac{(1-\nu_1{}^2)}{E_1} + \frac{(1-\nu_2{}^2)}{E_2}$	(9)
Rigid distance of approach of contacting bodies		$d_r = 0.655.n_d.(\frac{F^2.\gamma^2}{K})^{1/3}$	(10)
Ratio of b to a		$k = \frac{b}{a}$	(11)
The maximum pressure [7, 11]		$P_{Hertz} = \frac{3.F}{2.\pi.a.b}$	(12)

where: a - semi-major axis of elliptical contact; b - semi-minor axis of elliptical contact; d_r - rigid distance of approach of contacting bodies ; E - Young's modulus; υ - Poisson's ratio; P_{Hertz} - maximum pressure (or stress) at the center of an ellipsoidal; E(e) - elliptic Integral; R_1 - ball radius; R_2 - external race radius; R_3 - inner race radius; F - maximum load on a bearing ball, [10,11].

Table 2. Parameters of the contact between the balls and the race.

Parameters	Sphere	Circular Race		Unit
Poisson's ratio [*υ1, υ2*]	0.29	0.29		---
Elastic modulus [*E1,E2*]	210000	210000		MPa
Radius of objects	*R1, R2*	Ball radius = 7.54	8.01	mm
	R1, R3		7.665	mm
Force [*F*]		10		KN

2.1.Analytical results

Based on the parameters presented in table 2, the analytical calculations results are presented in the table 3.

IOP Conf. Series: Materials Science and Engineering **968** (2020) 012026 doi:10.1088/1757-899X/968/1/012026

Table 3. Analytical results.

Parameter		Unit
Maximum Hertzian contact pressure, [P_{Hertz}]	3110	MPa
Rigid distance of approach of contacting bodies, [d_r]	0.038	mm
Semi major axis of contact ellipse, [a]	0,312	mm
Semi minor axis of contact ellipse, [b]	4.930	mm

3. Simulation of static behaviour in ANSYS

ANSYS is commonly used and enjoy extended use in structural analysis. ANSYS consists of three main phases: pre-processing, conducting or importing of the solid model system that is to be analysed, solid meshing design in finite elements, implementation of boundary conditions and loads limit, processing, numeric solving of the characteristic equations of the system and getting the solution. Post-processing, viewing the results to analyse system reaction and identification of areas with critical applications [6].

In this paper it was considered a ball bearing according to the URB RULMENTI S.A. Barlad company catalogue [12], designation 6308 C3L, which have an inner diameter (d) of 40 mm, an outer diameter (D) of 90 mm, and a width (B) of 23 mm. The model is made in SolidWorks (CAD (*. SLDASM. format), figure 3.a, and the meshing are done after importing the model in ANSYS. Because of 3D geometry, the tetrahedron element was used, figure 3.b. In the present analysis, the effect of the cage is neglected.

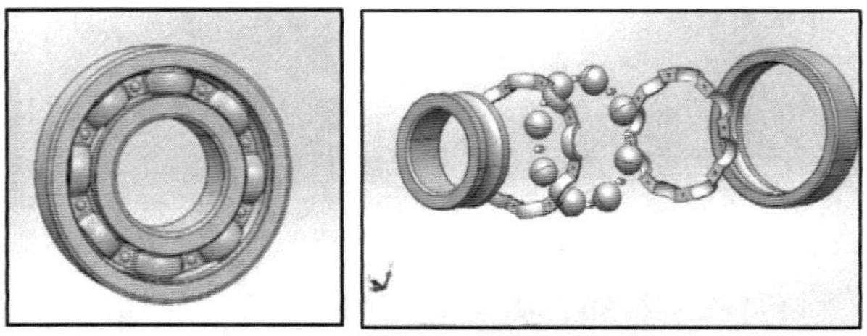

Figure 3.a SolidWorks model.

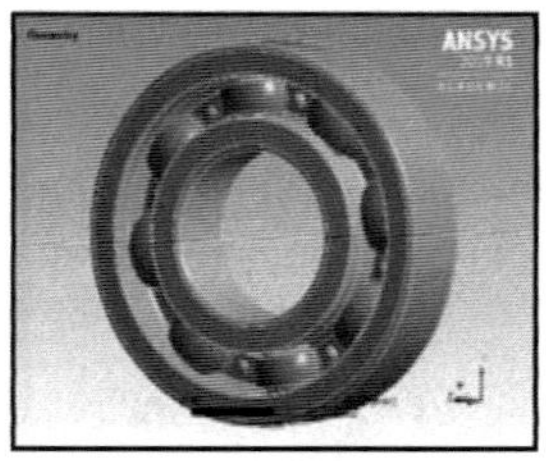

Figure 3.b ANSYS meshing model.

For both rings and balls the material used was ASTM 52100 Bearing Steel. The material properties are presented in table 4.

Table 4. ASTM 52100 Bearing Steel Properties.

Bulk modulus (typical for steel)	140 GPa
Shear modulus (typical for steel)	80 GPa
Elastic modulus	190-210 GPa
Poisson's ratio	0.27-0.30
Hardness, Rockwell C	60-67
Hardness, Vickers (converted from Rockwell C hardness)	848
Machinability (spheroidized annealed and cold drawn. Based on 100 machinability for AISI 1212 steel)	40

3.1. Boundary conditions

The value of external load is 10,000 N as Bearing Load in the negative y-axis direction with a bearing time of one second for simulation of impact load. (The value of the force is used in multiple papers and references in the literature [3;19])

The FEM analysis was performed in ANSYS Workbench. The simulation of the investigated bearing was solved as a linear isotropic material model with given material properties of linear structural steel E = 210,000 MPa, υ = 0.29.

The boundary conditions of the bearing, table 5, have been specified for each part as follows:

The outer ring and also inner ring displacement were removed in the X-axis direction and allowed in the direction of the Z and Y axes in the global Cartesian coordinate system. The coordinate systems are marked in figure 4.

- The first case (figure 4.a): the load is applied on the inner ring and the outer ring is fixed, as boundary condition
- The second case (figure 4.b): the load is applied on the outer ring; and the inner ring is fixed, as boundary condition

The contact between the elements, figure 5, was given as friction contact f=0.002 (the chosen of the friction contact coefficient for Deep Groove Ball Bearing (0.0010～0.0015))[13][18].

Table 5. Boundary conditions.

Displacement	X
Load	Y
Intensity of the load	10,000N
Friction Contact	f= 0.002

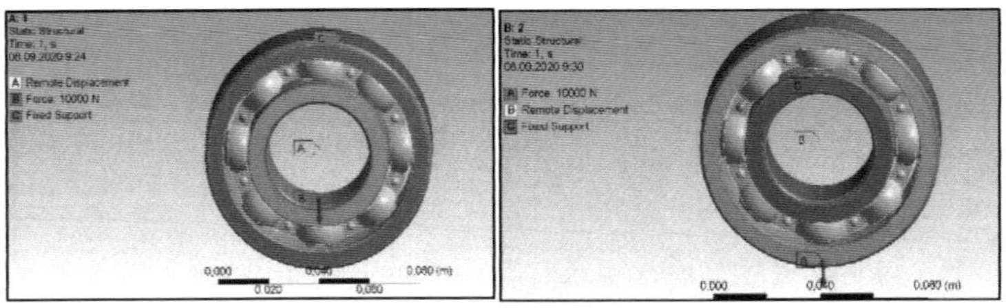

a) Case 1 b) Case 2

Figure 4. Boundary conditions.

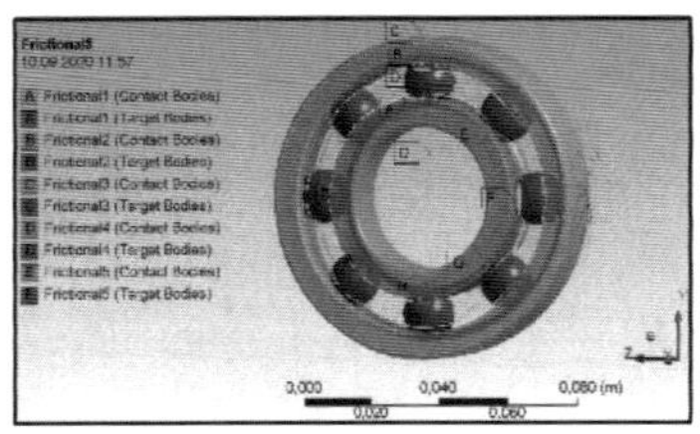

Figure 5. Contact Parameters.

3.2. *Meshing*

The mesh structure is the subdivision of the mathematical model which has a simple geometric form, and does not overlap, called a finite element. The answer to each finite element simulation is expressed on a finite number of freedom degrees that represent the values of unknown function (movement function) in several crucial points. Thus, the mathematical model will result in an approximation of the response obtained by assembling a discrete model with all other model elements [14]. Optimization of the mesh was carried out in order to obtain a continuous mesh and a correlation between the number of finite elements and the number of associated nodes. Choosing a simple type, here we used "Tetrahedrons", will result a mesh of top quality with 28716 nodes and 97422 elements in view of the structure and complexity, figure 6 and table 6. Also, for rings were established a custom mesh [15].

Table 6. Meshing.

Type	Tetrahedrons
Element	97422
Nodes	28716

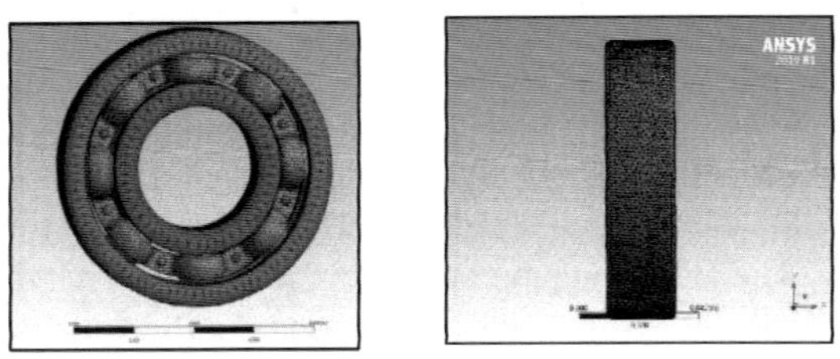

Figure 6. Meshing.

The simulation parameters are similar to those used for theoretical analysis shown in Table 2.

3.3. Simulation results

After running the model, we obtain important information about the contact pressure.

The results of the simulation are presented below:

The first case of loading:

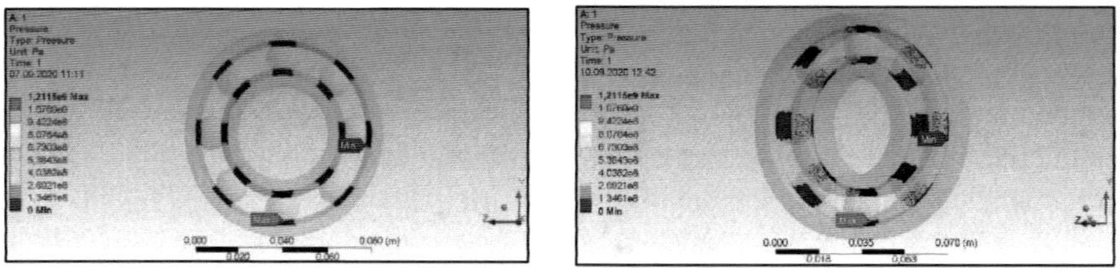

Figure 7. Contact pressure (case 1).

Table 7. Results for the first case of loading.

Object Name	Pressure
Minimum	0 MPa
Maximum	1211.5 MPa

The second case of loading:

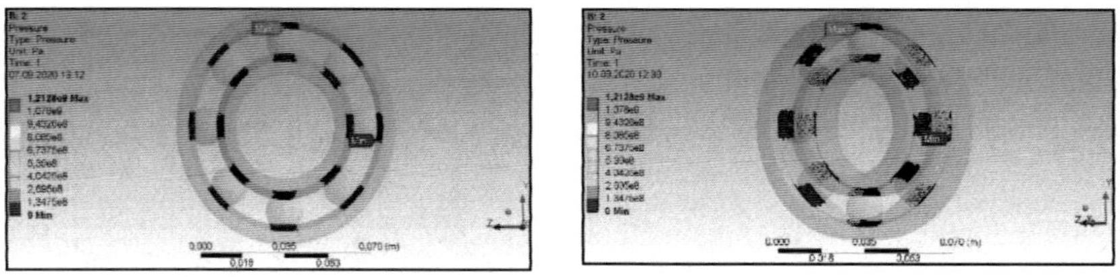

Figure 8. Contact pressure (case 2).

Table 8. Results for the second case of loading.

Object Name	Pressure
Minimum	0 MPa
Maximum	1212.8 MPa

In both cases, the biggest contact pressure appears at the contact point to correspond the line which radial force acts.

4. Simulation with MESYS

The rolling bearing analysis software MESYS calculates the life of a rolling bearings according to ISO/TS 16281 considering the inner geometry of the bearing, the load distribution within the bearing, the clearance and the tilting angle will affect the resulting bearing life. The inner geometry of the rolling bearing is provided by the user, but it can also be approximated from the load capacities by the software. The calculation returns the pressure distribution between the rolling elements and the reference life according to ISO/TS 16281 for a given loading (force and moment or tilting) [14, 16].

IOP Conf. Series: Materials Science and Engineering **968** (2020) 012026 doi:10.1088/1757-899X/968/1/012026

The input of the bearing geometrical parameters on MESYS and the loading F (0 N; 10,000 N; 0 N) are presented in figures 9 and 10.

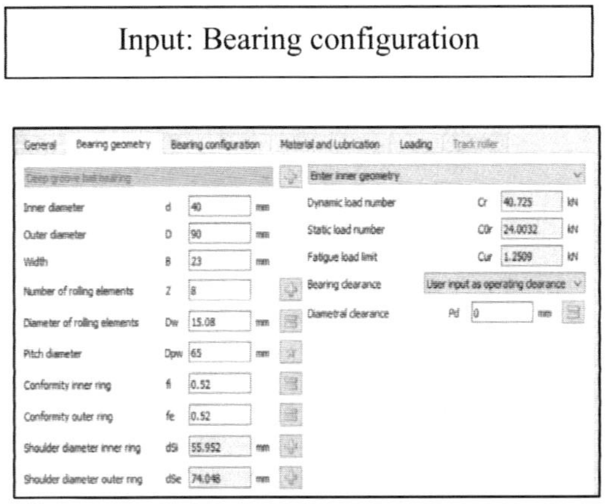

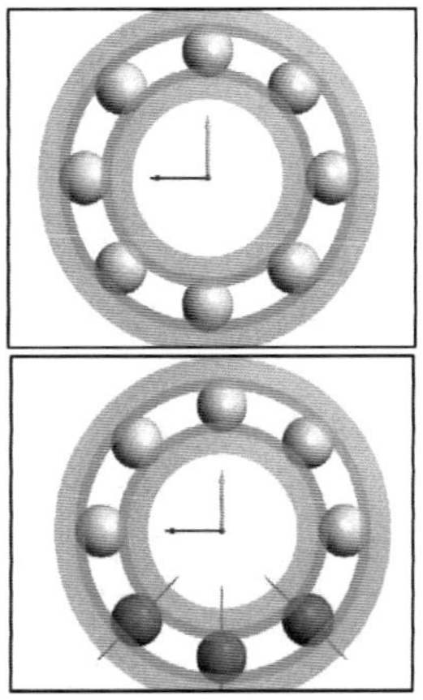

Figure 9. Application of the load in MESYS.

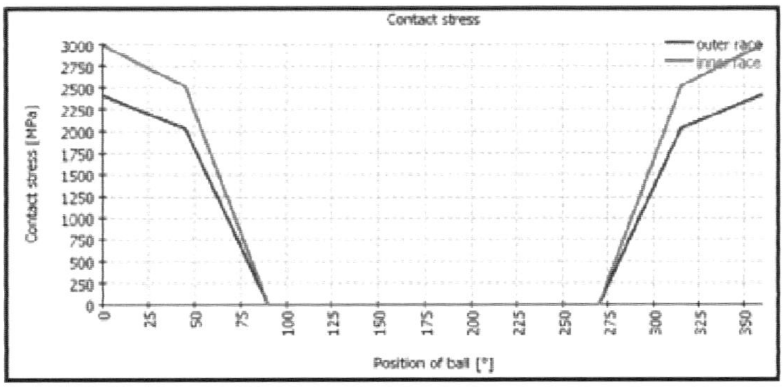

Figure 10. Contact stresses

In MESYS software the user could not choose where to apply the load or the external load, unlike Ansys software were the chosen of the point of the application of the load is easily configured. The user can only define the amplitude of the load. Results are provided as a small result overview directly in the software, a main PDF text report, a tolerances report, and a separate graphic window. In figure 11 are presented the results of simulations.

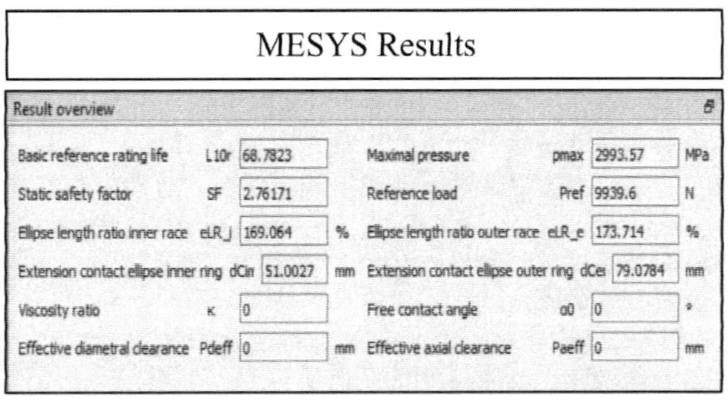

Figure 11. Results of MESYS simulations

5. Comparison between the results

Through the analysis, it was found that the value of the maximum contact pressure gets different size between the first case applying the load on the inner ring and fix the outer ring and the value of for the second case, applying the load on the outer ring and define the inner the ring as fixed support with a deviation of 0,001 %.

Table 9. Comparison between the obtained results.

		Analytical result contact pressure, MPa	ANSYS result contact pressure, MPa		MESYS result contact pressure MPa
Maximum pressure		3110.00	1211.5	1212.8	2993.57
Error MESYS result		1,6%	59,5%	59,4%	
Error ANSYS Result	Case 1	61,04%			
	Case 2	61,00%			

The results of the simulation by finite element method, and their comparison with the analytical results, in terms of contact pressure, see table 9, show a deviation 61% as percentage which is in accordance to the work of P. Šulka, A. Sapietová [3].

6. Conclusions

In the paper, we made a comparison between the results obtained from the analytical model, obtained with the ANSYS software and those obtained with the MESYS software Hertzian contact pressure.

The comparison between the MESYS results and the analytical calculation results are closer, because MESYS uses the same equations as the analytical ones to calculate the maximum Hertzian pressure.

Analysing the results between MESYS and ANSYS led to the following conclusions: the observed differences may be because MESYS does not take into account the presence of the cage; at the same time, ANSYS does not take into account the roughness of the surfaces.

The results and data obtained from this study can be used to improve the characteristics and geometries of the bearings of the next-generation products.

7. References

[1] T A Harris and M N Kotzalas 2006 *Essential concepts of bearing technology*: CRC press
[2] S Belabend 2019 *Maintenance based on simulation of static structural analysis for bearing behavior* Dunare de Jos Galati. Badji Mokhtar Annaba University, Galati,
[3] J-P Gerval, G Morel and C Querre, *Behaviour modelling and vibrations analysis applied to predictive maintenance: Different approaches leading to the same conclusions* **1** pp I/544-I/546
[4] P Šulka, A Sapietová, V Dekýš *et al. Comparison of analytical and numerical solution of bearing contact analysis* p 02022
[5] H Hertz 1987 *Le mémoire de hertz sur les contacts ponctuels*: Ecole nationale supérieure d'arts et métiers
[6] A M Puşcaşu, O Lupescu and A Bădănac *Study regarding the structural response of standard cylindrical roller bearings using ANSYS and MESYS* p 05012
[7] B Zengin, S Taşkaya and K Kaymaz 2018 *Investigation of force and moment effect of St 37 and St 70 roof lattice steels in Ansys program Middle East Journal Of Science,* **4**(1) pp 23-35
[8] P Šulka, A Sapietová, V Dekýš *et al. Static structural analysis of rolling ball bearing* p 01023
[9] A Jones 1960 *A general theory for elastically constrained ball and radial roller bearings under arbitrary load and speed conditions*
[10] K Chennakesavulu and D Rao 2015 *Structural analysis of ball bearings in ANSYS International Journal of Scientific Engineering and Technology Research,* **4**(35) pp 7086-90
[11] P Guay, and A Frikha *Ball bearing stiffness. A new approach offering analytical expressions* pp 23-25
[12] URB 2017 *URB Ultra Reliable* - General catalog

[13] V P Selma Belabend and R Khelif 2019 *Simulation Of Ball Bearing Behaviours Under Static Loading* Scientific Conference of Doctoral Schools SCDS-UDJG Perspectives and challenges in doctoral research, Galati,.
[14] Z Yongqi, T Qingchang, Z Kuo *et al.* 2012 *Analysis of Stress and Strain of the Rolling Bearing by FEA method Physics Procedia* **24** pp 19-24
[15] M Bucki, C Lobos, Y Payan *et al.* 2011 *Jacobian-based repair method for finite element meshes after registration Engineering with Computers,* **27**(3) pp 285-297
[16] I. O. 1990 Standardization, *Rolling Bearings: Dynamic Load Ratings and Rating Life*: International Organization for Standardization
[17] D Collins 2020 *What are Hertz contact stresses and how do they affect linear bearings?* https://www.linearmotiontips.com/what-are-hertz-contact-stresses-how-do-they-affect-linear-bearings/.
[18] *** 2020 *Selection of bearing type* https://koyo.jtekt.co.jp/en/support/bearing-knowledge/3-0000.html.
[19] W Jacobs, R Boonen, P Sas et al. 2014 *The influence of the lubricant film on the stiffness and damping characteristics of a deep groove ball bearing* Mechanical Systems and Signal Processing **42**(1-2) pp 335-350

Acknowledgements
This work was supported by Erasmus+ Programme, KA2 Capacity Building in Higher Education, project no.586035-EPP-1-2017-1-DZ-EPPKA2-CBHE-JP, entitled Algerian National Laboratory for Maintenance Education - ANL Med.

PLM as a sequential round of the technological revolution

E A Gromova
Peter the Great St. Petersburg Polytechnic University, Politechnicheskaya st., 29, St. Petersburg, 195251, Russia

E-mail: lizaveta-90@yandex.ru

Abstract. The fourth industrial revolution is actively developing in the world. Since 2011 it has become widespread. Affecting all areas of life activity, economy, which manifests itself in the qualitative reorganization of its mechanisms, creates new adapted management models, changes the nature of production processes. The digital revolution opens up new opportunities for using digital networks, agile production systems, and new business models for all stages of production. The faster the company makes this transition, the better its chances of succeeding and outperforming its competitors. Product lifecycle management (PLM) refers to the management of data and processes used in the design, engineering, manufacturing, sales, and service of a product across its entire lifecycle. The introduction of digital technologies occurs at all stages of the product lifecycle: product development, technology development, preparation and launch of production, product production, its operation and maintenance. PLM affects the ideology of creating a new product: to reduce the time to market and save resources, the system pushes to shift the labor intensity from the "production" stage to the "design" stage. The purpose of this study is to analyze product lifecycle management in agile management and to relate it to the current state of the Russian industrial development. Theoretical and practical aspects of this concept are given. The example of PLM implementation in the Russian industrial sector is presented. The Kalashnikov concern has built a product lifecycle management system. Today, the development of production of an industrial enterprise is impossible without a clear understanding of all stages of life cycle management of the product that it produces. Summarizing, PLM is offered as a promising concept for Russian economy.

1. Introduction

Industry 4.0 is not the near future, but a reality that introduces fundamental changes, especially evident in the development of the economy. The creation of a digital economy is a preference for the modern level of Russian industrial development. Industry 4.0 is making its own adjustments to the balance of power in the world market, and even the flagships of Russian engineering will not be able to ignore them. The old approaches to management and organization of production do not work, only digital technologies are able to ensure stable growth of efficiency and solution of strategic tasks, namely:

1. Fast product introduction to the market. Product innovation and the speed of its introduction to the market are the most important factors of competitiveness. At the same time, it is very difficult to provide them to manufacturers: the complexity and manufacturability of products is growing, the production chain is becoming more complex, and there is a need to process large amounts of data. Modern it solutions make it possible to solve these problems, providing the necessary level of interaction between the product and production.

2. Quality improvement. Without full compliance of the product with the consumer's expectations, the company cannot expect to be competitive. The key to quality is the ability to detect the root causes

of defects, rather than fighting their consequences, which in turn depends on the transparency of processes. Modern technologies of process monitoring and production management successfully solve this problem.

3. Improving efficiency. Production efficiency ensures the company's resource intensity – rational use of material, labor, and time resources and the ability to offer more competitive prices by reducing the cost of production. Optimization of production resources is also one of the consequences of effective implementation of digital technologies.

4. Increased agility. The need to increase the agility of production is caused by the variety of individual consumer requirements for the product, the instability of markets and the task of increasing labor productivity. Digitalization can provide production agility and product customization.

It is Industry 4.0 that opens up new opportunities for manufacturers to use digital networks, agile production systems and new business models for all stages of production. Development of industrial enterprise production is becoming impossible without a clear understanding of all stages of the product management life cycle. And the faster the company makes this transition, the better its chances of succeeding and beating the competition. So, product lifecycle management in agile management comes to the fore.

The theoretical and practical aspects of the modern management concepts are studied by many scholars. Focusing on agile management [1-8], it is important to select the following: P.T. Kidd (1994), S.L. Goldman, R.N. Nagel and K. Preiss (1995), A. Gunasekaran (1998, 2014, 2017), J. Sharp, Z. Irani, and S. Desai (1999), C. Larman (2004), J. Sutherland (2014). Noteworthy that Russian sector of economy has traditionally been conservative. So, the purpose of this research is to analyze PLM in agile management and to relate it to the current state of the Russian industrial development.

2. Materials and Methods

Product lifecycle management (PLM) [9] is the process of managing the entire lifecycle of a product from inception, through engineering design and manufacture, to service and disposal of manufactured products. PLM integrates people, data, processes and business systems and provides a product information backbone for companies and their extended enterprise. PLM is a completely consistent round of the technological revolution.

PLM influences the ideology of creating a new product: in order to reduce the time to market and save resources, the system pushes to shift the labor intensity from the "Production" stage to the "Design" stage. So, Professor G. Taguchi [10] believes that approximately 85% of the future success of products is laid before the start of production, and only 15% is accounted for directly by production. Figure 1 shows this schematically, and also shows the advantage of implementing digital design. Digital technologies are implemented at all stages of the product lifecycle: product development, technology development, preparation and launch of production, product production, operation and maintenance. Universal digitalization provides for the creation of a "digital product double", a "digital production process double", and a "digital equipment double".

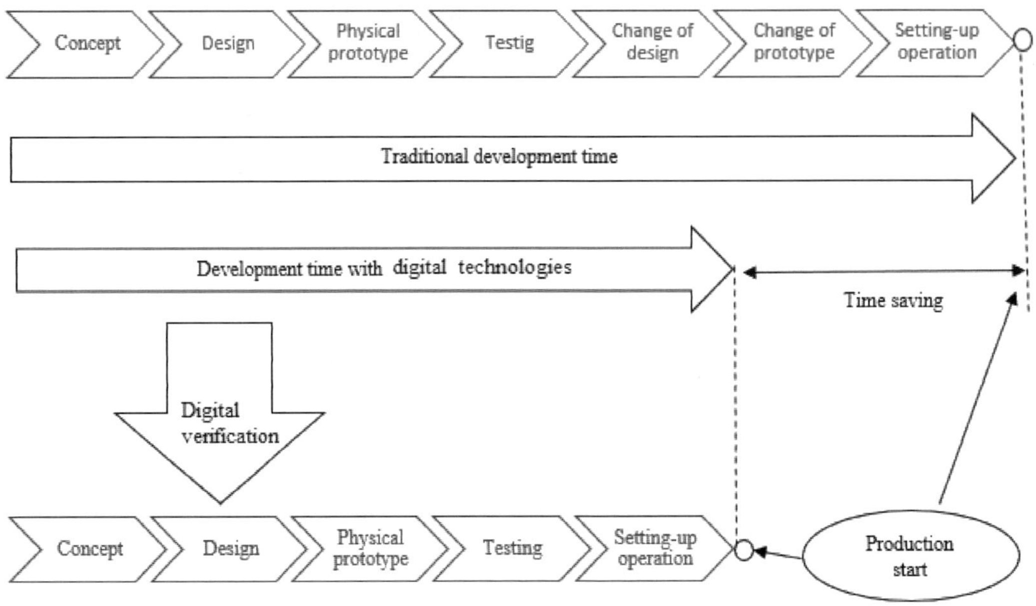

Figure 1. Comparison of the traditional design process and the design process with digital technologies, based on the [11].

The requirements for the technical task for products, registration of documents, and design development are increased. To meet these requirements, it is necessary to move from an ascending to a descending design method.

In the ascending development method, designers design individual small parts independently of each other, then combine them in the program. The first node of the product is obtained. The rest of the components are also designed, and at the end, all the parts are finally assembled. This approach is not ideal because unnecessary intersections of parts may appear in the final assembly, or, conversely, the gaps between them. Since the parts were designed by different people in different programs, there are no connections between the design elements. You need to manually change the parameters of the component that entered the assembly incorrectly, and all the surrounding parts. The disadvantage of the ascending method is that you may not notice an extra intersection. Due to the lack of connections between parts of the product, the program does not recognize where the intersection is really needed and where it is not. If the designer didn't see the error, the incorrect layout will be sent to production. Hence, there are defects, time to recheck the layout, and resources to restart production.

In descending design, the chief designer of the project creates a sketch of the product – a "skeleton", then splits it into nodes and distributes these nodes among the designers. It is important that all project designers work in the same program and see the same "desktop": the model frame, main surfaces, and axes. When a part or node is ready, you don't need to move it from program to program. They are already where they should be. If the design components overlap or do not fit together, the program will show this: the entire product was originally designed within the same it system, so it can recognize the connections between the design details. In descending design, the probability of transferring an incorrect layout to production is lower, which means that the risk of defects is lower.

Transferring a product to a virtual environment allows you to check its performance even before it appears in reality. Understand what will happen to each cog of the rocket during its launch and entry into orbit, conduct a crash test of a new car model, or calculate the strength of the bolt when fired from a machine gun-all this can be calculated thanks to PLM in a couple of days. The system includes integrated systems for modeling the operation of products. In terms of time and cost, virtual tests are more efficient than real ones.

A promising area of PLM development is interaction with customers. Instead of showing customers "live" samples, in theory it is possible to provide them with an electronic product layout. If customers want to upgrade something in the design or purchase a product with an individual configuration, the company will make changes to the digital design of the product. Production of the model will start only after the client approves its design and all technical characteristics. If the sample does not meet expectations, you can replace it with another one, without wasting months of work of all the plant's services.

3. Results and Discussion

Modern organizational concepts in production including agile manufacturing have been implemented sporadically in Russian industrial sector of economy [12-18].

JSC Kalashnikov Concern is the largest Russian manufacturer of a wide range of precision weapons. A large segment of civilian products includes hunting rifles, sporting rifles, machine tools and tools. 51% of the concern's shares belong to the State Corporation Rostec, 49% - to private investors. Kalashnikov concern products are delivered to more than 27 countries. The concern includes three product brands: Kalashnikov-military weapons, Baikal-hunting and civilian weapons, and Izhmash-sports weapons. Today, the group of companies is implementing a development strategy until 2020, the main priorities of which are to produce a wide range of products that are competitive on the world market, improve the efficiency of production processes, build an effective management structure and create comfortable working conditions for all employees of the enterprise.

Kalashnikov Concern took part in the First all-Russian conference "Digital industry of industrial Russia" (CIPR), which takes place from 7 to 10 June 2016 in Tatarstan [19]. The group's specialists shared their experience in building and developing smart manufacturing, as well as supply chain management (SCM), warehouse management (WMS), and product lifecycle management (PLM) systems. Today, the development of production of an industrial enterprise is impossible without a clear understanding of all stages of life cycle management of the product that it produces. Even the traditionally conservative arms industry can't do without dramatically speeding up the process of developing and bringing a product to market. This acceleration can be achieved through the use of a agile development methodology and an emphasis on quality control of the development process, as well as the introduction of a unified product lifecycle management environment for the designer, technologist and customer.

PLM combines all the enterprise it platforms used to create products. It all starts with a computer-aided design system, in which designers create an electronic 3D model of the product. Based on the ready-made layout, technologists develop control programs for processing parts on CNC machines (numerical control). Information about what tools and materials are needed for the production of each part of the product, production standards and deadlines, as well as other data is sent to KSAUP QC and MES&SCADA QC: systems for production planning. The first one plans globally - in the period from a month to a year, the second one is intended for forming a shift-daily task and monitoring its implementation at the level of production, shop, site and each employee. Every PLM project is unique because every enterprise is unique. All have their own internal regulations, project management methods, organization of processes, interaction between services, and a different it landscape.

The implementation process is very extensive: there are many information systems in the enterprise, and they are all different. According to the head of the Department for technological preparation of production and technological design of the Department of development and support of PLM A. Pleshkova: "Since the start of the project, the format of work has changed several times, and priorities in tasks have changed." As a result, the company switched to scrum and divided the team into two parts: design and technological preparation of production. This distinction is mandatory. If you consistently transfer designers to the system first, and then technologists, the latter will sooner or later find themselves in an information vacuum.

IOP Conf. Series: Materials Science and Engineering **968** (2020) 012027 doi:10.1088/1757-899X/968/1/012027

The first product that was developed in the electronic layout format was the new AK-12 assault rifle. It was a complex and lengthy project: the equipment was updated to meet the requirements of the system, the format of interaction between services was changed, people were trained to work in the program, and a new approach to design was tested. In the course of digitizing the product, many problems arose. The development and support team changed the system architecture several times based on feedback. Despite the fact that many employees were involved in the project, including specialists from the design and technology center, digitizing the product took quite a long time. And this is absolutely justified: together with the introduction of a new computer-aided design system, the process of creating a new product was completely rethought.

Before the introduction of PLM, it was only possible to mark and control the product path based on reports on paper and uploads from different programs. In an industrial environment, PLM combines all information about a product from the beginning of work on it, and provides data integration between engineering programs, planning systems, and production management. The relationship between multiple it platforms will help to avoid defects and to improve the quality of the product. Since all this happens in the same ecosystem, everyone who has access to the project will see any change that a colleague has made. No matter what distance the participants are from each other. PLM eliminates errors: designers, technologists, purchasing, and related departments are all located in the same information field.

Another factor that affects the speed of product creation is the complex geometry of product parts. When it is justified by the requirements for the technical characteristics of the product or its appearance, technologists look for production methods and methods of processing materials. It happens that the geometry can be simplified without losing the quality of the product, then technologists and designers discuss how to do it better. PLM includes a set of programs for calculating the parameters of design solutions. For example, the strength calculation system will show that the geometry of some part of the part does not meet the requirements. Because of this, the element may fail during operation of the product. After seeing this, the designer can change the geometry before passing the finished layout to the technologists.

4. Conclusions

According to the results of the study, the following conclusions can be drawn:

1) Industry 4.0 initiates a change in the strategic guidelines of industrial enterprises. Transformation of operational models into new digital models entails an identification of new effective models for the industrial development in new reality. Agile management is one of them, which basic idea is to adapt and rapid response to changes in the external environment, using their advantages as market opportunities;
2) currently, the development of industrial enterprise production is impossible without a clear understanding of all stages of the product management life cycle. So, PLM comes to the fore. PLM affects the ideology of creating a new product: to reduce the time to market and save resources, the system pushes to shift the labor intensity from the "production" stage to the "design" stage;
3) successful existing example of JSC Kalashnikov Concern of PLM implementation in the context of agile approach indicates the possibility of improving the state of the Russian industrial sector and, in general, the prospects of this concept in Russian economy. JSC Kalashnikov Concern can be competitive in the global industry;
4) when all stages of the product lifecycle are automated, engineers will be able to devote more time to the question "what do you need from the model?". The question "how to achieve this?" will take a back seat and will be provided with PLM. The lack of a framework in the thinking of inventors brings us closer to the era of the smartest and most complex machines in the history of mankind.

A special value of the work is the described aspects of JSC Kalashnikov Concern's case study. This example is significant in the context of the author's research of enterprises in the industrial sector of Russia. It makes possible to highlight the distinctive features of the development of Industry 4.0 in Russia. The study is a value for researchers in the field of digital transformation of industrial enterprises due to requirements of Industry 4.0 across the globe.

5. References

[1] Kidd P T 1994 *Agile manufacturing: forging new frontiers* Addison-Wesley
[2] Larman C 2004 *Agile and iterative development: A manager's guide* Addison-Wesley
[3] Sutherland J 2014 *SCRUM. The art of doing twice the work in half the time* Crown Business
[4] Goldman S L, Nagel R N and Preiss K 1995 *Agile competitors and virtual organizations: strategies for enriching the customer* Van Nostrand Reinhold
[5] Gunasekaran A 2017 *International Journal of Production Research* pp 1-13
[6] Dubey R and Gunasekaran A 2014 Agile Manufacturing: framework and its empirical validation *The International Journal of Advanced Manufacturing Technology* **76**(9-12) 2147-57
[7] Gunasekaran A 1998 Agile Manufacturing: Enablers and an Implementation Framework *International Journal of Production Research* **36**(5) 1223-47
[8] Sharp J, Irani Z and Desai S 1999 Working towards agile manufacturing in the UK industry *International Journal of Production Economics*, **62** pp 155-169
[9] Grieves M 2005 *Product Lifecycle Management: Driving the Next Generation of Lean Thinking* McGraw-Hill
[10] Adler Ju 2018 *Development of lean production systems in Russia: from history to the present*, Academic project
[11] Altintas Y, Brecher C, Weck M and Witt S 2005 *Virtual machine tool CIRP Annals,* **54**(2) pp 115-138
[12] Gromova E A 2019 *Digital economy development with an emphasis on automotive industry in Russia Espacios* **40**(6) p 27
[13] Rudskaya I and Rodionov D 2017 Econometric modeling as a tool for evaluating the performance of regional innovation systems (with regions of the Russian Federation as the example) *Academy of Strategic Management Journal* **16**(2)
[14] Pupentsova S V and Livintsova M G 2018 Qualimetric assessment of investment attractiveness of the real estate property *Real Estate Management and Valuation* **26**(2) pp 5-11
[15] Kalinina O and Valebnikova O 2018 Human Capital Management as Innovation Technologies for Municipal Organization *Advances in Intelligent Systems and Computing* pp 1315-1322
[16] Rodionov D G, Konnikov E A and Konnikova O A 2018 Approaches to ensuring the sustainability of industrial enterprises of different technological levels *Journal of Social Sciences Research* pp 277-282
[17] Glukhov V, Turichin G and Klimova-Korsmik O 2016 Quality management of metal products prepared by high-speed direct laser deposition technology *Key Engineering Materials* **684** pp 461-467
[18] Kuladzhi T, Babkin A and Murtazaev S A 2018 Matrix Tool for Efficiency Assessment of Production of Building Materials and Constructions in the Digital Economy *Advances in Intelligent Systems and Computing* **692**
[19] *Official site of the JSC "Kalashnikov Concern"* https://rostec.ru/news/4518354/ (accessed on Jul. 2020)

Acknowledgements
This research work was supported by the Academic Excellence Project 5-100 proposed by Peter the Great St. Petersburg Polytechnic University.

Tribological Performance of Graphene and Graphene Oxide Films as Solid Lubricant Layers on Tool Steel Surfaces

M Píška[1], J Sedláček[2], V Foksová[2] and P Sliwková[1]
[1] Brno University of Technology, Faculty of Mechanical Engineering, Institute of Manufacturing Technology, 616 69 Brno, Czech Republic
[2] MORAVIA CANS a. s., Tovární 532, 687 71 Bojkovice, Czech Republic

E-mail: piska@fme.vutbr.cz

Abstract. Some high productive processes induce a use of effective cooling and lubricating for forming and cutting tools today. Apart of various fluids, foams, oils and emulsions with EP (extreme pressures) additives some very effective means based on graphene or graphene oxide show excellent performance in their solid phase. In this study, a very effective way to enhance the tribological performance of graphene layers on tool steel surfaces is studied. The solid lubricants based on graphene and graphene oxide flakes showed a very good thermal stability, low coefficient of friction and high wear resistance. However, some technological conditions and topographies of surfaces for their successful applications should be made in advance as prerequisites. The research work deals with a high productive forming of metals and improving the mechanical and functional properties of special products. It contributes to finding the optimal internal varnishes, testing their resistance to stress and chemical resistance to the components, understanding all relevant mechanical properties especially of graphene (friction properties, adhesion and abrasion resistance). This research and development is concerned with a production of thin walled products, which will provide future customers with a more durable and safe products and also would have better properties compared to similar products already available on the market.

1. Introduction

Today´s scientific databases (WOS, Scopus) offer several thousand scientific works devoted to graphene and graphene oxide. About 1,000 papers deals with super friction properties, adhesion, superior strength, resistance to wear and the effect on mechanical properties of tribological systems. However, only a few of them with their behaviour at high speed deformation rates conditions used in industry, on real part surfaces. Right there the moving mechanical parts or work pieces undergoing sever plastic deformation consume a lot of energy also due to the friction. Therefore, investigations have been looking for innovative ways to minimize the friction and generation of heat via use of the advanced materials:

a. Graphene in general is a single-layered two-dimensional material, crystalline allotrope with a hexagonal lattice structure made from carbon atoms – Figure 1a. Graphene was first by the mechanical exfoliation of 3D graphite crystals. Nevertheless, the production costs for a mass use are still high.

b. Graphene-oxide (including the reduced graphene oxide) are also frequently referred to the perspective materials displaying extremal low friction mentioned also as super- lubricity, see Figure 1b. Graphene oxide (GO) is a similar material, made of carbon, hydrogen and oxygen molecules. GO seems to be a similar but cheaper product with some better mechanical properties compared to pure graphene. Through sonication, graphite adopts oxygen-containing functional groups that allow the material to be better dispersible in water that promote its use as a colloid. Despite all these research efforts and fundamental approaches, the super-lubricity has seldom been achieved at technological production systems. There are many theories and reasons for this fact, stemming from physical or chemical interactions that occur simultaneously at the sliding interfaces far away for hypothetical ideal planes, however, it is still a challenge for many engineers to realize them.

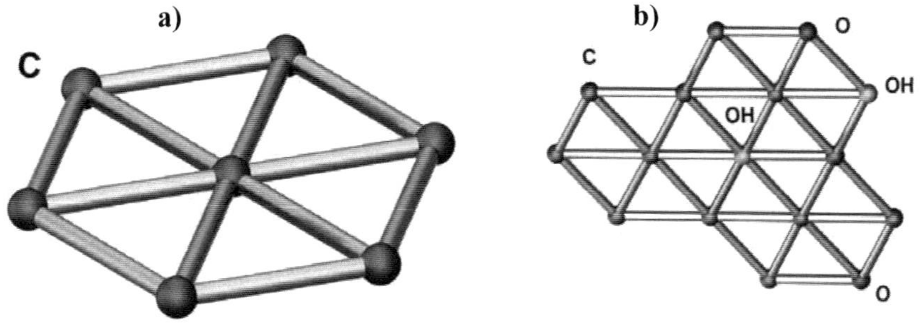

Figure 1. A single structure of graphene (a), and graphene oxide (b).

Graphene can have several morphological phases - spheres (hollow), of a size about 200 nm [1], flakes [2], wafers [3,4], ribbons [5], layers [6,7], nanotubes [8]. Its applications can take place on solid surfaces in the solid form, but also as a liquid (dispersed in water or oil) and even foam applications. In some cases, it can be mixed with MoS2 or nano-TiO2 [9], which are also excellent lubricants themselves. These nano-additives based on fullerenes (carbon beads) can be further filled inside with inorganic tungsten sulfide (IF-WS2) [10]. Such materials can be resistant to the extreme pressures, temperatures up to 500 °C. They have been developed for military purposes about 10 years ago in Israel and massively made in the U.S.A. later. The composition was based on IF-WS2/polymer and recommended for use in metalworking (drawing, forging, stamping, etc.) where provides excellent surface finishes also. The products were based on multifunctional closed spherical particles of inorganic tungsten bisulfide similar to fullerenes (IF-WS2), which provided unique properties due to exfoliation and tribological film formation on the metal surface. This product surpassed conventional synthetic and semi-synthetic fluids in its properties and did not contain chlorine, boron, formaldehyde or zinc.

Graphene and functionally deposited graphene are promising candidates as solid ultra-thin lubricants for solving adhesion and friction even in micro- and nano/electromechanical systems. Friction as a function of load shows a nonlinear characteristic [11] due to their strong adhesion. The phenomenon of increasing adhesion increases with increasing surface energy. Improvements in dynamic adhesion and friction are associated with associated hydrophilic properties. The beneficial effect of multilayer graphene in friction on M2 steel was demonstrated in [12], which confirmed the formation of shear layers with a low coefficient of friction.

Graphene is theoretically an excellent mechanical system with extreme flexibility [13], ultra-strong adhesion, and gas impermeability. For this two-dimensional material, the graphene interactions with the support substrate are unique. Amonton's first law states that macroscopic friction is proportional to the load applied, justified by the argument of increasing the load to increase the microscopic contact area between the two surfaces. However, due to its ultra-strong adhesion and low flexural stiffness, graphene does not require any higher loading to achieve an almost perfect fit to the nanoscale of the topography of its substrate. In contrast, the work [14] shows that damage to the graphene layers can lead to mutual sliding and up to several times higher deterioration of friction, although micrometric graphene residues can continue to reduce friction, even if the connection of the graphene layer is damaged. Friction on multilayer graphene is known to be dependent on the thickness of the layer [15]. The exact role that this interface plays, especially its size, contact and substrate roughness, in the graphene friction dependence on the layer has not yet been fully elucidated, which is the subject of further research.

2. Materials and method

This Alumina alloy samples - AlCu4Mg according to the standards EN AW 6082 T6, EN 753-3 EN 755-2, in dimensions 30x6.6x100 mm (3x3 pieces, each tested 6 times) - Figure 2 - have been machined precisely, ground, polished and covered with three sorts of an epoxy lacquer and the same thickness 5.0±0.2 μm (measured via kalotest) marked as:

a) EP – a standard epoxy lacquer (a company know-how),
b) HP – the standard epoxy lacquer enriched by 0.02 wt.% of Graphene GNP HP F8/1 - nanoplatelets of size up to 26.5 μm, 73% ≤ 10 layers,
c) RGO – the standard epoxy lacquer enriched by reduced graphene oxide (10% dispersion in isopropanolu, and the dose of 1.5 wt.% of the dispersion), with a low degree of oxidation, average size of particles 500 nm, 90% particles of dimensions less 800 nm with number of layers less 10.

All samples were set at the temperature 225 °C for 5 minutes in a laboratory heat oven Hoba 7940 (PPG, U.S.A.) The cylindrical tool was made from hardened steel 42CrMo4V - Table 1, turned precisely to a glossy surface with polycrystalline cubic boron nitride (PCBN) round cutting inserts SECO CCGW120408S-01020-L1-B CBN010 and ground by sapphire sandpapers of grit sizes FEPA 120, 240, 480 and 600 to suppress turning surface morphology without any releasing or re-fastening of the tool. The morphology of surfaces, roughness and other geometrical parameters were evaluated with Alicona IF G5 (Bruker Alicona, Austria), some selected parameters with Zeiss Stemi 2000C (Carl Zeiss AG, Germany) also. The surface temperature of the tool was measured with the thermal camera Flir 2000 (Flir, U.S.A.).

Table 1. Steel 42CrMo4V CSN EN 10083-1: 1991+A1: 1996; DIN 17200 – hardened state.

Chemical composition (weight %)							
C	Cr	Mo	V	Si	P	S	Fe
0.38	0.15	0.15	0.15	0.22	0.013	0.017	balance

Mechanical properties			
HRC Hardness*	Yield point Re [MPa]	Tensile strength Rm [MPa]	Young modulus [GPa]
35-36	920	1103-1137	224

* *Zwick/Roeller Rockwell hardness tester ZHR 4045/4150/8150, loading 150 kg, diamond indenter.*

After each test a re-polishing and cooling of the circumferential contact surface to 20-25 °C followed. Tribological tests have been running at a precise turning machine (SV18RD, TOS Trencin, a.s., overhauled by the M-MOOS Ltd., Lipnik nad Becvou in 2020) and use of a special friction facility that enabled many advantages – Figure 2:

- production of symmetrical and coaxial cylindrical steel surface aligned to the axis of its rotation, with high quality of the surface (Ra<0.4 μm) - Figure 3 and Figure 4,
- repeatable setting of the same loading conditions for the tribological tests,
- an adjustable sample positioning, dry and wet testing conditions etc.

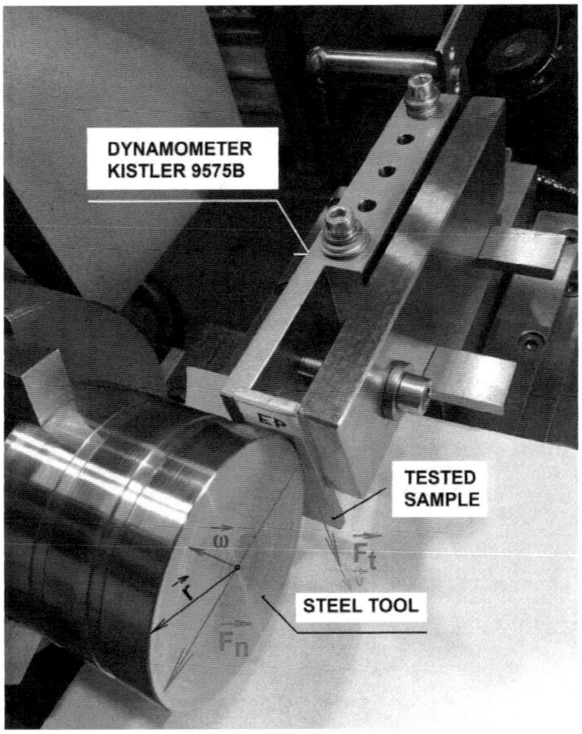

Figure 2. Principle of the friction test - measurement of normal force $\overrightarrow{F_n}$ and tangential force $\overrightarrow{F_t}$ for angular speed $\vec{\omega}$ and circumferential speed $\vec{v}$ according to tool radius $\vec{r}$.

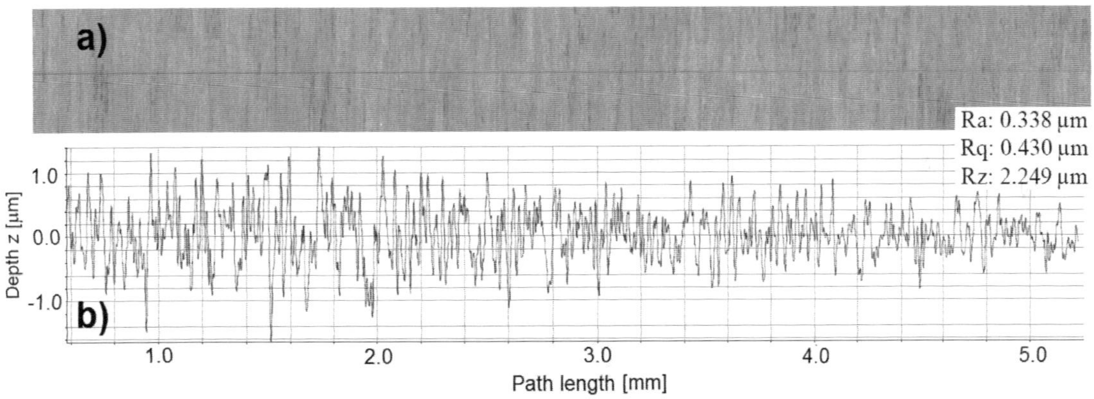

Figure 3. The surface of the steel tool and its quality; (a) – the morphology after the final polishing, (b) the typical selected surface amplitude characteristics.

The polished tool surface – see Figure 4 – exhibited a very low fraction of peaks above the core of material with surface texture aspect ratio over 88°, so a very good alignment of the circumferential speed to the machined surface was set to avoid a tangential removal of particles by micro-cutting due to a helix path that had been made by the cutting insert in the turning operation.

The laboratory tests were performed at a circumferential speed of 100 m/min (resp. 1.67 m/s), with a constant pressing force of 40 N, derived mechanically by means of a weight. The speed was set using a frequency converter and read by the digital chronometer (TESLA, a.s., Czech Republic). The piezoelectric system KISTLER 9575B (Kistler, Switzerland) was mounted on an axial bearing with

IOP Conf. Series: Materials Science and Engineering **968** (2020) 012028 doi:10.1088/1757-899X/968/1/012028

virtually zero friction along the motion axis. The basic scanning frequency was 100 Hz, with a long time constant and a low-pass filter of 10 Hz. The apparatus was calibrated using a laboratory mechanical weights in the scale 10–60 N with a high linearity (a statistical correlation over 98%). Testing was performed under dry conditions.

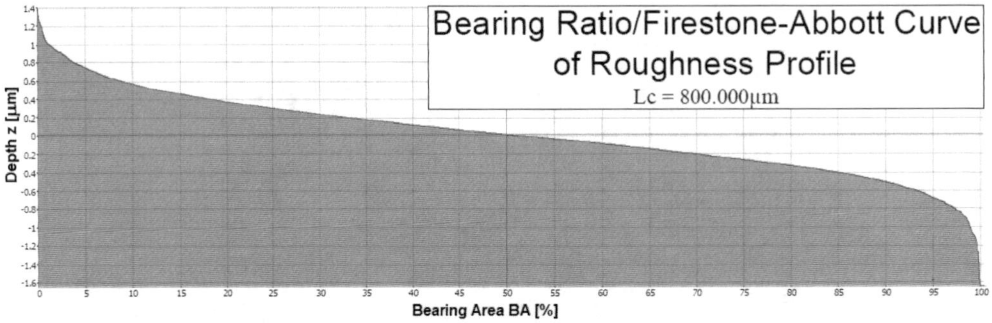

Name	Value	Unit	Description
Rk	1.077	μm	Core roughness depth, Height of the core material
Rpk	0.448	μm	Reduced peak height, mean height of the peaks above the core material
Rvk	0.453	μm	Reduced valley height, mean depth of the valleys below the core material
Rmr1	10.370	%	Peak material component, the fraction of the surface which consists of peaks above the core material
Rmr2	90.490	%	Peak material component, the fraction of the surface which will carry the load
l	5.800	mm	Profile Length
Lc	800.000	μm	LambdaC: cut off wavelength

Figure 4. The bearing ratios of the polished steel tool surface.

The actual course of a test started with a gradual motion of the sample forward the rotating tool until the full force thrust was reached. Then the constant engagement continued until the first traces of adhesion appeared on the surface, when the testing was stopped, and sample was released – Figure 5. All the time the force data were measured and downloaded. Their evaluation was performed using the Data Processing module in Microsoft Excel v. 10, U.S.A., which was based on a statistical evaluation of instantaneous Coulomb friction μ_i according to the relation of instantaneous tangential force Ft_i and normal force Fn_i:

$$\mu_i = \frac{Ft_i}{Fn_i} \tag{1}$$

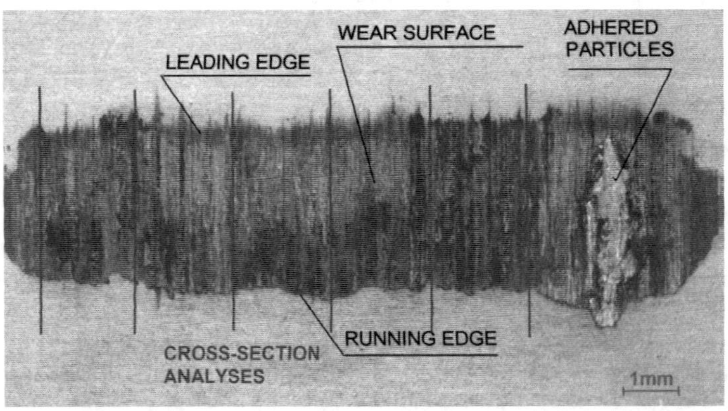

Figure 5. A typical pattern of wear at the end of testing (Zeiss Stemi 2000C).

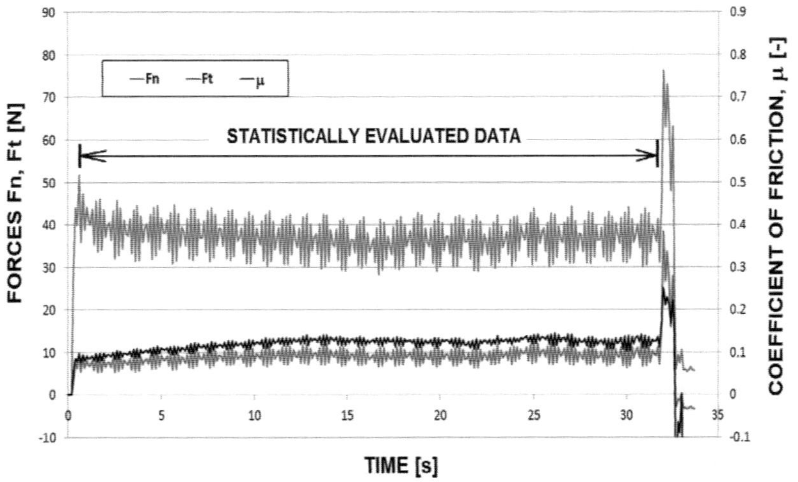

Figure 6. Selected measured force data from tribological tests, limits for a statistical assessment.

3. Results and discussions

This method of testing resulted in very consistent files for the friction evaluation with normal Gaussians distributions (assessed with the test $\chi 2$). The range of statistically processed data captured a steady, stable phase of the friction see Figure 6 and Figure 7 and did not include either the first moments of the tool contacts with the lacquers, or a formation of the metal contacts between the tool and the aluminum base of the samples in the end of the test.

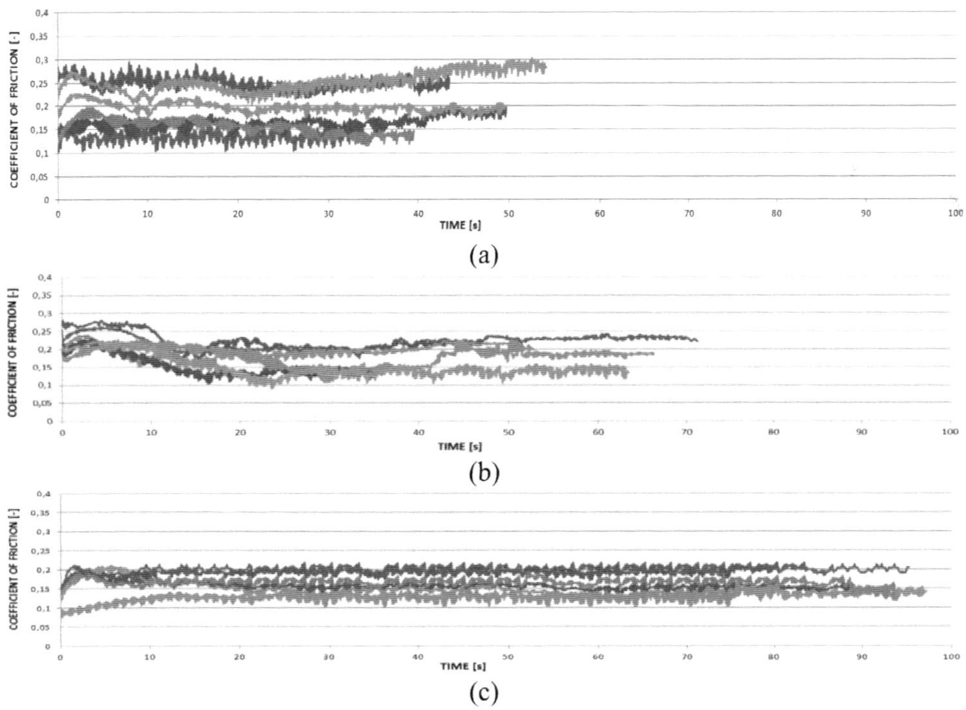

Figure 7. Time series of friction coefficients: (a) epoxy lacquer, (b) epoxy lacquer with graphene, (c) epoxy lacquer with reduced grahene oxide.

Table 2. Statistical evaluation of the data - mean and standard deviation of friction coefficient for all tested materials.

Statistics	EP	HP	RGO
mean [N]	0.1988	0.1853	**0.1648**
standard deviation [N]	0.0482	0.0404	0.0280

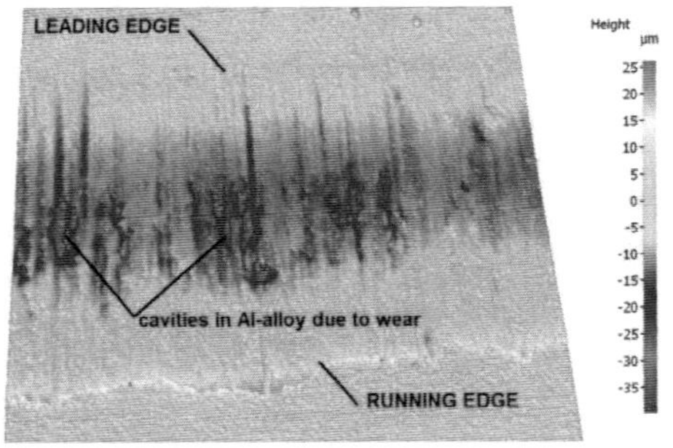

Figure 8. A typical pattern of the wear.

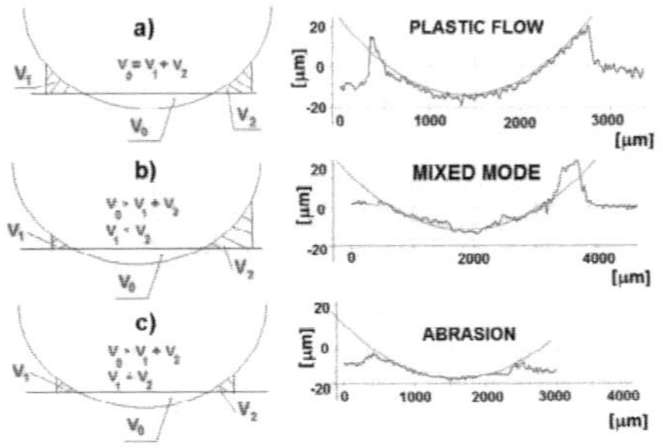

Figure 9. Three types of wear.

A typical pattern of the wear can be seen in Figure 8, where the morphology of a worn surface in the moment of tool penetration into the alumina surface is shown. In this case, the whole thickness of the lacquer was removed from the contact area in the deepest distance. Anyway, three types of wear have been watched at the cross-section of the surface and analysed - Figure 9, each with different material areas and volumes removed by the rotating tool. The prevailing mode of plastic flow was observed for the pure epoxy lacquer. The dominant abrasive mode was rare, but it has been found a few times for the lacquers enriched with graphene and graphene oxide. The dominant mechanism of wear for lacquers with graphene and graphene oxide resulted in the mixed mode.

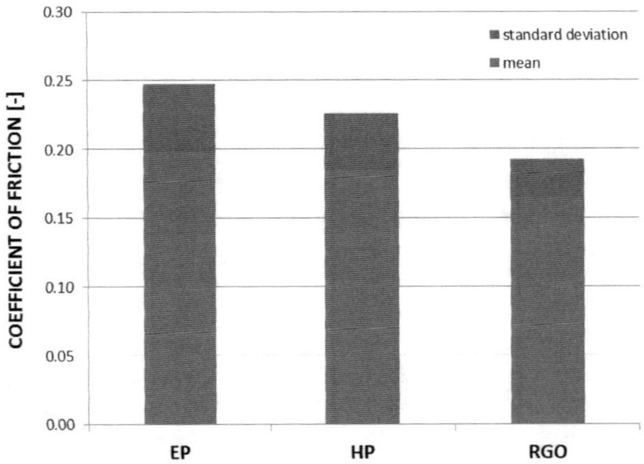

Figure 10. Statistical data of the coefficient of friction for individual coatings and given loading conditions.

All wear patterns and morphologies were analysed and classified in details. The analyses showed that the initial abrasion mechanism for all layers was similar. First, there was an abrasive removal of the contacting lacquer irregularities, then an increase in the contacting area and in temperature also due to friction and plastic flow of the loaded lacquer layer. The material flow at the leading edge was usually lower than the one at the running edge. However, it was found that:

- the intensity of wear for the various materials was different with respect to the time,
- the tests were, among other facts, sensitive to a corrugation or some irregularities of the samples at the microscopic level (a careful preparation of the flat surfaces had to be guaranteed),
- a partial problem can be un-known orientation of the graphene or graphene-oxide flakes exerted to the loading, so a little plastic flow or a suitable ordered particle orientation with a convenient sliding planes can help to low friction [17,18],
- the mechanical loading (40 N) seemed to be optimal, because higher loadings (50 N and more) caused a quicker testing, but followed with a higher dispersion of data and a lower loading (30 N) prolonged the testing and some "burnt" micro-particles begun to appear at the frictional areas.

Based on statistical processing of the results - Table 2, Figure 10 - the RGO type coating was found to be the superior coating for the given loading conditions, showing the lowest value of the friction, low data dispersion (expressed with standard deviations) and the longest time of its performance. There is the statistically significant difference between the mean values of the epoxy lacquer with reduced graphene oxide compared to the epoxy lacquer and epoxy lacquer with addition of graphene (α=0.05) also.

The effect of heating of the lacquer when loaded can be important, because it can soften the hardness and mechanical properties of the layer, its dynamic viscosity. Nevertheless, all tests started from the surface temperature 20±1°C and ended in the range of 50–62°C. However, the values are moreless informative, because it was difficult to define the surface emissivity exactly. The deeper analyses of the mechanisms continue.

4. Conclusions, future prospects

This application of graphene and graphene oxide can be beneficial for a wide range of applications in technical and medical practice. These phases, in a wide range of morphologies and dimensions, have been shown to reduce friction at the point of contact and protect technical components, forming and cutting tools from wear. These layers work theoretically from a single atomic layer, but provided perfectly flat and smooth contact surfaces. This is difficult to achieve in the common technical practice, and therefore it is necessary to determine the type of phase applied, its concentration and possibly the size of the layering. Experimental tests have shown very promising results for reduced graphene oxide, which could be used for application to the technological surfaces.

Next research with use of Raman spectroscopy would be oriented to a study of the induced dominant particle orientation and a technology of its deposition. A tiled composition of the flakes with very strong covalent bonds in the hexagonal structures, situated perpendicularly to the expected loading, and the weak Van der Waals bonds within the layers might yield in a significant lubrication effect and other benefits.

5. References

[1] Song H, Wang Z and Yang J 2016 Tribological properties of graphene oxide and carbon spheres as lubricating additives *Appl. Phys.* A **122** 933

[2] Merel S, van Wijk M, de Wijn A S and Fasolino A 2016 Collective superlubricity of graphene flakes *J. Phys.: Condens. Matter* **28** 13 p6

[3] Ma T B, Hu Y Z, Xu L, Wang L F and Wang H 2011 Shear-induced lamellar ordering and interfacial sliding in amorphous carbon films: A superlow friction regime *Chemical Physics Letters* **514** 4 pp325-329
[4] Porwal H, Tatarko P, Saggar R, Grasso S, Mani M K, Dlouhý I, Dusza J and Reece M J 2014 Tribological properties of silica–graphene nano-platelet composites *Ceramics International* **40**(8) pp12067–12074
[5] Fang Jun, Chen Bin and Pan H, Anomalous friction of graphene nanoribbons on waved graphenes *Theoretical and Applied Mechanics Letters* **5** pp212–215
[6] Wang L. and Duan F. Nanoscale wear mechanisms of few-layer graphene sheets induced byinterfacial adhesion *Tribology International* **123** pp266–272
[7] Zhang J, Lu W, Tour J M and Lou J 2012 Nanoscale frictional characteristics of graphene nanoribbons *Appl. Phys. Lett.* **101**, 123104
[8] Liu P and Zhang Y 2011 A theoretical analysis of frictional and defect characteristics of graphene probed by a capped single-walled carbon nanotube *Carbon* **49**(11) pp3687-3697
[9] Hu K H, Xu Y, Hu E Z, Guo J H and Hu X G 2016 Rolling friction performance and functional conversion fromlubrication to photocatalysis of hollow spherical nano-MoS2/nano-TiO2 *Tribology International* **104** pp131–139
[10] https://nanotech-europe.cz/prumyslove-oleje-a-mazadi-prisady-do-procesnich-kapalin/
[11] Ge X, Li J, Luo R, Zhang C and Luo J 2018 Macroscale Superlubricity Enabled by Synergy Effect of Graphene-oxide Nanoflakes and Ethanediol *ACS Appl. Mater. Interfaces* **10** pp8214–8224
[12] http://core-cms.prod.aop.campbridge.org/societes/materials-research-society
[13] Alexander N, Kitt L, Qi Z, Rémi S, Park H S, Swan A K and Goldberg B B 2013 How Graphene Slides: Measurement and Theory of Strain-Dependent Frictional Forces between Graphene and SiO2 *Nano Lett.* **13** pp2605–2610
[14] Marchetto D, Held C, Hausen F, Wahlisch F, Dienwiebel M and Bennewit R 2012 Friction and Wear on Single-Layer Epitaxial Graphene in Multi-Asperity Contacts *Tribol Lett* **48** 77-82
[15] Ye Z, Balkanci A, Martini A and Baykara M Z 2017 Effect of roughness on the layer-dependent friction of few-layer graphene *Physical Review* B **96** 115401
[16] Jianguo S et al. 2014 Preparation and Characterization of Graphene Oxide *Journal of Nanomaterials* 276143
[17] Berman D, Erdemir A, Sumant A V 2018 Approaches for Achieving Superlubricity in Two-Dimensional Materials *ACS Nano* **12**(3) pp 2122-2137
[18] Srivyas P D and Charoo MS 2019 Graphene An effective lubricant for tribological applications *Lecture Notes in Mechanical Engineering* pp 239-258

Acknowledgment
This work was supported by the grant „Application of the New Surface Treatment Technologies in Metal Packaging Industry“, PID: FV40313, Ministry of Industry and Trade, The Czech Republic.

A Python and Java software approach for 2.5 axes, self-adaptive stretch forming process and IoT solution

C C Grigoras, B Chirita and G Brabie
"Vasile Alecsandri" University of Bacau, Faculty of Engineering, Calea Marasesti 157, 600115, Bacau, Romania

E-mail: cosmin.grigoras@ub.ro

Abstract. Stretch forming is a metal forming process that implies bending and stretching a sheet of metal over a die, with the sheet being plastically deformed into the desired shape. The purpose is to obtain large parts. One of the main aspects is that the metal sheet is locked into position by gripping jaws. A hydraulic ram is raised into the metal sheet, therefore increasing the tensile forces. This process is used to draw intro shape materials like aluminium, magnesium, titanium alloys, stainless steel, Inconel. These materials have in common the fact that they have poor formability or the elastic spring-back has unacceptable values. Taking this into consideration, the nature of the materials along with the fact that the metal sheet can crack, due to excessive strain, we propose a self-adaptive stretch forming process. Underlining this process is the stress-strain curve, that has three inputs: force, sample section area and strain. For each of these factors, our method uses a particular approach, as in a Python-based software and Android-based IoT solution, that uses stress and strain data. Furthermore, it controls, in real-time the hydraulic press to the point at which the material is stretch close to its ultimate yield strength.

1. Introduction

While not so popular among the research topics, the stretch forming process is an essential part of the aeronautics industry, as it allows the making of large contoured panels in one piece. Investigations in this field are mainly conducted in R&D projects for the military sector or by major corporations like Stork Fokker or Dassault Aviation [1, 2] and access to this work is restricted. For other sectors, like automotive, stretch forming can be used for components like door panel or hoods but the process is less efficient and very cost-intensive [3].

From the analyzed researched papers, the key aspect of stretch forming is strain control [4-6]. Furthermore, a considerable number of finite element analysis were conducted to understand the material flow [2, 7-9]. Approaches of the stretch forming process include stretch forming under fluid pressure [10], electromagnetic incremental forming [11], axisymmetric stretch forming combined with deep drawing [12], hybrid incremental sheet forming combined with stretch forming [13] adaptive manufacturing using 3D printed dies, fuzzy logic fracture investigation [14-16] and determination of friction coefficient [17]. In present days concepts like digital strategy, process tracking, IoT platform, artificial intelligence, automation, process digitalization, data analytics, smart logistics, IT infrastructure, cloud & connectivity, data transparency, sustainability, augmented & virtual reality, expert networks or predictive maintenance are implemented with success by industrial and IT groups like Schaeffler, Trumpf, NEONEX, Siemens, BASF, Pia Automation, IoT One, Mercedes, own space, Bosch & Bosch Rexroth, MUNSCH, Caterpillar, Honeywell, Boeing, Airbus and many others. The integration of these smart solutions leads to improvements in manufacturing costs, operations agility, product innovation and improved client service [18].

Taking any industrial process to the next level of automation is one essential step in implementing the Industry 4.0 standard [19]. Therefore, this paper proposes a novel approach, using a complex Python-based algorithm. This takes advantage of the stress and strain developed in the material during the stretch forming process. Furthermore, an Android Java-based application was developed as an auxiliary tool for real-time data display and in-process control, thus closing one small gap between smart process implementation and stretch forming.

2. The methodology of the adaptive stretch forming process

The ASFP (Adaptive Stretch Forming Process) consists of implementing image processing, object detection and tracking algorithms along with dedicated scripting to retrieve force data from a dynamometer. The algorithms search the video feed for white on black 3/5 [mm] reference markers (figure 1, camera view), locking them into position and measuring the initial distance between them "L_o", as XY coordinates. The software tracks the markers and returns their position "L_f", calculating in real-time the elongation of the material. Predefined level of stresses is used as a threshold so that the algorithm switches between elastic and plastic deformation modes. The combination of these two input factors is necessary when working with materials like magnesium alloys, as previous research [14] revealed that the fracture limit is very hard to predict, using classic techniques.

2.1. Industrial equipment and hardware

The schematic representation of the industrial equipment is presented in figure 1. The hydraulic press (1) has an intermediary plate (2) that is software-controlled in the vertical direction, over which a Kistler 9272 type dynamometer is mounted. On top of it, a 150x150 [mm] steel alloy die is mounted, with a radius of 150 [mm]. The image acquisition hardware is provided by a USB camera (5), that is mounted on a light diffusion system (6), for shadow cancellation. A sheet of metal (7) is placed between the gripping jaws (8). The system is designed so that the hydraulic pistons (9) stretch the material in the horizontal plane, while the hydraulic pistons (10) assures that gripping jaws are on the same direction as the metal sheet.

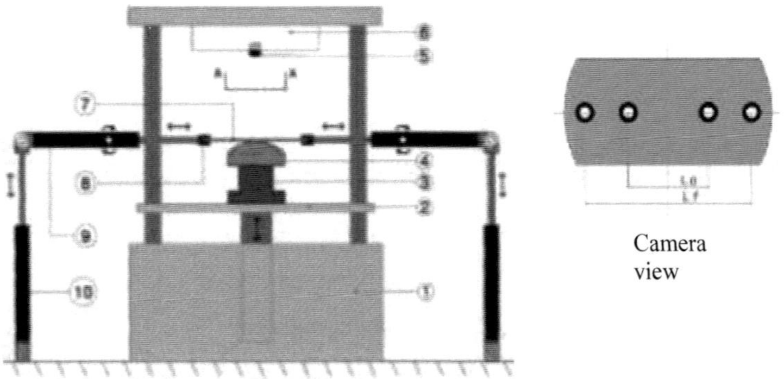

Figure 1. Schematic representation of the industrial equipment and camera view.

2.2. Software architecture

The software approach consists of two open-source applications, as shown in figure 2. A universal platform (Windows, Linux, OSX) Python-based software, which ensures data acquisition, data interpretation, decision making, and an IoT Android-based solution that returns real-time data from the stretch forming process and offers the possibility of in-process manual control. The software logic is based on the stress and strain input data. The input parameters are user defined (section area and modulus of elasticity) and measured (stress and strain). For this version of the algorithm some assumptions were made: a two-camera system is ideal for 3D measuring displacement, but it causes more problems than it solves in this actual stage, by means of camera calibration and intricate algorithms for 3D space motion, thus an approximation of the strain is made by a planar measurement and analytical approximation; this is made by knowing at each step the distance between the markers and the camera, this also allows to dynamically calculate the radius of the part, therefore knowing the distance between the markers as they on the radius of a circle; for the elastic region the stress is being calculated by force on section area while for the plastic region the stress is calculated using equation (1), where K and n are constant for each alloy, with n being the strain hardening exponent. The force is exported as a resultant of the normal and axial directions. The main goal is avoiding material failure. In the following, the paper presents how these values are obtained.

$$\sigma = K\varepsilon^n \tag{1}$$

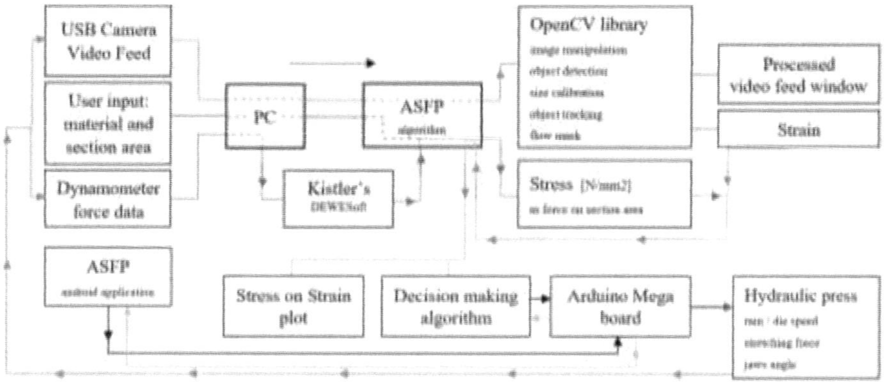

Figure 2. Simplified Adaptive Stretch Forming Process software architecture.

2.2.1. Python-based adaptive stretch forming software.

The ASFP application consist of eight Python-based individual modules joined together under a GUI (Graphical User Interface) highlighted in figure 3. The first two modules "Video calibration" and "Test video calibration" open the video feed, place a mesh over it, tracing the ideal position of the markers and indicating if the self-adaptive stretch forming process can start with no errors.

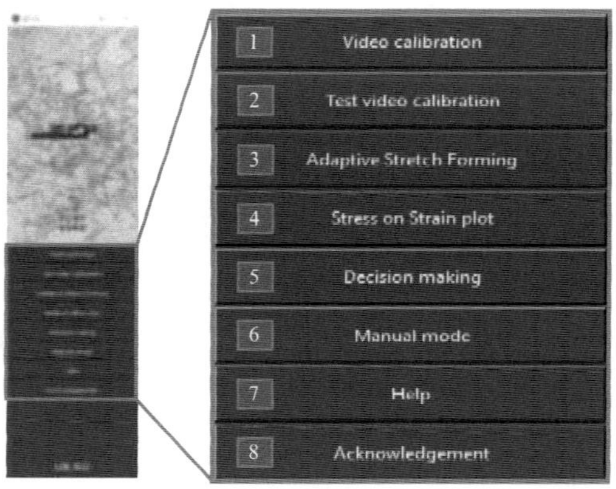

Figure 3. Adaptive Stretch Forming Process software GUI.

Module number three is a complex environment, composed of multiple scripts. The force values are returned from the dynamometer's official software (DEWESoft X3 SP10) via two python scripts and represent the resulting force developed on the vertical and horizontal directions. The processing speed is 200 Hz or 1 reading per 5 milliseconds. The data is timestamped and stored in a text file. The scripts were developed by DEWESoft technical team and adapted to this specific application; the communication, with the main application, is done by sockets and multiprocessing methods. The image processing is made in an OpenCV environment, that is represented by an open-source library. This includes several hundred computer vision algorithms, from which we have used image processing, video analysis, video capturing, camera calibration, object detection and object tracking. The script manages to detect the markers using the Shi-Thomas corner detector [20] and Hough Circle transformation [21] algorithms. The first method is rather a way of cancelling any additional point if they are found. For the tracking & tracing the Lucas-Kanade optical flow method is used [22]. It detects the outer shape of the black circles by RGB to GRAY colour conversion; it converts the shape to a circle and self-corrects the radius size; after this, it locks the circles centre with XY coordinates, as shown in figure 4. When the stretch forming process starts the algorithm traces those points, returning their position, thus the distance between them "L_f". Having these values, the algorithm calculates in real-time the elongation, by using equation (2). It does this at a maximum speed of 100 Hz or 1 frame per 10 milliseconds. If GPU processing is used, the algorithm can handle video and force data to speeds of 1000 Hz, thus a slow-motion camera can be connected. As the force and elongation data are recorded at different rates, the plot script uses timestamps and positions the data from the same timeframe in one line, that represent the real strain obtained at a specific stress. Furthermore, in this script, the decision making takes place via an "if-else goal achieve" algorithm. This controls an Arduino Mega, that is connected to the hydraulic press main panel. This way the intermediary plate (2) speed, the gripping jaws angle and stretching force in the hydraulic pistons (9 and 10) are controlled.

$$\varepsilon = \frac{\Delta L}{L_0} \Leftrightarrow \varepsilon = \frac{L_f - L_0}{L_0} \tag{2}$$

The video feed can be provided by USB camera, IP camera, smartphone camera or video file. The last one was needed as the software records the process from figure 4, and can be analysed afterwards. The "Decision making" module highlights the command given from module 3 to the Arduino Mega board. Connecting a 3D-connection space mouse allows working in module 6 in manual mode, useful for in-process user intervention or setup purpose.

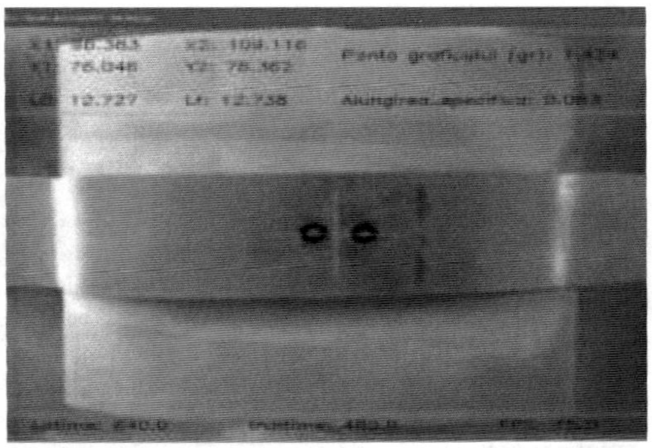

Figure 4. ASFP video feed analyser GUI.

2.2.2. Android IoT application.

This IoT solution was designed as an auxiliary tool and comes as an important step in linking this process to the Industry 4.0 standard. Having Bluetooth capabilities, it can connect wirelessly to the Arduino Mega board. The application is divided into the MainScreen.java (left side), MainScan.java (not shown) and the MainApp.java (right side) modules, as shown in figure 5. The main screen shows a list of the Bluetooth devices recently connected and can access the network scan module if the device is not listed. This interface also handles Bluetooth on-off switch and user permissions. After selecting the equipment, the main application screen appears. In the upper part of the screen information regarding the process elapsed time is displayed along with a "stop the connection" button. The next widget can be customized to show the video feed or it can display images taken from figure 4. The two seeker bars (stretch forming force and ram speed) can override the command given by the python ASFP algorithm and set the values as the user requested. To have full control of the two hydraulic pumps four-button were placed; the green ones turn on the hydraulic system for the ram and gripping system while the red one stops them. A scroll view widget is placed to return process information from the Arduino Board as text messages. In the lower part of the screen a manual input and send button was added, for custom control.

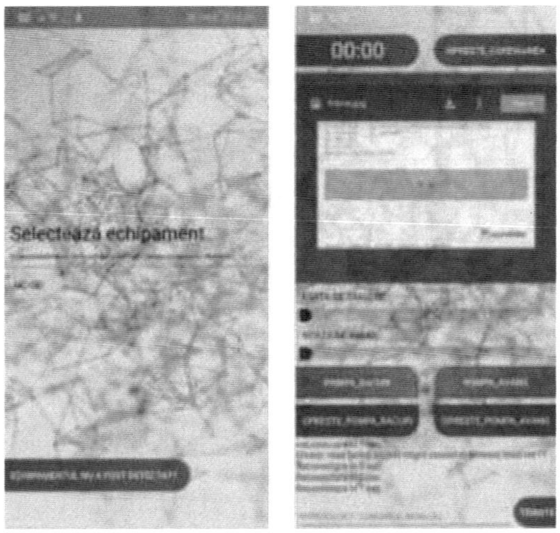

Figure 5. Adaptive Stretch Forming Process Android IoT software graphical user interface.

3. Conclusions

The applications ware developed by using two frameworks with more than 40 libraries. The precision of the camera measurements is within ±20 [μm]. The speed from data acquisition to decision making to process control if very fast. The software can handle as much as 50 complete set of instructions per second, but it's limited to 10 due to the Arduino board.

Although permanent software development is made, the application, at its actual version, proves that this solution is a useful resource as it can control the stretch forming process in real-time.

4. References

[1] Wisselink H H and van den Boogaard AH 2005 Finite Element Simulation of the Stretch-forming of Aircraft Skins *Numisheet 2005* ed. Smith L M et al. (Detroit: American Institute of Physics) pp 60-65

[2] Kurukuri S et al. 2011 Simulation of stretch forming with intermediate heat treatments of aircraft skins *Int J Mater Form* **4** pp 129–140

[3] Vlahovic D and Liewald M 2008 Benchmarking Methods for Short Cycle Stretch-Forming *International Journal of Material Forming* **1**(1) pp 193-196

[4] *** 1990 Scientific Technical Information Program NASA Center for AeroSpace Information *Scientific and Technical Aerospace Reports. NASA, Office of Scientific and Technical Information* **28** p 2573

[5] Cubberly W H and Bakerjian R 1989 Society of Manufacturing Engineers, Tool and Manufacturing Engineers Handbook Desk Edition *Society of Manufacturing Engineers* pp 33-46

[6] Hardt D E, Norfleet W A, Valentin V M and Parris A 2000 In Process Control of Strain in a Stretch Forming Process *Journal of Engineering Materials and Technology* **123**(4) pp 496-503
[7] Tasan C C, Diehl M, Yan D, Zambaldi C, Shanthraj P, Roters F and Raabe D 2014 Integrated experimental–simulation analysis of stress and strain partitioning in multiphase alloys *Acta Materialia* **81** pp 386-400
[8] Yan A M and Klappka I 2008 Springback in stretch forming process of aeronautic panel production by finite element simulation *International Journal of Material Forming* **1**(1) pp 201-204
[9] Zhang H, Diehl M, Rotersa F, Raabe D 2016 A virtual laboratory using high resolution crystal plasticity simulations to determine the initial yield surface for sheet metal forming operations *International Journal of Plasticity* **80** pp 111-138
[10] Mellor P B 1956 Stretch forming under fluid pressure *Journal of the Mechanics and Physics of Solids* **5**(1) pp 41-56
[11] Cui X, Mob J, Lib J, Xiaoa X, Zhoub B and Fangb J 2016 Large-scale sheet deformation process by electromagnetic incremental forming combined with stretch forming *Journal of Materials Processing Technology* **237** pp 139-154
[12] Lim TC, Ramakrishna S and Shang H M 1999 Axisymmetric sheet forming of knitted fabric composite by combined stretch forming and deep drawing. Composites Part B *Engineering* **30**(5) pp 495-502
[13] Araghi B T, Manco G L, Bambach M and Hirt G 2009 Investigation into a new hybrid forming process: Incremental sheet forming combined with stretch forming *CIRP Annals* **58**(1) pp 225-228
[14] Ciofu C, Grigoraș C C, Chiriță B, Iancu C and Brabie G 2020 Fracture Investigation in Draw Bending of AZ31B Sheets using Fuzzy Logic Prediction *Procedia Manufacturing* **47** pp 1462-67
[15] Grigoraș C, Chiriță B and Brabie G 2019 Additive manufacturing of a stretch forming die using 3D printing technology *IOP Conference Series: Materials Science and Engineering* **564** p 012017
[16] Grigoras C, Chiriță B, Brabie G and Ciofu C 2019 Experimental analysis of AZ31B magnesium alloy sheet failure using punch stretching *IOP Conference Series: Materials Science and Engineering* **682** p 012009
[17] Kaftanoğlu B 1973 Determination of coefficient of friction under conditions of deep-drawing and stretch forming *Wear* **25**(2) pp 177-188
[18] Commision E 2017 Digital Transformation Monitor. Industry 4.0 in Aeronautics: IoT applications Available from: https://ec.europa.eu/growth/tools-databases/dem/monitor/sites/default/ files/DTM_Aeronautics%20-%20IoT%20Applications%20v1.pdf.
[19] Saurabh V, Prashant A and Santosh B 2018 Industry 4.0 – A Glimpse *Procedia Manufacturing* **20** pp 233-238
[20] Mordvintsev A, Abid K and Tomasi S 2013 Corner Detector & Good Features to Track Available from: https://opencv-python-tutroals.readthedocs.io/en/latest/py_tutorials/py_feature2d/py_shi_tomasi/py_shi_tomasi.html.
[21] Mordvintsev A and Abid K 2013 Hough Circle Transform Available from: https://opencv-python-tutroals.readthedocs.io/en/latest/py_tutorials/py_imgproc/py_houghcircles/py_houghcircles.html
[22] Mordvintsev A and Abid K 2013 Optical Flow Available from: https://opencv-python-tutroals.readthedocs.io/en/latest/py_tutorials/py_video/py_lucas_kanade/py_lucaskanade.html

Acknowledgements
This work was supported by a grant of the Romanian Ministry of Research and Innovation, CCCDI – UEFISCDI, project number PN-III-P1-1.2-PCCDI-20170446 / 82PCCDI / 2018, within PNCDI III.

IOP Conf. Series: Materials Science and Engineering **968** (2020) 012030 doi:10.1088/1757-899X/968/1/012030

Algorithm for screws profiling of a conical trilobed compressor - analytical solution

N Baroiu, G A Costin, V G Teodor and N Oancea
"Dunarea de Jos" University of Galati, Romania, Department of Manufacturing Engineering, Domneasca street, No. 111, 800201, Galati, Romania

E-mail: Alexandra.Costin@ugal.ro

Abstract. Screw compressors, as rotary machinery, are capable of working with high compression ratios for gas compression. They transform the mechanical work done by an electric motor or turbine into the potential energy of a gas, by reducing its volume and increasing the pressure. The essential elements of these compressors are the two helical rotors with lobes, mounted in a housing. Existing rotors in the construction of compressors are working bodies whose composite profiles include circle arcs, line segments and cycloidal curves, while also often including non-analytical curves. Such geometry of profiles gives the gear processes superior characteristics, in the sense of reducing friction and improving sealing conditions. This paper analyses, by analytical approach, the possibility of constructing a multilobated conical screw compressor with a number of lobes Z, Z + 1, of the male and female rotor. The parametric equations of the contained worm and the kinematics of the working process of the conical screw compressor shall be defined. The theory of surface wrapping comprises analytical methods of study, which were developed on the basis of the fundamental theorems on wrapping of known surfaces in differential geometry and the kinematics laws on rotor movements. Thus, this paper proposes an algorithm based on the approach to the problems analyzed under the "virtual focal points theorem" between the complex surfaces of the conical surface lobes, which constitute the active lobes of the male and female rotors. Normal to active surfaces are defined, and also the speed in relative movement, in the "gear" process of the conical screw compressor. Regarding rotors in conical screw compressors, the circular pitch of the teeth is constant. Also, the screw pitch of the conical compressor rotors is a constant value. In these conditions, the rotors of the screw compressors are acknowledged as active surfaces, conical screw surfaces with constant axial screw pitch. The cross section (in the normal plane to the driving rotor axis) determines a plane gear. This defines the specific winding condition, and the shape of the conical screw surface of the female rotor.

1. Introduction

Essentially, conical screws compressors were analytically developed and later, through industrial applications by some authors for concrete cases of the conical helical compressor with the driver screw with 2 lobes and the wrapper screw with 3 lobes [1-4].

The newly proposed constructions have special industrial applications, such compressors having various applications in different types of industries, including the automotive and aeronautical industries [5].

The functional principle differs from the traditional one, in the sense that instead of cylindrical rotors with parallel axes [6-9], the constructions from conical rotary compressors place a rotor inside another rotor.

This approach was obtained through a combination of complex algorithms, computing power given by the modern software and advanced technological processing [10-12], one of the practical results being the decrease of the constructive volume given by the progressive change of the rotor diameter.

This paper proposes an algorithm that was developed based on the „*virtual focal points*" theorem between reciprocal enveloping surfaces – in this case – the driver screw (wrapped screw) and the driven screw (wrapper screw) [13-15].

An analysis of a conical helical compressor of *z/(z+1)* type is proposed for a driver screw (wrapped screw). The flank of the helical surface is analyzed in analytical mode. A different constructive form is proposed, admitting that the contained rotor of the compressor has the shape of the cross section multilobated profile (3 or 4 lobes) in a circle arc. An example of application is proposed, based on dedicated programmes, which could allow the development of various constructive geometric shapes for the conical rotors of the compressor.

2. Frontal generatrix of the conical screw

The frontal profile of the wrapped (driver) conical screw consists, Figure 1, in a set of circular profiles, of radii r_1 and r_2, having the centers on the circle of radius R_{rp} – circle belonging to the conical axode, having in the frontal plane, plane perpendicular to the conical axode axis, in the points O_1, respectively O_2.

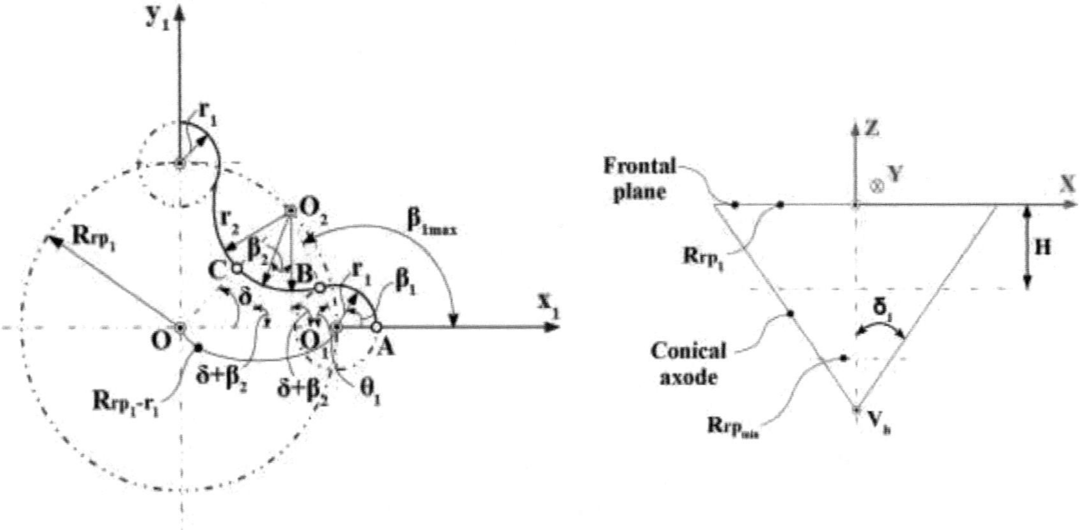

Figure 1. *a* – frontal profile of the driver conical screw; *b* – conical axode.

In connection with *AB*, *BC* profiles, the profiles radii sizes are defined, Figure 1: r_1 – circle arc with the center in O_1 on the centrode of radius R_{rp_1}; r_2 – circle arc with the center in O_2 on the centrode of radius R_{rp_1}. In this sense, it is observed that ΔOO_1O_2 is an isosceles triangle and, consequently:

$$\overline{OO_1} = \overline{OO_2} = R_{rp_1}. \tag{1}$$

Profiles are defined:

- for AB arc of equations in the X_1Y_1 plane:

$$\widehat{AB}: \begin{vmatrix} X_1 = R_{rp_1} + r_1 \cdot cos\,\beta_1\,; \\ Y_1 = r_1 \cdot sin\,\beta_1\,, \end{vmatrix} \tag{2}$$

with β_1 – variable size between the limits:

$$\beta_{1_{min}} = 0^{\circ}; \qquad \beta_{1_{max}} = \delta_1 - \theta_1. \tag{3}$$

- for $\widehat{BC}$ arc:

$$BC: \begin{vmatrix} X_1 = R_{rp_1} - (r_1 + r_2)\cdot cos\,\theta_1 - r_2 \cdot cos(\delta + \beta_2); \\ Y_1 = (r_1 + r_2)\cdot sin\,\theta_1 - r_2 \cdot sin(\delta + \beta_2), \end{vmatrix} \tag{4}$$

with β_2 variable angular size:

$$\beta_{2_{min}} = 0^{\circ}; \qquad \beta_{2_{max}} = \delta_1 - \delta - \theta_1. \tag{5}$$

The angular sizes are defined:
- $\delta = 45^{\circ}$, for conical screw with 4 lobes;
- $\delta = 60^{\circ}$, for conical screw with 3 lobes;

- $\theta_1 = \dfrac{\pi - \delta}{2};$ (6)

- $r_1 = \lambda \cdot R_{rp_1}$ - with λ constructive scalar, $\lambda<1$; (7)

- $r_2 = R_{rp_1} \cdot \dfrac{sin\,\delta}{sin\,\theta_1} - \lambda \cdot R_{rp_1}.$ (8)

Along the conical axis, the size of the rolling circle in the plane $H = i, i = 0, 1, 2 \ldots n$, is, see figure 2:

$$R_{rp_i} = R_{rp_1} - p \cdot \varphi_1 \cdot tg\,\delta_1, \tag{9}$$

where p is the axial helical parameter [mm] (constructive); φ – variable angular parameter; δ_1 – the angle of the conical axode generatrix, Figure 1b.

3. Helical flanks of the conical screw

In the $X_1Y_1Z_1$ system (Z_1 is the conical screw axis), left conical flanks equations, described by the AB, BC generatrices see (2) and (6), are given by the transformation:

$$\begin{pmatrix} X_1 \\ Y_1 \\ Z_1 \end{pmatrix} = \omega_3^T(\varphi) \cdot \begin{pmatrix} X_{1_{AB,BC}} \\ Y_{1_{AB,BC}} \\ Z_{1_{AB,BC}} \end{pmatrix} - \begin{pmatrix} 0 \\ 0 \\ p \cdot \varphi \end{pmatrix}, \tag{10}$$

with p – constructive helical parameter [mm].

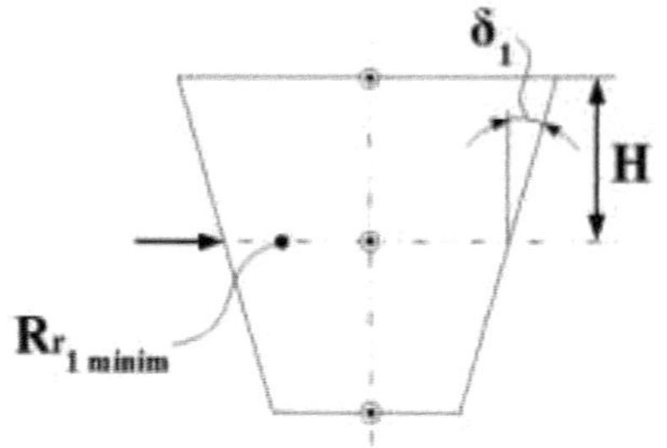

Figure 2. Angle of the conical axode generatrix, δ_1.

It is noted that the R_{rp_1} size varies along the screw axis, see (9) and also the sizes of r_1 and r_2 radii, see (7), (8) and (9), are variable sizes depending on the angular parameter φ, Figure 3.

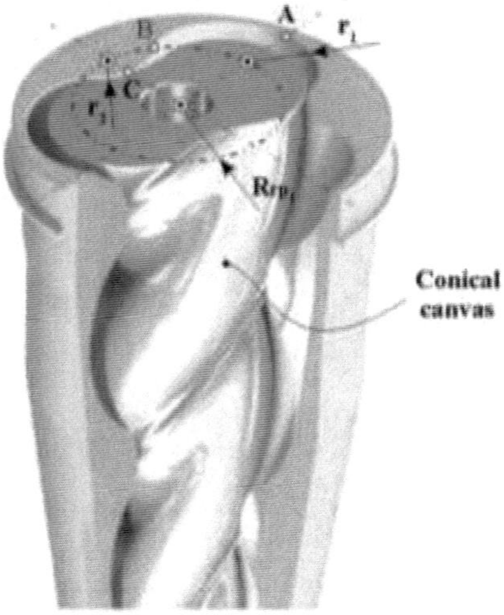

Figure 3. Conical canvas of rotor lobe [5].

The helical flanks are determined, see (2) and (4):

$$\Sigma_{AB}:\begin{pmatrix}X_1\\Y_1\\Z_1\end{pmatrix}=\begin{pmatrix}\cos\varphi_1 & -\sin\varphi_1 & 0\\ \sin\varphi_1 & \cos\varphi_1 & 0\\ 0 & 0 & 1\end{pmatrix}\cdot\begin{pmatrix}R_{rp_i}+r_1\cdot\cos\beta_1\\ r_1\cdot\sin\beta_1\end{pmatrix}+\begin{pmatrix}0\\0\\-p\cdot\varphi_1\end{pmatrix}, \tag{11}$$

with variable parameters φ_1 and β_1;

$$\Sigma_{BC}:\begin{pmatrix}X_1\\Y_1\\Z_1\end{pmatrix}=\begin{pmatrix}\cos\varphi_1 & -\sin\varphi_1 & 0\\ \sin\varphi_1 & \cos\varphi_1 & 0\\ 0 & 0 & 1\end{pmatrix}\cdot\begin{pmatrix}R_{rp_i}+(r_1+r_2)\cdot\cos\theta_1-r_2\cdot\cos(\delta+\beta_2)\\ (r_1+r_2)\cdot\sin\theta_1-r_2\cdot\sin(\delta+\beta_2)\\ 0\end{pmatrix}+\begin{pmatrix}0\\0\\-p\cdot\varphi_1\end{pmatrix}, \tag{12}$$

with variable parameters φ_1 and β_2.

After developments it results:

- Helical flank surface Σ_{AB}, from (11):

$$\Sigma_{AB}:\begin{vmatrix} X_1 = \left(R_{rp_i} + r_1 \cdot cos\,\beta_1\right) \cdot cos\,\varphi_1 - r_1 \cdot sin\,\varphi_1 \cdot sin\,\beta_1; \\ Y_1 = \left(R_{rp_i} + r_1 \cdot cos\,\beta_1\right) \cdot sin\,\varphi_1 + r_1 \cdot cos\,\varphi_1 \cdot sin\,\beta_1; \\ Z_1 = -p \cdot \varphi_1. \end{vmatrix} \tag{13}$$

- Helical flank surface Σ_{BC}, from (12):

$$\Sigma_{BC}:\begin{vmatrix} X_1 = \left[R_{rp_i} + (r_1 + r_2) \cdot cos\,\theta_1 - r_2 \cdot cos\left(\delta + \beta_2\right)\right] \cdot cos\,\varphi_1 - \\ -\left[(r_1 + r_2) \cdot sin\,\theta_1 - r_2 \cdot sin\left(\delta + \beta_2\right)\right] \cdot sin\,\varphi_1; \\ Y_1 = \left[R_{rp_i} + (r_1 + r_2) \cdot cos\,\theta_1 - r_2 \cdot cos\left(\delta + \beta_2\right)\right] \cdot sin\,\varphi_1 + \\ +\left[(r_1 + r_2) \cdot sin\,\theta_1 - r_2 \cdot sin\left(\delta + \beta_2\right)\right] \cdot cos\,\varphi_1; \\ Z_1 = -p \cdot \varphi_1. \end{vmatrix} \tag{14}$$

Σ_{AB}, Σ_{BC} are „left" conical helical surfaces.

Finally, the parametric equations of the conical screw flanks can be brought at the form:

$$\Sigma_{AB}:\begin{vmatrix} X_1 = R_{rp_i} \cdot cos\,\varphi_1 + r_1 \cdot cos\left(\varphi_1 + \beta_1\right); \\ Y_1 = R_{rp_i} \cdot sin\,\varphi_1 + r_1 \cdot sin\left(\varphi_1 + \beta_1\right); \\ Z_1 = -p \cdot \varphi_1, \end{vmatrix} \tag{15}$$

and for the $\widehat{BC}$ flank:

$$\Sigma_{BC}:\begin{vmatrix} X_1 = R_{rp_i} \cdot cos\,\varphi_1 + (r_1 + r_2) \cdot cos\left(\varphi_1 + \theta_1\right) - r_2 \cdot cos\left(\delta + \beta_2 + \varphi_1\right); \\ Y_1 = R_{rp_i} \cdot sin\,\varphi_1 + (r_1 + r_2) \cdot sin\left(\varphi_1 + \theta_1\right) - r_2 \cdot sin\left(\delta + \beta_2 + \varphi_1\right); \\ Z_1 = -p \cdot \varphi_1. \end{vmatrix} \tag{16}$$

It is specified that the R_{rp_1}, r_1 and r_2 sizes are variable depending on φ_1, see the definitions (7), (8) and (9):

$$\begin{aligned} &R_{rp_i} = R_{rp_1} - p \cdot \varphi_1 \cdot tg\,\delta_1; \\ &r_1 = \lambda \cdot R_{rp_i}; \\ &r_2 = R_{rp_i} \cdot \frac{sin\,\delta}{sin\,\theta_1} - \lambda \cdot R_{rp_i}, \end{aligned} \tag{17}$$

R_{rp_1} is the size of the primitive circle radius in $Z_1 = 0$ plane.

4. Kinematics of the axode rolling process, associated with the conical compressor rotors

The two rotors of the compressor are associated with two conical canvases - the primitives of the helical conical surfaces, of radius R_{rp_1}, respectively R_{rp_2}. The conical surfaces have a common generatrix, along which the two conical canvases are tangent. The common generatrix to the two conical canvases - the A_1 and A_2 axodes, has the length:

$$L = \frac{R_{rp_1}}{\sin\delta_1} = \frac{R_{rp_2}}{\sin\delta_2}. \tag{18}$$

If it is noted with ϕ_1, respectively ϕ_2 the angles in the rotational movements, of the same direction, around $\vec{V}_1$, $\vec{V}_2$ axes, then the transmission ratio of the movement is:

$$i_{1,2} = \frac{\phi_1}{\phi_2} = \frac{z_{1\ \mathrm{lob}}}{z_{2\ \mathrm{lobi}}}, \tag{19}$$

where $z_{1\ lob}$ and $z_{2\ lob}$ are the number of lobes of the driver (inner) and driven (outer) conical screws.

The reference systems are defined:

- $x_1y_1z_1$ - fixed system, joined with the axis of the wrapped cone (driver rotor);
- $x_2y_2z_2$ - fixed system, joined with the axis of the wrapper cone (driven rotor);
- $X_1Y_1Z_1$ - mobile system, joined with the surface of the driver screw;
- $X_2Y_2Z_2$ - mobile system, joined with the surface of the driven screw (wrapper screw).

The two conical surfaces execute rotational movements around the Z_1 and Z_2 axes - absolute movements:

$$\begin{aligned} x_1 &= \omega_3^T(\phi_1)\cdot X_1 \\ x_2 &= \omega_3^T(\phi_2)\cdot X_2 \end{aligned}. \tag{20}$$

The relative position of fixed systems, which have a common origin:

$$x_2 = \theta\cdot x_1, \tag{21}$$

with the θ matrix of form:

$$\theta = \begin{pmatrix} \cos\theta & 0 & \sin\theta \\ 0 & 1 & 0 \\ -\sin\theta & 0 & \cos\theta \end{pmatrix}. \tag{22}$$

Obviously, the position angular parameter of the $\vec{V}_1$ and $\vec{V}_2$ axes, θ, is:

$$\theta = \delta_2 - \delta_1, \tag{23}$$

with δ_1, δ_2 the semiangles of the primitive conical canvases associated with the driver, respectively driven screw.

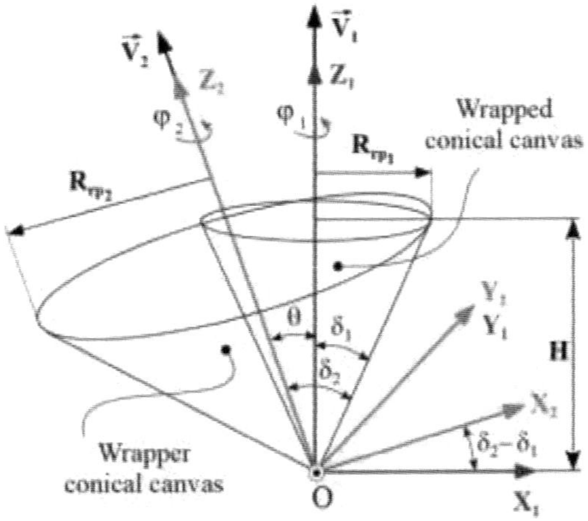

Figure 4. Primitive conical canvases of the rotors.

4.1. The relative movement of the wrapped cone towards to the wrapper cone
The primitive cones have the z_1 and z_2 axes fixed.

The reference systems to which the generating movements are defined, are as follows:
- $x_1y_1z_1$ - fixed system, joined with the axis of the wrapped cone;
- $x_2y_2z_2$ - fixed system, joined with the axis of the wrapper cone;
- $X_1Y_1Z_1$ - mobile system, joined with the surface of the wrapped primitive cone;
- $X_2Y_2Z_2$ - mobile system, joined with the surface of the wrapper primitive cone.

The fixed systems were represented only by the positions of the z_1 and z_2 axes, figure 4.

The absolute movements, rotations around the z_1 and z_2 axes, are given by the transformations:

$$x_1 = \omega_3^T(\phi_1)\cdot X_1; \tag{24}$$

$$x_2 = \omega_3^T(\phi_2)\cdot X_2. \tag{25}$$

The $x_2 = \theta\cdot x_1$ matrix represents the coordinates transformation between the fixed systems, with:

$$\theta = \begin{pmatrix} \cos\theta & 0 & \sin\theta \\ 0 & 1 & 0 \\ -\sin\theta & 0 & \cos\theta \end{pmatrix} \quad \text{and} \quad \theta = \delta_2 - \delta_1 . \tag{26}$$

5. The family of surfaces

From (20) and (21), results the relative movement:

$$X_2 = \omega_3(\phi_2)\cdot\theta\cdot\omega_3^T(\phi_1)\cdot X_1. \tag{27}$$

Basically, if the X_1 matrix represents the conical surfaces Σ_{AB} and Σ_{BC} of the driver rotor, in the form (see (15) and (16)):

$$\Sigma_{AB}\begin{vmatrix} X_1 = X_1\cdot(\beta_1,\varphi_1) \\ Y_1 = Y_1\cdot(\beta_1,\varphi_1) \\ Z_1 = p\cdot\varphi_1 \end{vmatrix}, \tag{28}$$

of variables β_1 and φ_1, respectively:

$$\Sigma_{BC}\begin{vmatrix} X_1 = X_1 \cdot (\beta_2, \varphi_1) \\ Y_1 = Y_1 \cdot (\beta_2, \varphi\ ) \\ Z_1 = p \cdot \varphi_1 \end{vmatrix}, \tag{29}$$

of variables β_2 and φ_1.

By the transformation (30), representing the relative movement between the mobile systems $X_1Y_1Z_1$ towards $X_2Y_2Z_2$ (movement of the driver rotor in relation with the driven rotor), by transformations (27), (28) and (29), it determines the surface families Σ_{AB}, Σ_{BC}, in the $X_2Y_2Z_2$ system:

$$\begin{pmatrix} X_2 \\ Y_2 \\ Z_2 \end{pmatrix} = \begin{pmatrix} \cos\phi_2 & \sin\phi_2 & 0 \\ -\sin\phi_2 & \cos\phi_2 & 0 \\ 0 & 0 & 1 \end{pmatrix} \cdot \begin{pmatrix} \cos\theta & 0 & \sin\theta \\ 0 & 1 & 0 \\ -\sin\theta & 0 & \cos\theta \end{pmatrix} \cdot \begin{pmatrix} \cos\phi_1 & -\sin\phi_1 & 0 \\ \sin\phi_1 & \cos\phi_1 & 0 \\ 0 & 0 & 1 \end{pmatrix} \cdot \begin{pmatrix} X_{1_{AB,BC}} \\ Y_{1_{AB,BC}} \\ p \cdot \varphi \end{pmatrix}. \tag{30}$$

Thus, the surface families $\Sigma_{\widehat{AB},\widehat{BC}}$ in mobile system $X_2Y_2Z_2$, of the wrapper screw, see (30), can be written in the form:

$$\begin{pmatrix} X_2 \\ Y_2 \\ Z_2 \end{pmatrix} = T \cdot \begin{pmatrix} X_{1_{AB,BC}} \\ Y_{1_{AB,BC}} \\ p \cdot \varphi \end{pmatrix}. \tag{31}$$

The T-matrix is the multiplication, see (27):

$$T = \begin{pmatrix} cos\,\theta \cdot cos\,\varphi_2 & sin\,\varphi_2 & sin\,\theta \cdot cos\,\varphi_2 \\ -cos\,\theta \cdot sin\,\varphi_2 & cos\,\varphi_2 & -sin\,\theta \cdot sin\,\varphi_2 \\ -sin\,\theta & 0 & cos\,\theta \end{pmatrix}. \tag{32}$$

Thus, the $\Sigma_{\widehat{AB},\widehat{BC}}$ surface families from the $X_1Y_1Z_1$ system of the driver screw in the $X_2Y_2Z_2$ system of the driven screw can be determined from (15), (16), (31) and (32):

- for the Σ_{AB} flank (15) it results:

$$\begin{pmatrix} X_2 \\ Y_2 \\ Z_2 \end{pmatrix} = (T) \cdot \begin{pmatrix} R_{rp_i} \cdot cos\,\varphi_1 + r_1 \cdot cos(\varphi_1 + \beta_1) \\ R_{rp_i} \cdot sin\,\varphi_1 + r_1 \cdot sin(\varphi_1 + \beta_1) \\ -p \cdot \varphi_1 \end{pmatrix} \tag{33}$$

- for the Σ_{BC} (16) it results:

$$\begin{pmatrix} X_2 \\ Y_2 \\ Z_2 \end{pmatrix} = (T) \cdot \begin{pmatrix} R_{rp_i} \cdot cos\,\varphi_1 + (r_1 + r_2) \cdot cos(\varphi_1 + \theta_1) - r_2 \cdot cos(\delta + \beta_2 + \varphi_1) \\ R_{rp_i} \cdot sin\,\varphi_1 + (r_1 + r_2) \cdot sin(\varphi_1 + \theta_1) - r_2 \cdot sin(\delta + \beta_2 + \varphi_1) \\ -p \cdot \varphi_1 \end{pmatrix}, \tag{34}$$

where the sizes r_1, r_2, R_{rp_i} are defined by (17) and β_1, β_2 and φ_1 are variable parameters.

Essentially, the matrix equations (33) and (34) determine in the $X_2Y_2Z_2$ system, the flanks families generated in the relative movement of the surfaces $\Sigma_{AB,BC}$, see (15), respectively (16), in the system associated with the wrapper screw, of equations:

$$\left(\Sigma_{AB}\right)_{\phi_1} \left| \begin{array}{l} X_2 = X_2\left(\varphi_1, \beta_1, \phi_1\right); \\ Y_2 = Y_2\left(\varphi_1, \beta_1, \phi_1\right); \\ Z_2 = -p \cdot \varphi_1. \end{array} \right. \tag{35}$$

$$\left(\Sigma_{BC}\right)_{\phi_1} \left| \begin{array}{l} X_2 = X_2\left(\varphi, \phi_1, \beta_2\right); \\ Y_2 = Y_2\left(\varphi, \phi_1, \beta_2\right); \\ Z_2 = -p \cdot \varphi_1. \end{array} \right. \tag{36}$$

the relation between ϕ_2 and ϕ_1 given by (19).

The enveloping of the surface families (35) and (36), represents, in the $X_2Y_2Z_2$ system, the peripheral surface of the conical screw conjugated to the wrapped screw (the driver screw) of the conical helical compressor.

The proposed issue - the determination of the enveloping of the surfaces families known in analytical form, of type $\left(\Sigma_{AB}\right)_{\phi_1}$ and $\left(\Sigma_{BC}\right)_{\phi_1}$, can be approached by the virtual focal points theorem between the two surfaces of the driver screw, see (15) and (16), associated with conical axode A_1, figure 3 and their enwrapping, associated with A_2 axode, in the rolling process of the two axodes A_1 and A_2.

6. Olivier theorem [14]

All the points belonging to the Σ_{AB} and Σ_{BC} surfaces represent, in the sense of the method, virtual points of the contact line at a certain time, between the enwrapping surfaces: the flanks of the driver screw and the homologous ones of the driven screw.

The selection condition from the totality of the virtual focal points, meaning what is currently called - the enveloping condition, the totality of the points of the future characteristic curve, assumes that the normal at the current point of the surfaces Σ_{AB}, Σ_{BC} is orthogonal to the „*velocity*" in relative movement of them in relation with the conjugated surfaces, in the process of rolling the conical axodes, A_1 and A_2.

In figure 5, a contact point is defined between the enwrapping surfaces: Σ, associated with the A_1 centrode of the wrapped screw (driver screw) and S, associated with the A_2 axode of the wrapper screw (driven screw). The normal $\vec{N}_{\Sigma}$, identical with the normal $\vec{N}_S$ at $\Sigma_{AB,\ BC}$ and $S_{AB,\ BC}$ surfaces – conjugated surfaces in the rolling movement of the two centrodes A_1 and A_2. The totality of the points M thus determined, define the characteristic curve $C_{\Sigma,S}$, the curve along which the two surfaces Σ and S are tangent. The Σ surface is known, and the S surface follows, in order to be determined. Clearly, the Σ and S surfaces can be seen as being generated by the characteristic curve $C_{\Sigma,S}$.

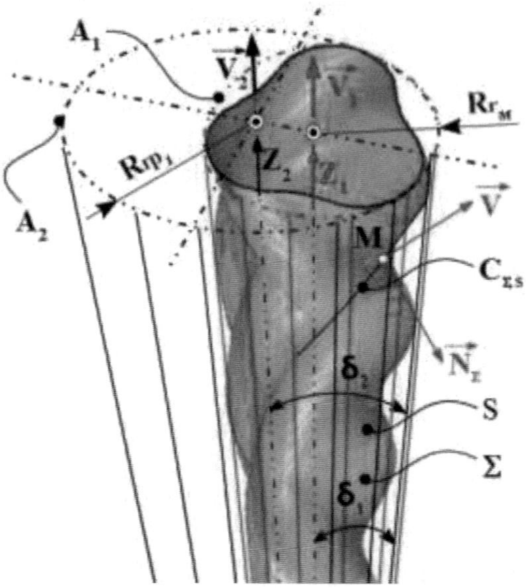

Figure 5. Primitive conical canvases of the rotors.

Thus, the normals at the $\Sigma_{AB,BC}$ surfaces must be determined, defined in $X_1Y_1Z_1$ system and subsequently transferred to $X_2Y_2Z_2$ system of the wrapper screw, by relation (27).

The normals at the helical surfaces Σ_{AB}, $_{BC}$, see (15) and (16):

$$\vec{N}_{\Sigma_{AB}} = \begin{pmatrix} \vec{i} & \vec{j} & \vec{k} \\ \dot{X}_{1\phi_1} & \dot{Y}_{1\phi_1} & \dot{Z}_{1\phi_1} \\ \dot{X}_{1\beta_1} & \dot{Y}_{1\beta_1} & \dot{Z}_{1\beta_1} \end{pmatrix}, \tag{37}$$

where, from (15), it results:

$$\begin{aligned} \dot{X}_{1\varphi} &= -(R - r_1)\cdot cos(\alpha-\varphi) - r_1\cdot sin(\beta+\varphi); \\ \dot{Y}_{1\varphi} &= (R - r_1)\cdot sin(\alpha-\varphi) + r_1\cdot cos(\beta+\varphi); \\ \dot{Z}_{1\varphi} &= p, \end{aligned} \tag{38}$$

and

$$\begin{aligned} \dot{X}_{1\beta} &= r_1\cdot sin(\beta+\varphi); \\ \dot{Y}_{1\beta} &= -r_1\cdot cos(\beta+\varphi); \\ \dot{Z}_{1\beta} &= 0. \end{aligned} \tag{39}$$

Therefore, the $\vec{N}_{\Sigma_{AB}}$ vector is defined:

$$\vec{N}_{\Sigma_{AB}} = -p\cdot r_1\cdot\cos(\beta+\varphi)\cdot\vec{i} + p\cdot r_1\cdot\sin(\beta+\varphi)\cdot\vec{j} + r_1\cdot\left[(R-r_1)\cdot\cos(\alpha+\beta)\right]\cdot\vec{k}\,. \tag{40}$$

From (16) the partial derivatives of the normal vector at Σ_{BC} are defined:

$$\vec{N}_{\Sigma_{AB}} = \begin{pmatrix} \vec{i} & \vec{j} & \vec{k} \\ \dot{X}_{1\varphi} & \dot{Y}_{1\varphi} & \dot{Z}_{1\varphi} \\ \dot{X}_{1\theta_2} & \dot{Y}_{1\theta_2} & \dot{Z}_{1\theta_2} \end{pmatrix}, \tag{41}$$

and from here the vectorial form of the vector $\vec{N}_{\Sigma_{BC}}$:

$$\vec{N}_{\Sigma BC} = \begin{pmatrix} \vec{\imath} & \vec{\jmath} & \vec{k} \\ \dot{X}_{1\phi} & \dot{Y}_{1\phi} & \dot{Z}_{1\phi} \\ \dot{X}_{1\theta_2} & \dot{Y}_{1\theta_2} & \dot{Z}_{1\theta_2} \end{pmatrix}, \quad (42)$$

$$\vec{N}_{\Sigma BC} = -p \cdot \dot{Y}_{1\theta_2} \cdot \vec{\imath} + p \cdot \dot{X}_{1\theta_2} \cdot \vec{\jmath} + [r_2 \cdot (R - r_1) \cdot cos(\theta_2 - \alpha) - r_2 \cdot (r_1 + r_2) \cdot sin(\theta_2 - \alpha)] \cdot \vec{k}, (43)$$

with partial derivatives:

$$\begin{aligned} \dot{X}_{1\varphi} &= (R - r_1) \cdot sin(\alpha - \varphi) - (r_1 + r_2) \cdot cos(\alpha - \varphi) - r_2 \cdot cos(\theta_2 - \varphi); \\ \dot{Y}_{1\varphi} &= -(R - r_1) \cdot cos(\alpha - \varphi) + (r_1 + r_2) \cdot sin(\alpha - \varphi) - r_2 \cdot sin(\theta_2 - \varphi); \\ \dot{Z}_{1\varphi} &= p, \end{aligned} \quad (44)$$

and

$$\begin{aligned} \dot{X}_{1\theta_2} &= r_2 \cdot cos(\theta_2 - \varphi); \\ \dot{Y}_{1\theta_2} &= r_2 \cdot sin(\theta_2 - \varphi); \\ \dot{Z}_{1\theta_2} &= 0. \end{aligned} \quad (45)$$

The normal forms $\vec{N}_{\Sigma_{AB}}$ and $\vec{N}_{\Sigma_{BC}}$ are transferred from $X_1Y_1Z_1$ system (see Figure 4) of the wrapped rotor to $X_2Y_2Z_2$ system of the wrapper rotor, by the transformation:

$$\begin{pmatrix} N_{X_2,\Sigma_{AB,BC}} \\ N_{Y_2,\Sigma_{AB,BC}} \\ N_{Z_2,\Sigma_{AB,BC}} \end{pmatrix} = (T) \cdot \begin{pmatrix} N_{X_1,\Sigma_{AB,BC}} \\ N_{Y_1,\Sigma_{AB,BC}} \\ N_{Z_1,\Sigma_{AB,BC}} \end{pmatrix}, \quad (46)$$

with the (T) matrix defined by (32).

Therefore, considering the vectors $\vec{N}_{X_1,\Sigma_{AB}}$ definitions from (40) and (43), corresponding to the conical helical surface corresponding to the AB arc of the cross profile of the screw, respectively BC, see $\vec{N}_{X_1,\Sigma_{BC}}$ - (42), the forms of these vectors are obtained in the $X_2Y_2Z_2$ system, by the (46) transformation - the system associated with the wrapper screw.

The vectors equations (43) and (46) have in $X_2Y_2Z_2$ system the forms:

$$\vec{N}_{X_2\Sigma_{AB}} = N_{X_2\Sigma_{AB}} \cdot \vec{\imath} + N_{Y_2\Sigma_{AB}} \cdot \vec{\jmath} + N_{Z_2\Sigma_{AB}} \cdot \vec{k}; \quad (47)$$

$$\vec{N}_{X_2\Sigma_{BC}} = N_{X_2\Sigma_{BC}} \cdot \vec{\imath} + N_{Y_2\Sigma_{BC}} \cdot \vec{\jmath} + N_{Z_2\Sigma_{BC}} \cdot \vec{k}, \quad (48)$$

where, the directors parameters $N_{X_2\Sigma_{AB}}$ ... $N_{X_2\Sigma_{BC}}$ are functions of the type:

$$N_{X_2\Sigma_{AB}} = N_{X_2\Sigma_{AB}}(\varphi, \beta, \phi_1); \quad N_{Y_2\Sigma_{AB}} = N_{Y_2\Sigma_{AB}}(\varphi, \beta, \phi_1); \quad (49)$$

$$N_{X_2\Sigma_{BC}} = N_{X_2\Sigma_{BC}}(\varphi,\ \theta_2,\ \phi_1);\quad N_{Y_2\Sigma_{BC}} = N_{Y_2\Sigma_{BC}}(\varphi,\ \theta_2,\ \phi_1). \tag{50}$$

The functions (49) and (50) are directors parameters of normals at the Σ_{AB}, Σ_{BC} flanks in the $X_2Y_2Z_2$ system.

The „*velocity vector*" $\vec{V}$ of the current point on the driver screw lobes, associated with the conical axode A_1, meaning the Σ_{AB} and Σ_{BC} surfaces in the relative movement towards to the $X_2Y_2Z_2$ system, must be defined.

The $X_1Y_1Z_1$ system movement towards $X_2Y_2Z_2$ and, joined with it, the conical helical flanks of the driver screw, meaning Σ_{AB}, Σ_{BC}, see (15) and (16), in relation to the wrapper (driven) screw of the compressor is described by the transformation (27) and (30).

Therefore, resuming the matrix equation (27), the velocity vector $\vec{V}$ has the direction defined by $\vec{R}_{\phi_1}$:

$$\vec{R}_{\phi_1} = \dot{\omega}_3(\phi_2)\cdot\frac{d\phi_2}{d\phi_1}\cdot\theta\cdot\omega_3^T(\varphi_1)\cdot X_1 + \omega_3(\phi_2)\cdot\theta\cdot\dot{\omega}_3^T(\varphi_1)\cdot X_1, \tag{51}$$

in which the transmission ratio is considered:

$$I = \frac{d\phi_2}{d\phi_1}, \tag{52}$$

I being a constant size.

Evidently, the following form can be written:

$$R_{\phi_1} = [I\cdot\dot{\omega}_3(\phi_2)\cdot\theta\cdot\omega_3^T(\varphi_1) + \omega_3(\phi_2)\cdot\theta\cdot\dot{\omega}_3^T(\varphi_1)]\cdot X_1, \tag{53}$$

where X_1 is the matrix formed with the coordinates of the current point on Σ_{AB}, respectively, Σ_{BC} see (15) and (16).

The development of the vector (53) leads to the form:

$$R_{\phi_1} = \left\{ I\cdot\begin{pmatrix} -\sin\phi_2 & \cos\phi_2 & 0 \\ -\cos\phi_2 & -\sin\phi_2 & 0 \\ 0 & 0 & 0 \end{pmatrix}\cdot\begin{pmatrix} \cos\theta & 0 & \sin\theta \\ 0 & 1 & 0 \\ -\sin\theta & 0 & \cos\theta \end{pmatrix}\cdot\begin{pmatrix} \cos\phi_1 & -\sin\phi_1 & 0 \\ \sin\phi_1 & \cos\phi_1 & 0 \\ 0 & 0 & 1 \end{pmatrix} + \right.$$
$$\left. + \begin{pmatrix} \cos\phi_2 & \sin\phi_2 & 0 \\ -\sin\phi_2 & \cos\phi_2 & 0 \\ 0 & 0 & 1 \end{pmatrix}\cdot\begin{pmatrix} \cos\theta & 0 & \sin\theta \\ 0 & 1 & 0 \\ -\sin\theta & 0 & \cos\theta \end{pmatrix}\cdot\begin{pmatrix} -\sin\phi_1 & -cos\phi_1 & 0 \\ \cos\phi_1 & -\sin\phi_1 & 0 \\ 0 & 0 & 0 \end{pmatrix}\right\}\cdot\begin{pmatrix} X_1 \\ Y_1 \\ Z_1 \end{pmatrix}. \tag{54}$$

It is noted that $\phi_2 = I\cdot\phi_1$ and, consequently, the directors parameters of the $\vec{R}_{\phi_1}$ vector basically, are by the form:

$$\vec{R}_{\phi_{1AB}} = R_{\phi_{1X_{2AB}}}\cdot\vec{i} + R_{\phi_{1Y_{2AB}}}\cdot\vec{j} + R_{\phi_{1Z_{2AB}}}\cdot\vec{k} \tag{55}$$

and similarly:

$$\vec{R}_{\phi_{1BC}} = R_{\phi_{1X_{2BC}}}\cdot\vec{i} + R_{\phi_{1Y_{2BC}}}\cdot\vec{j} + R_{\phi_{1Z_{2BC}}}\cdot\vec{k}, \tag{56}$$

representing the velocity vectors of the points on $\Sigma_{AB,\ BC}$ in $X_1Y_1Z_1$ system, in relative movement towards to $X_2Y_2Z_2$ system, see (33) and (34) and the condition for identifying the contact points between the conjugated flanks.

The flanks family, see (35) and (36):

$$\left(\Sigma_{AB}\right)_{\phi_1} \begin{vmatrix} X_2 = X_2\left(\varphi, \beta, \phi_1\right) \\ Y_2 = X_2\left(\varphi, \beta, \phi_1\right) \\ Z_2 = p \cdot \varphi \end{vmatrix} . \tag{57}$$

The normal at the conical helical flank is:

$$\vec{N}_{X_2\Sigma_{AB}} = N_{X_2\Sigma_{AB}} \cdot \vec{i} + N_{Y_2\Sigma_{AB}} \cdot \vec{j} + N_{Z_2\Sigma_{AB}} \cdot \vec{k} \tag{58}$$

and similarly, for BC flank, see (49) and (50):

$$\vec{N}_{X_2\Sigma_{BC}} = N_{X_2\Sigma_{BC}} \cdot \vec{i} + N_{Y_2\Sigma_{BC}} \cdot \vec{j} + N_{Z_2\Sigma_{BC}} \cdot \vec{k} . \tag{59}$$

The velocity vector, see (55) is:

$$\vec{R}_{\phi_{1AB}} = R_{\phi_{1X_{2AB}}} \cdot \vec{i} + R_{\phi_{1Y_{2AB}}} \cdot \vec{j} + R_{\phi_{1Z_{2AB}}} \cdot \vec{k} \tag{60}$$

and similarly, $\vec{R}_{\phi_{1BC}}$:

$$\vec{R}_{\phi_{1BC}} = R_{\phi_{1X_{2BC}}} \cdot \vec{i} + R_{\phi_{1Y_{2BC}}} \cdot \vec{j} + R_{\phi_{1Z_{2BC}}} \cdot \vec{k} . \tag{61}$$

The enwrapping condition:

$$\vec{N}_{X_2\Sigma_{AB}} \cdot \vec{R}_{\phi_{1AB}} = K_1\left(\varphi, \beta, \phi_1\right) = 0 \tag{62}$$

and similarly, for BC flank:

$$\vec{N}_{X_2\Sigma_{BC}} \cdot \vec{R}_{\phi_{1BC}} = K_2\left(\varphi, \phi_2, \phi_1\right) = 0 . \tag{63}$$

The equations assembly (35) and (36) and the conditions (61), (63) allow, by the elimination of one of the variable parameters $\left(\varphi, \theta_2, \phi_1\right)$ or $\left(\varphi, \beta, \phi_1\right)$, the determination of the conical helical compressor screw flank form.

7. Graphical solution in CATIA for representation of the driver screw and the driven screw

The graphical method developed in CATIA filled out the algorithm that was developed based on the „virtual focal points" theorem using the facilities of the *Generative Shape Design*, *Assembly* and *Kinematics* modules. It started from the hypothesis that the inner screw has a rotational movement inside the outer screw, with decreasing diameter, which, from a practical point of view, allows much higher compression ratios in a single work cycle. As a functional principle, the wrapper screw rotates around a central axis, while the wrapped screw rotates along an inclined axis, coplanar with that of the wrapper screw, the two screws having a common generatrix.

In CATIA design environment, with the help of specific commands from Generative Shape Design, the 3D model of the inner screw was obtained, the parameters of the graphic model: the outer diameter of the inner screw: 21.48 mm; outer diameter of the outer screw: 28.68 mm; rotor length: 180 mm; screw propeller pitch: 90 mm; maximum eccentricity between axles: 3,629 mm, Figure 6.

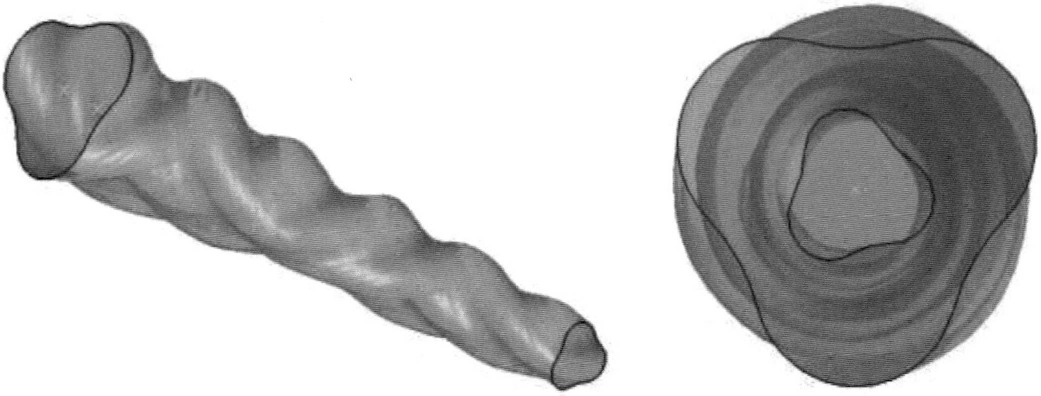

Figure 6. 3D model of the wrapped conical rotor.

The graphical method developed in the CATIA graphic environment, starts from the 2D design of the front profiles of the rotors by defining the *XYZ* reference system (*Z* is the axis of the helical surface). These approximate profiles are curves of spline type, with which the helical surface of the driver rotor is determined (*Multi Sections Volume* command) and the 3D model of the helical parameter screw *p* is generated. In table 1 and Figure 7, the points coordinates are specified and the 2D model of the cross profile of the wrapped screw respectively of the wrapper screw is presented.

Tabelul 1. The points coordinates on the front profile of the wrapped and wrapper conical rotor.

Wrapper screw				*Wrapped screw*			
No crt.	*X [mm]*	*Y [mm]*	*Z [mm]*	*No crt.*	*X [mm]*	*Y [mm]*	*Z [mm]*
1.	22.32	0	0	1	18.75	0	0
2.	21.88	2.86	0	2	18.22	3.19	0
3.	20.62	5.50	0	3	16.70	3.06	0
4.	18.61	7.70	0	4	14.34	8.28	0
5.	16.00	9.29	0	5	11.40	9.64	0
6.	14.56	10.02	0	6	9.78	10.16	0
7.	13.32	10.92	0	7	8.21	10.81	0
8.	11.99	12.01	0	8	6.70	11.60	0

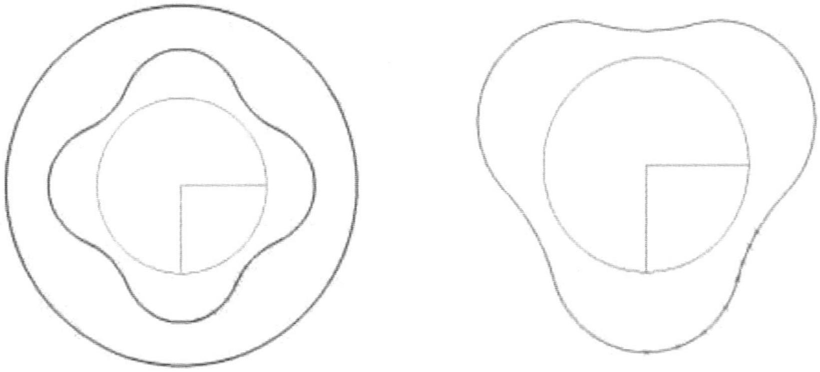

a. *b.*

Figure 7. Points on the front profile of the wrapper (a) and wrapped (b) conical rotor.

The facilities offered by CATIA graphic environment allowed the 2D and 3D modeling of the front sections of the rotors, as well as the obtaining of their solid models, with the advantage that changes of the geometric parameters could lead to other constructive forms of the compressors rotors of 3/4 type with lobes with circle arc profile, first of all, by decreasing the length of the rotors, as well as their outer diameters.

8. Conclusions

Compressors with conical rotors are considered much more compact, due to the driver screw located inside the driven one. These have a higher efficiency compared to an equivalent compressor with cylindrical rotors, being able to develop higher pressure ratios.

This paper presents an algorithm developed in analytical form for profiling the wrapper screw of the conical helical compressor, conical gearing $z/z+1$.

The algorithm was developed based on the Olivier theorem, between the constant pitch conical helical surfaces, associated with conical axodes of concurrent axes. The theorem presented is a complementary variant of the Gohman fundamental theorem.

A graphic solution of the 3D model of the driver rotor of the compressor, developed in the CATIA graphic design environment, was also presented. The CATIA design environment allowed, through its software capabilities, to obtain an intuitive and faster 3D CAD solution, with precise results for defining the helical surface that represents the driver screw.

The driver (inner) screw is considered a screw with connected circle arc generatrices, the example presented referring to a driver with three starts.

9. References

[1] Dmitriev O, Tabota E and Arbon I 2015 A miniature Rotary Compressor with a 1:10 compression ratio *IOP Conference Series: Materials Science and Engineering* **90** 012055

[2] Dmitriev O and Arbon I 2017 Comparison of energy-efficiency and size of portable oil-free screw and scroll compressors *IOP Conference Series: Materials Science and Engineering* **232** 012057

[3] Baroiu N, Păunoiu V, Teodor V G, Susac F and Oancea N 2018 Geometrical analysis, for rapid prototyping, of the compressor's helical conic rotor model *MATEC Web Conf.* **178**

[4] Baroiu N, Teodor V G, Păunoiu V and Oancea N 2018 Comparative study regarding two constructive solutions for conical worm from pump construction *The Annals of "Dunărea de Jos" University of Galati, Fascicle V* pp 41-48

[5] *** 2020 http://vertrotors.com/, accessed 20.02.2020

[6] Basha N, Kovacevic A and Rane S R 2019 Analysis of Oil-Injected Twin-Screw Compressor with Multiphase Flow Models *Designs* **3**(54)
[7] Husak E, Kovacevic A and Karabegovic I 2019 Calculation of clearances in twin screw compressors *IOP Conf. Series: Materials Science and Engineering* **604** 012001
[8] Lu Y, Kovacevic A and Read M 2019 Normal Rack Grid Generation Method for Screw Machines with Large Helix Angles *IOP Conf. Series: Materials Science and Engineering* **604** 012011
[9] Kovacevic A, Stosic N, Smith I K 2010 CFD Analysis of a Screw Compressor Performance *Advances of CFD in Fluid Machinery Design* (London: Purdue University) pp 1-9
[10] Stosic N, Smith I K and Kovacevic A 2005 Screw Compressors: Mathematical Modeling and Performance Calculation (Berlin: Springer Verlag)
[11] Teodor V G, Baroiu N and Susac F 2018 The Synthesis of New Algorithms for CAD Profiling of Cutting Tools (Beau Bassin: Lambert Academic Publishing)
[12] Rane S R 2015 Grid Generation and CFD Analysis of Variable Geometry Screw Machines. *Doctoral thesis* (London: City University)
[13] Oancea N 2004 Surfaces generation by winding vol. I-III (Galaţi: University Press)
[14] Radzevich S P 2018 Theory of Gearing: Kinematics, Geometry, and Synthesis *CRC Press*
[15] Radzevich S P 2007 On the Accuracy of Precision Involute Hobs: An Analytical Approach *Journal of Manufacturing Processes* **9**(2) pp 121-136

Acknowledgments
This work was supported by a grant of the Romanian National Authority for Scientific Research and Innovation, CNCS-UEFISCDI, project number PN-III-P1-1.2-PCCDI-2017-0446 - TFI PMAIAA, within PNCDI III.

AUTHOR INDEX